AGRICULTURAL BIODIVERSITY AND BIOTECHNOLOGY IN ECONOMIC DEVELOPMENT

NATURAL RESOURCE MANAGEMENT AND POLICY

Editors:

Ariel Dinar
Rural Development Department
The World Bank
1818 H Street, NW
Washington, DC 20433

David Zilberman
Dept. of Agricultural and
Resource Economics
Univ. of California, Berkeley
Berkeley, CA 94720

EDITORIAL STATEMENT

There is a growing awareness to the role that natural resources such as water, land, forests and environmental amenities play in our lives. There are many competing uses for natural resources, and society is challenged to manage them for improving social well being. Furthermore, there may be dire consequences to natural resources mismanagement. Renewable resources such as water, land and the environment are linked, and decisions made with regard to one may affect the others. Policy and management of natural resources now require interdisciplinary approach including natural and social sciences to correctly address our society preferences.

This series provides a collection of works containing most recent findings on economics, management and policy of renewable biological resources such as water, land, crop protection, sustainable agriculture, technology, and environmental health. It incorporates modern thinking and techniques of economics and management. Books in this series will incorporate knowledge and models of natural phenomena with economics and managerial decision frameworks to assess alternative options for managing natural resources and environment.

While there are significant trends in use of genetic resources in agricultural research and its application, heated public debate has evolved around the biotechnology, biodiversity, and biosafety aspects. Acknowledging the potential of biotechnology tools and products for ending hunger and poverty in the developing world, it is also contended that genetic modification may poses unacceptable risks for human health and the environment. "*Agricultural Biodiversity and Biotechnology in Economic Development*" attempts at addressing such issues in a structured approach, using economic, legal and institutional frameworks.

The Series Editors

Recently Published Books in the Series

Haddadin, Munther J.
Diplomacy on the Jordan: International Conflict and Negotiated Resolution
Renzetti, Steven
The Economics of Water Demands
Just, Richard E. and Pope, Rulon D.
A Comprehensive Assessment of the Role of Risk in U.S. Agriculture
Dinar, Ariel and Zilberman, David
Economics of Water Resources: The Contributions of Dan Yaron
Ünver,.I.H. Olcay, Gupta, Rajiv K. IAS, and Kibaroğlu, Ayşegül
Water Development and Poverty Reduction
d'Estrée, Tamra Pearson and Colby, Bonnie G.
Braving the Currents: Evaluating Environmental Conflict Resolution in the River Basins of the American West

AGRICULTURAL BIODIVERSITY AND BIOTECHNOLOGY IN ECONOMIC DEVELOPMENT

by

Joseph Cooper

Leslie Marie Lipper

David Zilberman

Library of Congress Cataloging-in-Publication Data

Agricultural biodiversity and biotechnology in economic development / [edited] by Joseph Cooper, Leslie Marie Lipper, David Zilberman.
p. cm. – (Natural resource management and policy ; 27)
Includes bibliographical references and index.
ISBN-10: 0-387-25407-2 (alk. paper)
ISBN 13: 978-0387-25407-4 (HC)
ISBN-10: 0-387-25408-0 (SC)
ISBN-13: 978-0387-25408-0 (SC),
E-ISBN-10: 0-387-25409-9
E-ISBN-13: 978-0387-25409-8
1. Agrobiodiversity. 2. Agricultural biotechnology. 3. Agrobiodiversity—Economic aspects. 4. Agricultural biotechnology—Economic aspects. I. Cooper, Joseph. II. Lipper, Leslie. III. Zilberman, David, 1947-IV. Series.

S494.5.A43A43 2005
333.95—dc22 2005049009

Printed in the United States of America.

9 8 7 6 5 4 3 2 1 SPIN 11395324

springeronline.com

Contents

Foreword

The topics addressed in this book are of vital importance to the survival of humankind. Agricultural biodiversity, encompassing genetic diversity as well as human knowledge, is the base upon which agricultural production has been built, and protecting this resource is critical to ensuring the capacity of current and future generations to adapt to unforeseen challenges. Agricultural biodiversity underpins the productivity of all agricultural systems and is particularly important for poor and food-insecure farmers, who maintain highly diverse production systems in response to the marginal and risky production conditions they operate under. Understanding the importance of agricultural biodiversity in the livelihoods of the food insecure and enhancing its performance through the use of a variety of tools, including biotechnology, is a critically important issue in the world today, where over 800 million people have insufficient food to meet minimum needs. A strong theme that runs throughout the book is the importance of good public policy interventions to promote the provision of public goods associated with agricultural biodiversity conservation and directing biotechnology development to meet the needs of the poor. The book's primary innovation is that it describes the relationship between biotechnology and plant genetic diversity and puts these in the context of agricultural development. Both the conservation of plant genetic diversity and agricultural biotechnology have received extensive examination, but thc linkages between the two have not, despite the apparently obvious relationship between the two. Biotechnologies, which cover a wide range of techniques and products, represent a valuable new tool for utilizing genetic resources. If applied with due precaution and risk analysis, they can increase the value of maintaining genetic diversity by reducing uncertainty about the characteristics and values of genetic resources. Biotechnology allows greater precision in the human manipulation of plant genetic resources, and even transfers of individual traits between species. However there are several potential risks associated with this technology and its application in agricultural development, which form an important part of the analysis presented in this volume.

One controvcrsy addressed in the book is the potential of genetically modified organisms (GMOs) to benefit poor agricultural producers in developing countries. Various aspects of the debate are covered in this book, many of which were also discussed in the 2004 State of Food and Agriculture Report, which focused on the potential of biotechnology to meet the needs of the poor. Concerns about the risks associated with the technology are both technical and socioeconomic. There is uncertainty about the long-term impact of releasing transgenic species into existing gene pools and concern that irreversible and ultimately negative impacts may ensue. This uncertainty gives rise to the need for biosafety regulations which can be expensive and difficult to implement, particularly in developing countries with limited regulatory capacity. Increased privatization and concentration of agricultural research associated with the development of biotechnologies has also been raised as a potential problem, with fears of a loss of control of genetic resources on the part of farmers and developing countries. Against these concerns is weighed the evidence that transgenics can provide an effective means of addressing some of the most difficult and persistent problems in increasing agricultural productivity in developing countries, as well as a means of significantly reducing other environmental problems, especially those associated with pesticide use. Another controversial issue raised in this book is how best to approach the *in situ* conservation of plant genetic resources for food and agriculture. FAO estimates that about three-quarters of the genetic diversity found in agricultural crops have been lost over the last century. Of 6,300 animal breeds, 1,350 are endangered or already extinct (Scherf, 2000).

This rapidly diminishing gene pool is cause for great concern and a pressing need to design effective conservation strategies. Defining what should be conserved, its value to various groups in society as well as future generations, how much conservation is needed, where it should take place, and the most effective means of attaining it, are all controversial topics which are being debated today, and which several chapters in this book shed light on. A key issue raised is the relationship between *in situ* conservation and agricultural development. At present, the primary providers of *in situ* conservation are developing country farmers located in areas of high native diversity and who, in many cases, do not have the opportunity to adopt more homogenous utilization patterns of crop genetic resources because no suitable modern varieties have been developed to meet their conditions. These producers are likely to be the least cost sources of *in situ* conservation at present. However, is it fair and appropriate to rely upon their lack of access to improved genetic materials to provide cheap conservation in the future—at the expense of their own potential for productivity increases and improvements in their welfare? Developing strategies, which rely on diversity to achieve productivity and livelihood improvements, is one way to avoid this dilemma. Possible candidates for such strategies have been identified in the book; examples include broadening the genetic base of modern breeding programs, participatory plant breeding, and using biotechnology to insert important traits into traditional varieties.

A third controversial issue the book covers is the justification for, and means of, sharing benefits from the utilization of plant genetic resources for food and agriculture. Questions such as: the amount of compensation due to farm communities that have preserved traditional varieties, which embody the crop genetic resources upon which new varieties are built and sold for profit, are addressed in several chapters. The kinds of mechanisms can be established to facilitate benefit-sharing without reducing access to the resources or incentives to develop new means of utilizing them is another important issue examined. The International Treaty on Plant Genetic Resources, which came into force in June, 2004, is an important step forward in resolving these questions. The Treaty brings governments, farmers, and plant breeders together and offers a multilateral framework for accessing genetic resources and sharing the benefits derived from them. Farmers' rights are an important concept under the Treaty, recognizing that farmers around the world, particularly those in low-income countries, have developed and conserved plant genetic resources over the millennia. There are moral, political, and economic justifications for rewarding farm communities that provide *in situ* conservation. These rights do not conflict with other forms of intellectual property rights such as patents; rather, they are complementary. There are still many issues to resolve in the design of mechanisms to implement the Treaty, many of which have been raised and addressed in this volume.

One single book cannot give a satisfactory reply to all these questions, but the authors in this book have raised a wide range of provocative thoughts and proposed potential solutions and ways to move forward on several aspects of these important questions. Stimulating the ongoing debate on these topics is a critical part of identifying solutions, and this is our key purpose in promoting efforts such as this book.

HARTWIG DE HAEN
Assistant Director General
Economic and Social Department
Food and Agriculture Organization of the UN

Scherf, B., 2000, *World Watch List for Domestic Animal Diversity*, Part 1.9 http://dad.fao.org/en/Home.htm - databases. FAO/UNDP, 3rd ed., Rome, Italy.

Acknowledgment

This book is an outcome of a FAO research project on agricultural biotechnology and biodiversity that was initiated by Joseph Cooper and managed under his leadership while at FAO, and then by Leslie Lipper after his departure. This book would not have been possible without the continuing support of Kostas Stamoulis, the Chief of ESAE Service at FAO, for whose patience the editors of this book are very grateful. The editors gratefully acknowledge the support and input of Hartwig de Haen, Prahbu Pingali, Willie Meyers of the FAO, Robbin Shoemaker, Kitty Smith, and Keith Wiebe of ERS. We would like to acknowledge the encouragement and support we received from Vittorio Santaniello, of the University of Rome "Tor Vergata," and the input and inspiration we got from the members of the International Consortium on Agricultural Biotechnology Research (ICABR) during their meetings in Rome and Ravello.

We would like to acknowledge the financial support of the Department of Agricultural and Resource Economics and the Giannini Foundation at the University of California at Berkeley, and thank wholeheartedly Laurie Lyser for her assistance in formatting the chapters and Amor Nolan for her tireless editorial effort.

Disclaimer

The views presented herein as those of the authors, and do not necessarily represent the views of the Economic Research Service, the United States Department of Agriculture, or the Food and Agriculture Organization of the United Nations.

PART I.
Overview and Key Issues

Chapter 1

INTRODUCTION~ AGRICULTURAL BIODIVERSITY AND BIOTECHNOLOGY: *ECONOMIC ISSUES AND FRAMEWORK FOR ANALYSIS*[*]

Joseph C. Cooper,[1] Leslie Lipper,[2] and David Zilberman[3]

[1]*Deputy Director for Staff Analysis, Resource Economics Division, Economic Research Service (United States Department of Agriculture), 1800 M Street, NW, Washington, DC 20036-5831;* [2]*Economist, Agricultural and Development Economic Analysis Division, Food and Agriculture Organization of the U.N., Viale delle Terme di Caracalla 00100, Rome, Italy;* [3]*Professor, Department of Agricultural and Resource Economics, University of California, 207 Giannini Hall, University of California, Berkeley, CA 94720*

Abstract: This chapter provides an overview of the book, *Agricultural Biodiversity and Biotechnology: Economic Issues and Framework for Analysis*. The book presents the results of three years of collaborative research in which the authors aimed to develop a coherent and economics-based approach to policy-making in the management of biotechnology and biodiversity. Namely, it explores the economics of both the conservation of plant genetic resources for food and agriculture and the adoption of molecular biotechnology, the economics of whether or not their respective policies should be linked, and, if so, how. This book begins with a section containing chapters overviewing the global setting in which the management of biotechnology and biodiversity are taking place, including an analysis of major socioeconomic trends and institutional developments and their potential impacts. The next section provides an analysis of the current and potential value of biotechnology in developing countries and the types of institutional reforms needed to realize this potential. The book is then concluded with a summary chapter that integrates the policy implications drawn from earlier sections on biodiversity and biotechnology in the context of development.

Key words: adoption; agriculture; biodiversity; biotechnology; conservation; developing countries; economic analysis; plant genetic resources; policy implications.

---

* The views contained herein are those of the authors and do not necessarily represent policies or views of the Economic Research Service or United States Department of Agriculture.

1. OBJECTIVES AND CONTRIBUTIONS

The emergence of biotechnology has expanded the human capacity to take advantage of genetic resources and manipulate biological material to obtain food, medicine, and other valuable substances. Biotechnology products generally contain a large intellectual component requiring significant up-front investment, and they have a highly valuable commercial potential, which has created an impetus for the privatization of the knowledge input to their production. The establishment of intellectual property rights (IPRs) for knowledge about biological processes and properties has increased the value and importance of maintaining biodiversity, as genetic resources are a key input to biotechnology production. The management of genetic resources and biotechnology has created new policy challenges in the attempt to attain a socially optimal allocation of costs and benefits between the public and private sectors.

The gene revolution originated in the developed world, but much of its promise may lie in addressing production and consumption problems in developing countries, with significant potential for alleviating hunger and poverty. However, there is a great deal of criticism and uncertainty about the capacity of the current institutional framework governing access to biotechnology to facilitate the transfer of technologies controlled by the private sector in developed countries to benefit a broad range of producers and consumers in developing countries. At the same time, most of the world's biodiversity resources are located in developing countries. Thus, strategies for their conservation and utilization in sustainable economic development need to be considered in the context of generating equitable access to the benefits from the management and development of genetic resources, as well as the need for efficient approaches to their conservation under conditions where economic development is imperative. The study of the policy nexus of managing biodiversity and biotechnology, especially within the context of the developing world, is an intellectual challenge which is highly relevant to current policy debates. This books aims to provide a state-of-the-art summary of knowledge and policy debate in this critical area.

The book presents the results of three years of collaborative research in which the authors aimed to develop a coherent and economics-based approach to policymaking in the management of biotechnology and biodiversity. Leading experts in various aspects of this policy debate were asked to contribute chapters on specific issues to which they could apply their unique expertise. By integrating the continuous effort of the editors with the insight of the other authors, we hope to have a fluid augmentation that is rich with insight and unique knowledge.

Target audiences for this book include agricultural economists who are working on technology and resource management issues, and especially on biotechnology and biodiversity, development economists addressing issues of resources and agricultural sector in developing countries, and environmental and resource economists. These individuals may be in academia, in government, in nongovernmental organizations, and in private companies. Another target audience is policy scholars in government, schools of public policy or schools of environment that are interested in issues of biotechnology policy, IPRs, and biodiversity, as well as the interaction between developed and developing nations regarding these issues.

A third target audience is scholars in both development studies and resource management studies. By largely de-emphasizing technical presentation in the main text and emphasizing conceptual and policy issues, we believe that we will reach scholars whose aim is to analyze these major issues of development and resource management without heavy emphasis on economics.

Finally, interest in these topics presented in the book is strong among scholars and policymakers both in the developed and developing world, and in international organizations such as The World Bank and United Nations agencies. Thus, policymakers throughout the world who are addressing the issues of this book are an important target audience.

While most of the chapters rely upon economic analysis and tools, most of the book is written in a manner that aims to reach a broad range of experts interested in the topic, including noneconomists. It also contains contributions by noneconomists who are experts in biotechnology and biodiversity. It aims to familiarize the reader with some of the major debates, policy options associated with management of biotechnology and biodiversity in the developing world, and conceptual approaches that aim to identify policies and management schemes that will lead to strategies that will improve social welfare and reduce poverty.

This book begins with a section containing chapters overviewing the global setting in which the management of biotechnology and biodiversity are taking place, including an analysis of major socioeconomic trends and institutional developments and their potential impacts. This section is followed by one containing chapters summarizing the major issues in the management of agricultural and wild biodiversity, including valuation and incentives for conservation. Equity concerns and their implications for the distribution of costs and benefits associated with the conservation and use of genetic resources are the subject of the chapters in the following section.

The next section provides an analysis of the current and potential value of biotechnology in developing countries and the types of institutional reforms needed to realize this potential. The book is then concluded with a summary chapter, which integrates the policy implications drawn from earlier sections on biodiversity and biotechnology in the context of development. In the remainder of this chapter, we provide a general introduction to the links between agricultural biodiversity and biotechnology, based upon issues and findings of the chapters as summarized below.

2. LINKAGES BETWEEN AGRICULTURAL BIODIVERSITY AND BIOTECHNOLOGY: OVERVIEW

Between 1961 and 1999, global per capita cereal production increased by 22% while total acreage devoted to cereals increased by only 4.9%.[4] This increase in productivity is partially attributable to an increase in fertilizer, pesticide, and water use. However, in a recent study, Evenson and Gollin (2003) show that the development and adoption of improved genetic materials were a significant and large part of the increase in agricultural productivity over this period. They estimate that between 1961 and 1980, 21% of the growth in yields in food production in developing countries was attributable to the adoption of modern varieties among farmers, as was 50% of the yield growth experienced between 1981 and 2000. Modern production systems are frequently characterized by their domination by monoculture, the adoption of which can lead to decreased genetic diversity, at least by some measures of diversity. The loss of genetic diversity generates costs in terms of reduced resilience of farming systems and reduced options for future crop and variety development.

The concern over the erosion of genetic resources may be linked in part to the increasing globalization of the economy, which has created pressures and conditions for the increasing intensification of agriculture, leading to the adoption of modern plant varieties around the world and, in turn, possible loss of traditional plant varieties. At the same time, some developing countries perceive that major international corporations primarily from developed countries are likely to earn much income through the utilization of genetic materials that have been conserved mainly by farmers in developing countries. The desire to maintain national sovereignty over their genetic resources has led to at least a dozen countries establishing controls over access to their genetic resources, and an equal number of nations developing such controls.

Enough international concern has developed over the need to conserve agricultural genetic resources to lead to the establishment of a multilateral system of access and benefit-sharing for key crops, the International Treaty

---

[4] "Cereals" include wheat, rice, barley, maize, millet, sorghum, as well as other grains. These figures are derived from the FAOSTAT database (www.fao.org).

on Plant Genetic Resources for Food and Agriculture—hereafter denoted as the International Treaty. The International Treaty is considered a major step towards guaranteeing the future availability of the diversity of plant genetic resources for food and agriculture (PGRFAs) on which farmers and breeders depend, as well as a fair and equitable sharing of benefits. This treaty entered into force on June 29, 2004.

Of course, PGRFAs are the basic biological input into the breeding of new crop varieties.[5] Molecular biotechnology is increasingly at the forefront of modern crop breeding techniques. However, as applying such techniques is costly, modern crop varieties tend to be produced with developed country conditions and markets in mind, thereby limiting the extent of their relevance to developing country conditions and likely adoption rates. This situation is unfortunate as biotechnology can potentially be of great use to developing countries in helping them meet the demands of feeding their populations. At the same time, to the extent that farmers in developing countries are adopting modern crop varieties, their adoption may be coming at the expense of traditional farmers' varieties, or landraces, concentrations of which tend to be in developing countries. In the process, it is possible that PGRFAs of potential future value in crop breeding may be lost. However, as the cost of biotechnology applications fall, and consequently, biotechnology transfers to developing countries increase, agricultural biodiversity in these countries could be increasingly threatened. In sum, PGRFA conservation and the promotion of biotechnology applications in developing countries may be strongly linked. If so, policy mechanisms addressing each would be more efficient if they were linked. This book explores the economics of both the conservation of PGRFAs and adoption of molecular biotechnology and the economics of whether or not their respective policies should be linked and, if so, how.

3. OVERVIEW OF THIS BOOK

This section provides a summary of the contents of the rest of the chapters in this book. Chapter 2 is an overview of the processes of globalization (particularly trade liberalization), environmentalism, consumerism, and the rise of the information economy, all of which are key factors that shape the evolution of agriculture, biotechnology, and biodiversity.

---

[5] PGRFAs consist of the diversity of genetic materials contained in all domestic cultivars as well as wild plant relatives and other wild plant species and plant matter (germplasm) that are used in the breeding of new varieties—either through traditional breeding or through modern biotechnology techniques.

Chapter 3 covers the evolution of plant improvement research, focusing on changes in the research process from the "green revolution" to the "gene revolution." Among the chapter's key points are that the green revolution was largely based in the public sector and involved crops and varieties that were suited for developing countries with highly productive farming areas. Varietal adoption patterns were also very much conditioned upon the presence of local breeding capacity. With the subsequent gene revolution, agricultural research and development (R&D) are now largely based in the private sector. The result of a shift in center of research innovation from the public to the private sector is that the focus of R&D will be on seeds or varieties with significant commercial value, which tend not to be seeds or varieties adopted for specific developing country conditions. Given the focus of biotechnology applications on varieties intended for profitable developed country conditions, the ability of developing countries to benefit from biotechnology will depend on their local breeding capacity. Of three breeding options examined in the chapter—local breeding with local varieties, regional breeding with adapted varieties, and adoption of seeds produced elsewhere—the first option is most expensive and last is the least. However, the first option is more likely to produce higher benefits in terms of biotechnology adoption as well as biodiversity conservation, as local varieties will be used in breeding.

Part II of this book covers valuation and conservation issues for genetic resources and biodiversity. Chapter 4 discusses the economic value of maintaining crop diversity as insurance against vulnerability to disease and pests. Based on an empirical assessment of the change in welfare resulting from a marginal change in number of potential parents, the author finds little value overall in maintaining a large number of potential parents in breeding lines. On the other hand, while noting that this chapter does not cover exactly the same subjects as Chapters 5, 6, and 7, these latter chapters argue that *in situ* conservation is an important means of conserving a valuable aspect of plant genetic diversity: The evolutionary process which occurs as a result of both human and natural selection pressures.

Chapter 5 tackles the economic incentives for conserving crop genetic diversity on farms. The chapter starts off with an assessment of the market failure that arises from the public good nature of *in situ* conservation, in which farmers bear the cost of conservation but perhaps a small share of the benefits to society of such conservation. The chapter argues that there is a greater harmony between public and private values in terms of managing biodiversity for reduction in vulnerability to pests and diseases (i.e., a form of portfolio diversification at the farm level), but not for reducing genetic erosion, which has public good aspects. Rural populations depend to some extent on diversity in the genetic base, particularly in areas with isolated markets, as a form of insurance. However, such a dependence is not necessarily sufficient to promoting socially optimal levels of *in situ* conservation. Policies to promote conservation *in situ* include promotion of demand for products of diverse (landrace) varieties, e.g., building niche markets, labeling, and raising public awareness. Other methods include changes in plant-breeding methods, such as participatory plant breeding, community seed banks, seed registers, and protection of farmers' varieties through "farmers' rights." The argument made in this chapter—that *in situ* conservation is cheapest where opportunity costs associated with the adoption of modern varieties are highest—also comes out in other conservation chapters in this section.

Chapter 6 focuses on *in situ* conservation methods and their costs, but in a less micro–oriented fashion than the previous chapter. Like Chapters 5 and 7, Chapter 6 asserts that the cheapest means of promoting *in situ* conservation is to look for such conservation situations with the lowest opportunity costs for maintaining diversity *in situ*. However, the chapter provides more analysis of the processes required to keep conservation incentives in place even while promoting economic development, given that the cheapest conservation possibilities tend to be in areas with relatively low levels of economic development. The chapter differs from all other chapters on *in situ* conservation in that it addresses the consequences of having more PGRFA conservation than is optimal for society.

Chapter 7 uses empirical evidence from Mexico to investigate the factors driving on-farm diversity in PGRFAs. It argues that we need better information on what factors determine the selection of particular varieties for adoption by farmers, what impacts the process of selection has on genetic populations and what (e.g., trading networks, markets, seed exchange networks) determines genetic flows in and out of PGRFA populations. Of all the chapters that address conservation issues, this chapter goes into the most detail about how human selection of PGRFAs interacts with natural selection in determining patterns of diversity.

Chapter 8 talks about the backbone of all regional and international collaboration for PGRFA conservation, namely, the presence of reliable national conservation programs. International funding does not remove the need for domestic funding. Emphasis must be on measures that improve the efficiency of conservation, and measures to be targeted for improvement include regional and international collaboration, data and information management, and over-duplication of samples. For example, conservation efficiency can be raised through the creation of a multilaterally accessible database with information on the *ex situ* and *in situ* germplasms that are available in the regions from which the germplasm is drawn.

The author of the chapter asserts that a final prerequisite for any collaboration on the regional or international level is the maintenance of national sovereignty of those countries involved. Namely, only with their sovereign rights maintained over materials such as germplasm are countries willing to place such materials in secure storage facilities outside their borders.

Part III of this book covers distributional issues in the management of genetic resources. Chapter 9 discusses the sharing of benefits derived from the utilization of PGRFAs in the breeding of new varieties. From an economic efficiency as well as equity standpoint, it seems reasonable to tie a country's contribution to a benefit-sharing fund (such as that envisioned under the auspices of the new International Treaty) to the benefits it receives from its use of PGRFAs. Every country benefits from utilization of PGRFAs in the production of new goods, but some countries may benefit more than others. Unfortunately, as discussed in this chapter, these benefits cannot be quantified, except perhaps in limited case studies. Hence, given that political considerations dictate that a benefit-sharing fund be created, an alternative can be to appeal to indicators that take equity and development considerations into account in determining contributions, and that acknowledge at least some of the characteristics of the benefits of PGRFAs. Thirteen potentially feasible indicators are examined in this chapter. All the feasible indicators are deficient in some way.

While Chapter 9 discusses who should contribute to a benefit-sharing fund, and how much, Chapter 10 examines potential economic criteria for distributing money from the conservation fund for the conservation and sustainable development of plant genetic resources. However, benefits accruing from the distribution of these funds for conservation activities are almost impossible to ascertain. The question then becomes what is the most economically efficient method of distributing the funds among countries or throughout the world, given the available data. This chapter describes a proxy indicator for the importance of a region as a primary center of diversity. It then goes on to rank their importance to the global community and to OECD countries based upon the consumption of crops originating from various centers of diversity.

Chapter 11 extends the institutional discussion in the previous chapter with a demonstration of how cooperative game theory can be applied to determining the "fair and equitable sharing" of the benefits arising from the use of PGRFAs, using as a starting point the regional allocations from the previous chapter. Using this approach, the impacts of the players' (e.g., countries') bargaining power on the resulting allocations can be empirically assessed. Furthermore, the approach allows us to explicitly account for potentially competing interests of the players, thereby introducing some equity to the allocation. The implications of three different allocation regimes are modeled. One of these assumes that funds will be distributed by the International Treaty only to world regions, which then will be responsible for allocating the funds within their regions. This scenario was found to be particularly appropriate as part of a flexible mechanism for biodiversity conservation as it allows the use of different types of control mechanisms at different levels of negotiation processes.

Part IV of this book includes chapters that address biotechnology concepts, economic valuation of biotechnology, and management of biotechnology production and processes. Chapter 12 gives an overview of evolution of agricultural biotechnology concepts and applications. While this chapter provides a conceptual overview of the present state of biotechnology applications to agriculture, the first section of Chapter 13 provides a technical overview that includes examples of specific products.

Specifically, Chapter 13 provides a relatively detailed summary of existing biotechnology applications and provides details on second-generation biotechnologies that are being developed and that may be of relevance to developing countries. The chapter also presents data on the adoption of genetically modified organisms (GMOs) in developing countries. It also discusses GM products that are further down the production pipeline, and does so by country, crop, and trait and, for livestock, by country, species, and trait.

Chapter 14 examines how differing IPR regimes, states of development of the seed industry, and agricultural R&D capacities will affect the nature of biotechnology adoption in developing countries. For example, with strong IPRs, a strong breeding sector, but high transaction costs in trading IPRs, the most likely outcome is that the biotechnology company will directly introduce GM varieties that are not locally adapted, resulting in a loss of *in situ* diversity in PGRFAs. With weak IPRs and a strong breeding sector, every breeder or seed company can use commercialized GM varieties in order to cross-breed the technology into their own germplasm. Thus, many different GM varieties will be available on the market, although in the long run there may be less access to technology, due to developer's inability to capture rents.

Chapter 15 utilizes the example of biotechnological innovation in the global canola sector to identify some lessons for how developing countries might participate and benefit from this innovation. Developing countries are facing ever-rising technical, economic, and political barriers that limit their capacity to use biotechnology in their fight against hunger. Developing countries require functioning economic markets, physical and scientific infrastructure, and political and legal capacities. In many cases, these countries will need to create the appropriate input and output market conditions for the new technology to be disbursed. Firms will only go where there is supporting infrastructure, research collaborators, functioning labor markets, competent regulators, and markets that are accepting of GM products. Some developing countries, such as China, India, and Brazil, have the prospect of assembling institutions adequate to promoting adoption of GM crops.

Like the previous chapter, Chapter 16 addresses the economics of the adoption of biotechnology and the constraints to its adoption, but does so from the farm level. The chapter argues that divisible technologies that are simple to use and that have limited fixed costs (e.g., GM seed varieties and tissue culture technologies) hold the most promise for adoption by small, poor farmers. The fact that biotechnology varieties do not require high inputs of human capital—in fact, they often result in reduced management requirements—also means they may be well suited for adoption among low-income farmers. Nevertheless, adoption may be constrained by several farm-level factors including farm size, agroecological conditions, availability of credit, and risk. These factors have been shown to be important in the adoption decisions among smallholders in developing countries. Adoption levels also depend on macro-level factors including a country's research capacity and characteristics of its input and output markets. China has shown the greatest success with GM crop adoption—where farmers have benefited instead of foreign firms due to a combination of weak IPRs and significant government involvement in biotechnology research. In Latin America, a growing gap between small, poor farmers and large multinational cooperation, as well as negative public perception of GM crops, continue to be significant constraints to farmer adoption of GM crops.

Part V, the final section, draws policy implications by identifying and expounding on the themes that cut across the biodiversity, biotechnology, and development issues raised in this book. Chapter 17 examines the potential of biotechnology for poverty alleviation and is the only chapter to consider indirect impacts (via labor markets and food prices) of biotechnology adoption on the poor. The chapter raises questions about biotechnology as a means of poverty alleviation: (1) Do faster and cheaper means of economic development exist than through agricultural technology change; (2) are there faster and cheaper means of agricultural technology change than through biotechnology; (3) do many market failures (e.g., in the provision of credit) exist that may prevent agricultural biotechnology from being effective; and (4) do other basic needs of the poor need to be addressed before biotechnology adoption would be effective? After raising these questions, the chapter provides many policy recommendations for how to get biotechnology to work as a tool for poverty alleviation.

Chapter 18 describes a possible mechanism for reconciling the economic tension that exists between the public and private economic forces that drive agricultural research. That mechanism is the establishment of an intellectual property clearinghouse for agricultural biotechnology. This clearinghouse would provide three essential functions: (1) identification of all relevant intellectual property that exists over a given technology and what properties are available and how they could be accessed; (2) the establishment of a pricing scheme and terms of contract that depend on the identity of the buyer; and (3) the establishment of an arbitration mechanism for monitoring and enforcement of the contracts made through the clearinghouse. The purpose of the clearinghouse would be to reduce market failures in agricultural biotechnology markets. It would also increase access to agricultural biotechnology in the National Agricultural Research Systems (NARS) in the developing countries, the Consultative Group on International Agricultural Research (CGIAR) system, universities, and, ultimately, farmers in developing countries.

Chapter 19 picks up on and amplifies policy themes in agricultural biodiversity conservation and sustainable use that were raised in earlier chapters. The chapter discusses the effectiveness of various types of payment mechanisms for conservation. It identifies the wide range of actors who are, or potentially could become, involved in conservation through the use of a wide range of mechanisms that go well beyond the traditional concepts of conservation activities. A key theme throughout the discussion in the chapter is the importance of recognizing human knowledge as a key component of agricultural biodiversity and the necessity of incorporating means for knowledge preservation as much as the physical conservation of agricultural biodiversity. Chapter 20 provides a detailed history and description of the International Treaty on Plant Genetic Resources for Food and Agriculture. Finally, Chapter 21 is a synthesis that attempts to identify and reconcile the common themes across the chapters and draws some major economic conclusions and policy recommendations from this synthesis.

REFERENCES

Evenson, R. E., and Gollin, D., 2003, Assessing the impact of the Green Revolution, 1960 to 2000, *Science* **300**:758.

Chapter 2

MAJOR PROCESSES SHAPING THE EVOLUTION OF AGRICULTURE, BIOTECHNOLOGY, AND BIODIVERSITY

David Zilberman[1] and Leslie Lipper[2]
[1]Professor, Agricultural and Resource Economics, 207 Giannini Hall, University of California, Berkeley, CA, 94720; [2]Economist, Agricultural and Development Economic Analysis Division, Food and Agriculture Organization of the U.N., Viale delle Terme di Caracalla 00100, Rome, Italy

Abstract: The paper identifies five major global trends that are likely to impact agricultural biodiversity conservation and the adoption of agricultural biotechnologies. The trends covered include trade and capital market liberalization, the rise of the environmental movement, consumerism, privatization and devolution of government services, and the emergence of the information age. We find that trade liberalization is likely to lead to increased incentives and capacity for biotechnology adoption, with unclear but potentially negative impacts on agricultural biodiversity. Environmentalism has generated a system of environmental governance and regulation, which may come into conflict with those established under global trade agreements. However, the way in which these disputes will be resolved is still unclear, but it will likely have important implications for both agricultural biotechnology and biodiversity. The rise in consumer power associated with increased incomes and the expansion of markets will affect biotechnology adoption through two opposing effects: the expression of consumer concerns about environmental and food safety, balanced against the delivery of quality characteristics that biotechnology can deliver. Privatization in the agricultural research and development sector increases incentives for the development of agricultural biotechnologies, but may create barriers to their adoption in developing countries, while the privatization of environmental services generates increased incentives for biodiversity conservation. Rapid improvements in information technologies increase the capacity for effective biodiversity conservation and are fundamental components of the development of biotechnologies.

Key words: agricultural biodiversity; agricultural biotechnology; environmental treaties, globalization, information technologies; privatization.

Over the past 20 years, several global trends have been unfolding which have implications for the evolution of agricultural biotechnology and the conservation and sustainable use of agricultural biodiversity. These trends are interlinked and in some cases have opposing effects, and their final outcomes are yet to be determined. In this chapter we provide a short survey of these developments together with an analysis of their potential implications for the use of agricultural biotechnology and the management of agricultural biodiversity. The trends covered include trade and capital market liberalization, the rise of the environmental movement, consumerism, the privatization and devolution of government services, and the emergence of the information age.

Both biotechnology and the concept of biodiversity are fairly recent arrivals onto the human scene, and their management has raised several controversies. For example, biotechnology is a product that is comprised of a large intellectual component, e.g., it represents the culmination of a process of research. This research has mostly been carried out in the private sector, although it also often involves the use of genetic resources which originated in the public domain. There is considerable disagreement on the best means of protecting the property rights to the intellectual component embodied in biotechnology, while recognizing both the private and public contributions to the end product. In addition, agricultural biotechnology products are the result of a major scientific advance and have only very recently become available. Due to their novelty, there is only limited information on the long-run impacts they might have on environmental and food safety. A great deal of uncertainty exists on how much risk such products entail, as well as much controversy on how it should be measured and how much is socially acceptable.

Considerable uncertainty and conflict exist over the conservation of agricultural and wild biodiversity as well. Assigning values to biodiversity conservation is fraught with uncertainty. One of the most significant values associated with biodiversity is preserving potential future options for the use of the genetic resources maintained—and this is very difficult to assign value to. There is even considerable uncertainty with determining the use values of agricultural biodiversity, which ostensibly is easier to measure.

Uncertainty over values leads to controversy over conservation strategies: how much and what should be preserved. Controversy is particularly sharp when conservation conflicts with economic development (see Chapter 19).

These controversies are currently under discussion and negotiation in a variety of formal and informal forums, and they are being shaped by the global trends, which we identified in the first paragraph. In the discussion which follows below, we discuss how these global processes are shaping the ongoing debates in various contexts and draw conclusions as to their potential implications for the management and use of agricultural biotechnology and biodiversity in developing countries. Our discussion is kept to a fairly general level, which does not fully capture the tremendous variation that exists among developing countries in terms of their endowments and capacities. More specific analyses related to the management of biotechnology and agricultural biodiversity in the varied context of developing countries are given in later chapters of this book.

1. GLOBALIZATION OF TRADE AND CAPITAL MARKETS

Over the last 20 years, the volume of trade between countries has expanded remarkably as a result of the reduction of trade barriers, as well as decreasing costs in transport and communications and the increased mobility of capital across international boundaries. International and regional trade agreements have been the primary mechanism by which trade barriers have been lowered, such as the General Agreement on Trade and Tariffs (GATT), and subsequently the World Trade Organization (WTO) at the global level, and North American Free Trade Agreement (NAFTA), the European Community, and MERCOSUR[3] as examples of regional blocs.

Liberalization has also occurred in agricultural trade markets, although this is one of the most contentious areas of international trade policy and one where significant distortions still exist, particularly among developed countries. Indeed it was deadlock over agricultural trade which caused the breakdown of negotiations at the 2003 WTO meeting in Cancun. Nonetheless, there has been considerable movement towards the liberalization of agricultural trade markets, and more is expected in the future. In the United States, there is a move towards converting commodity support programs towards "green payments," e.g., paying for environmental services. In Europe, the expansion of the European Union is creating pressures to reform the Common Agricultural Policy (CAP) and reduce production supports. Farmers are increasingly expected to rely on insurance instruments provided by the private sector and sometimes subsidized by the government for the management of production and revenue risk. Future markets and forward contracts are also likely to play a major role in reducing risk in agriculture.

In basic grain markets, the impact has been a shift in production from high cost to a few lower cost producers such as the United States, Argentina, and Australia, as well as Thailand and Vietnam. At the same time many developing countries as well as the transition economies of Eastern Europe have become net importers of grain, and this trend is expected to continue with liberalization (Bruinsma, 2003).

---

[3] El Mercado Común del Sur, includes Brazil, Argentina, Paraguay, and Uruguay.

If indeed agricultural support prices in developed countries are reduced, producer prices for some agricultural commodities are likely to increase in developing countries and new market opportunities created. One impact of these changes may be increased incentives for the adoption of new yield-increasing biotechnologies. Agricultural trade liberalization increases competitive pressures among producers and creates incentives for increasing yields and reducing costs in agriculture. It also exposes producers to the demand requirements of a larger group of consumers. This may expand the demand for both yield-increasing and pest-controlling biotechnology products. For example, the ability to export to markets in Japan, Canada, and other countries may be determined by the ability to control pest problems with minimal or no chemical residues. Concern about ozone depletion is leading to regulations banning the use of methyl bromide and other chemicals. These measures provide increased incentives for the adoption of pest-controlling biotechnology products.

By reducing investment barriers, trade liberalization creates the potential for investors such as multinational companies to invest in both production and marketing infrastructure in developing countries with promising commercial market potential, or which establish incentives to attract mobile capital. Profound changes are occurring in the organization of the food sector in developing countries due to globalization, as well as urbanization, increasing incomes, and the opportunity costs of food purchasing and preparation. The rise of multinational retail chains, supermarkets, fast food chains, and other forms of pre-prepared foods are manifestations of this change. The developments in the structure of food markets raise challenges and opportunities for local and global suppliers, and have implications for both agricultural biotechnology and biodiversity.

On the one hand, food producers can potentially take advantage of the income-earning opportunities created in a dynamic and rapidly expanding market. This could increase the demands for agricultural biotechnology and incentives to adopt among producers in order to remain competitive. On the other hand, small producers unable to adapt to the required institutional and organizational changes, and the technology and management requirements that they entail, risk marginalization in terms of market participation. Some evidence of this trend is available with concentration in the food supply chain linked to increased farm consolidation and reduced market participation among small producers (Reardon et al., 2003; Berdegué et al., 2003). It is not clear what impact this will have on either agricultural diversity or biotechnology, although it is likely to lead to a higher demand for biotechnology products from both the commercial farm sector and the food processing industries, but will reduce demand from small farmers

While agricultural trade liberalization may result in increased incentives for producers in developing countries to adopt agricultural biotechnology, the extent to which adoption actually will occur depends on the types of innovations biotechnology delivers, and the degree to which these substitute for scarce factors of production and address key production and consumption constraints (see Chapters 13 and 17). At present, agricultural biotechnology innovations are being developed primarily to reduce production costs or increase yields under conditions present in developed countries, which constitute the main market for these products. In many developing countries, production constraints are of a different nature than those in developed countries; barriers to productivity increases are often more related to controlling for the incidence of drought, poor soil quality, and high rates of pest and disease, whereas in developed countries reducing management costs and pesticide use are more important concerns. Trade liberalization may exert some positive influence on the commercial attractiveness of developing innovations to address these needs through its impact on the global demand for inputs; however, this will only apply to technologies that have the potential for a significant commercial market.

Even where technologies are suitable for the production conditions in developing countries, it will be necessary for countries to have in place an adequate level of research, extension, and regulation to achieve dissemination and adoption of such technologies (see Chapters 3, 14, 15, and 16) The institutional requirements are significantly higher and more sophisticated than has been the case in the past for the adoption of improved agricultural technologies. Issues such as biosafety regulation, the negotiation of intellectual property rights (IPRs), and the technological capacity to modify technologies to suit local conditions place fairly significant burdens on the research and development (R&D) infrastructure of developing countries, and the capacity to meet such demands varies widely among them.

An important effect which the liberalization of trade may have on both agricultural biotechnology adoption and the management of biodiversity is the degree to which consumer concerns for the environment and food safety are allowed to be manifested through trade regulations and labeling (Anderson and Nielsen, 2001). The key principle of the WTO is nondiscrimination among member states, e.g., a standardization of product definition and treatment. However, consumer preferences for the environmental and health attributes of agricultural products are heterogeneous across national boundaries and could potentially be manifested in trade regulations. The ability of countries to regulate trade based on environmental and food safety concerns and specifically the degree to which countries may apply their own standards to reflect such concerns are governed by two agreements made under the WTO: the Agreement on Sanitary and Phytosanitary Measures (SPS) and the Agreement on Technical Barriers to Trade (TBT) (Anderson and Nielsen, 2001; Josling, 1999). These agreements allow members to impose restrictions on trade based on environmental and food safety concerns, but they also seek to ensure that such regulations are no more trade restrictive than necessary by imposing restrictions on the use of such "nontariff barriers to trade." In addition, they do not apply to the processes by which agricultural and other goods are produced, but only to the products themselves, which limits the degree to which environmental and food safety concerns can be used to establish trade barriers (Anderson and Nielsen, 2001). Nonetheless, consumer concerns over the environment and food safety could potentially impact the production practices in exporting countries. Ultimately, this impact will depend on the type of specific attributes that are demanded, the willingness to pay among consumers for such attributes, the capacity to distinguish such characteristics in products (e.g., labeling), and the degree to which the expression of such preferences is allowed under the WTO regulations.

The WTO includes another important agreement that has major implications for the dissemination of biotechnology and the management of biodiversity: the Agreement on Trade-Related Aspects of Intellectual Property Rights (TRIPS). The main thrust of this agreement is to facilitate trade in products that have a high intellectual property content. The agreement mandates a minimum standard for IPRs among member states, but leaves them free to determine the appropriate method of implementing them under their own legal system. Article (27.3(b)) of the agreement explicitly refers to the protection of plant varieties and stipulates that new varieties need to be protected either by patents or an "effective sui generis" system such as that of the International Union for the Protection of New Varieties of Plants (UPOV). Under the UPOV system of plant protection, plant breeders' and farmers' rights may be recognized; e.g., breeders have the right to use protected genetic materials in the development of new varieties, and farmers may have the right to save and re-use seeds from protected varieties for their own use (Helfer, 2002). The TRIPS agreement also allows members to exclude from patentability inventions whose use would seriously prejudice the environment. Implementation of this agreement is likely to increase the incentives for the developers of biotechnology innovations to expand into new markets, due to the increased protection it provides for their investment into the technology. Since agricultural biotechnology innovations are being produced mostly by the private sector, this protection is critically important for creating incentives among the suppliers of the technology for its dissemination.

The TRIPS agreement also has implications for the conservation and sustainable use of agricultural biodiversity. Agricultural biodiversity is maintained through systems of access and exchange from the farm to the international level, and property rights to plant genetic resources are likely to effect current patterns of exchange. There are several options for property rights over plant genetic resources, and their impacts on diversity are expected to be varied (Correa, 1999). Property rights and their degree of enforcement are also likely to impact the extent and nature of transgenic crop adoption, which will have implications for both spatial and temporal agricultural diversity (see Chapter 3; also Wright, 1998) The increased value of plant genetic resources as an input to breeding under private breeding programs may lead to increased demand for diversity (see Chapter 19). Concern that the establishment of property rights will lead to reduced levels of access and exchange of plant genetic resources and thus reduced levels of agricultural biodiversity have also been raised (FAO, 1998; Crucible Group, 1994). This includes concerns about the potentially negative impacts on access imposed by farmers' rights mechanisms (Gollin, 1998).

The agreements made under the WTO are not the only international agreements which drive the way the globalization of trade networks proceeds and impacts on biotechnology and biodiversity; there are several environmentally related conventions and agreements which are discussed in the following section and which may have an impact on the ways the WTO agreements are interpreted and implemented. However, the framework laid out under the WTO is the most important in determining what the potential impacts of trade liberalization on biodiversity and biotechnology will be, as this agreement has wide and expanding membership and its signatories include some of the key national players in this arena, which is not the case with many of the environmentally related agreements discussed below.

2. ENVIRONMENTALISM

Environmentalism has arisen from two main motivations: (1) the interest in preserving species, environmental quality, and ecosystems and (2) the concern about environmental and health side effects of agricultural practices. The 1957 publication of Rachel Carson's book, *Silent Spring,* was a major benchmark in the evolution of the environmental movement. It raised awareness about the negative side effects of pesticides and other agricultural practices. Over the last 30years, with a growing availability of information on the incidence and costs of environmental degradation, concerns over the necessity and means for controlling and reversing the process have become manifested in governmental policies from the international to the local level, as well as through activities in civil society. A key thrust of the environmental movement is the promotion of awareness of the nonmarket as well as market values of environmental goods and services and pressures to account for this value through government regulations as well as consumer behavior. Specific manifestations of the impacts of the environmental movement are considered in the next few paragraphs.

2.1 Establishment of environmental protection legislation and agencies

Since the late 1960s, most countries have established national agencies of environmental protection that are at the ministerial level and an increasing body of environmental regulations at all levels of government. However, in many cases the implementation of environmental regulation has been hampered because of political economic constraints of information about the processes that drive environmental degradation and the means to control them. There is a large body of evidence (Damania, Fredriksson, and List, 2003; Deacon 1999) showing that higher income countries attain higher standards of environmental quality and that corruption and flawed governance reduce the effectiveness of environmental policy.

The primary means of environmental regulation have been through the implementation of "command and control" measures, which are fairly blunt and achieve environmental objectives at excessive costs (Oates and Baumol, 1996). However, at present, there is gradual transition to financial incentives (payment for environmental services) and market-based mechanisms (trading in water rights or pollution permits). The regulation of chemical pesticides and drugs consists of strict preregistration testing and "learning by doing" once a product is released. The regulating authorities establish applications, standards, and tests for efficacy and side effects before registration. Products are recalled once a sever defect (carcinogenicity) is detected. The high cost of registration may be a barrier to entry, but it serves to address concerns about product safety and environmental impacts (National Research Council, 2000). Cropper et al. (1992), in an analysis of the regulations of pesticides in the United States, suggest that the Environmental Protection Agency is capable of weighing benefits and costs when regulating environmental hazards; however, the implicit value placed on health risks—$35 million per applicator cancer case avoided—may be considered high by some people. The same regulatory approach is used for genetically modified (GM) varieties. The effectiveness of this regulatory approach depends on quantitative understanding of the processes through which biotechnology affects the environment. For example, concerns about the buildup of pest resistance have led to the establishment of *refugia* requirements (demanding allocation of some land to nonmodified varieties) with *Bacillus thuringiensis* (Bt) cotton. The challenges of establishing and implementing these regulations are apparent from a growing body of literature on their evaluation (Laxaminarayan, 2002). Performance measures are also very difficult to establish for the conservation and sustainable use of agricultural biodiversity. There is uncertainty on the status, measurement, and value of biological diversity, both for wild and agricultural biodiversity.

Chapter 19 the irreversibility of the loss of genetic resources also creates difficulties in assigning performance measures.

Until recently, agricultural biodiversity conservation policies have focused primarily on the *ex situ* preservation of genetic resources associated with economically important crops. At present, the portfolio of policies includes *ex situ* gene banks, the establishment of botanical gardens and experiment stations, and various forms of incentive measures to promote *in situ* conservation. The former are mechanisms for preserving genetic resources, while the latter conserve evolutionary processes and human knowledge in addition to genetic resources.

2.2 International agreements on global environmental problems

Increasing concerns about global environmental problems and the need for international coordination in addressing them have given rise to a proliferation of international agreements. At the U. N. Conference on Environment and Development (UNCED) held in Rio de Janeiro in 1992, a basis was laid for several international agreements in the areas of biodiversity preservation, climate change, desertification control, and others. Of direct relevance to the management of biodiversity and biotechnology is the Convention on Biological Diversity (CBD). The CBD is an intergovernmental convention that entered into force in 1993, which has now been ratified by 180 parties with the aim to achieve three main goals: (1) the conservation of biodiversity; (2) sustainable use of the components of biodiversity, and (3) sharing the benefits arising from the commercial and other utilization of genetic resources in a fair and equitable way.

In January, 2000, a supplementary agreement to the CBD, known as the Cartagena Protocol on Biosafety, was adopted. This agreement seeks to protect biological diversity from the potential risks posed by living modified organisms resulting from modern biotechnology. The two cornerstones of the Protocol are the concepts of Advance Informed Agreement (AIA) and the Precautionary Approach.[4] The AIA enables importing countries to subject all imports of Living Modified Organisms

[4] The CBD website on the Cartagena Protocol describes the Precautionary Approach: "One of the outcomes of the United Nations Conference on Environment and Development in 1992 . . . was the adoption of the Rio Declaration on Environment and Development, which contains 27 principles to underpin sustainable development. One of these principles is Principle 15 which states that 'In order to protect the environment, the precautionary approach shall be widely applied by States according to their capabilities. Where there are threats of serious or irreversible damage, lack of full scientific certainty shall not be used as a reason for postponing cost-effective measures to prevent environmental degradation.'"

(LMO's) to a risk assessment before allowing its entry, and such risk assessments may be made using a precautionary approach. This could have implications for the adoption of agricultural biotechnology, as this agreement could allow countries to block imports of seeds of GM plant varieties in the absence of sufficient scientific evidence about their safety. The agreement entered into force in September, 2003.

It is important to note that the members of the CBD and the Cartagena Protocol differ from the members of the WTO. Notably, the United States has not ratified the CBD (although it is a signatory) and is not a signatory to the CP and, as the primary developer and exporter of GM products, this is likely to have major implications for how these agreements are implemented. How these differences in legally binding commitments among countries will be resolved in international fora is still not clear, and there are attempts to try to harmonize any conflicting provisions (Josling, 1999). It is also not clear how varying standards for risk assessment allowed under the WTO and multilateral environmental agreements will be resolved. This will most likely emerge through dispute resolution and arbitration in international bodies (Anderson and Nielsen, 2001).

2.3 Proliferation of environmental groups in civil society

Public support for the environmental movement has been manifested by the establishment of nongovernmental organizations that emphasize various aspects of environmentalism. Some, like the Nature Conservancy, are engaged in the purchase of valuable environmental resources (mostly land and water), and others (e.g., Greenpeace) are engaged in political activism. Other key players include the World Conservation Union (IUCN) and the World Resources Institute (WRI), which play a role of information provision and policy support, and the World Wildlife Fund (WWF), which is engaged in the implementation of conservation-related projects. A key activity of many environmental groups is educating consumers on the environmental implications of various goods and services offered in the marketplace and the mobilization of pressure from consumers on producers through their purchasing decisions.

Several studies have found that the demand for environmental quality is related to income. The demand for environmental services and goods varies across income groups, with higher income categories being more likely to focus on conservation, while for lower income groups the sustainable utilization of natural resources is a more pressing concern. In developing countries major environmental concerns are related to problems of water quality, waste management, and sanitation, particularly in urban areas, as well as the sustainable use of natural resources in the process of economic development. Countries with higher income levels are more concerned with natural resource preservation, such as the preservation of open space, and the protection of endangered species. In general, concerns about global environmental goods and services, such as biodiversity and climate change, have been driven by developed countries, although there is increasing awareness and concern of the importance of these issues among developing countries.

3. CONSUMERISM

As income increases, consumer rights and preferences for improved quality have become the major determinant of economic activities. In most developed countries, the primary potential for revenue generation is through enhancing the value-added of food products. Indeed, in developed countries, sectors in the agricultural economy (e.g., poultry) that have been able to provide a wider variety of quality choices and extend their product mix have been very successful. Becker (1965) provided a conceptual framework to analyze consumer choices for improved product quality. They suggest that consumers derive enjoyment from the characteristics of market goods that they consume, and that consumption activities may entail some effort. For example, the value of a meal to a consumer may be comprised of the value of its nutritional content, its taste, its safety to consume, and the degree of effort that its preparation requires. Economic factors are a major determinant of food quality preferences. Some characteristics, such as convenience in preparation, exhibit higher elasticities of income. Cultural factors may also influence the values assigned to various food characteristics. Thus, one of the challenges of agricultural industries is to economically produce products that contain the food characteristics desirable in their target markets.

As income in developing countries rises, the demand for improved food quality is likely to increase significantly. In the next 50 years, we expect that vast populations in Asia and South America will reach income levels that will enable them to pursue improved food quality. Projections made by the FAO indicate that by 2015 rises in income will translate into consumption of an average level of over 3000 kcals/day/person by 54% of the world's population (Bruinsma, 2003). This increase in caloric intake will stimulate a transition in food consumption patterns as well, from starchy staples toward "luxury" goods such as dairy products, fish, and meat. The demand for food characteristics associated with a high elasticity of income, such as food safety, nutritional content, and convenience is thus also likely to increase.

According to Welch and Graham (2002), "Micronutrient malnutrition (e.g., Fe, Zn and vitamin A deficiencies) now afflicts over 40% of the world's population and is increasing especially in many developing nations. Green revolution cropping systems may have inadvertently contributed to the growth in micronutrient deficiencies in resource-poor populations. Current interventions to eliminate these deficiencies that rely on supplementation and food fortification programs do not reach all those affected and have not proven to be sustainable." They argue that one approach to the micronutrient deficiency problem is enhancing the nutritional content of staple food products.

One of the major promises of biotechnology is its potential to enhance food characteristics. Biotechnology may be used to extend shelf life, modify size and shape, and enhance flavors and nutritional content. Parker and Zilberman (1993) have shown that improved food quality may more than double the retail price of peaches, and quality-enhancing biotechnology may be a major source of income for agriculture in the long run. Environmental preferences are also manifested through consumer behavior. One dimension that may enhance the demand for biotechnology products is the desire to consume pesticide-free food.

At the same time, consumer concerns over the health and environmental impacts of biotechnology products is resulting in a slower rate of their adoption in agricultural production. On the health side, concerns over the potential for increased levels of allergic reactions from consuming foods generated through biotechnology have been raised. Environmental concerns have also been raised regarding the potential for genetically modified organisms (GMOs) to escape into the larger gene pool, resulting in an irreversible change in the composition of genetic resources and the potential for the spread of undesirable organisms such as "super weeds" (Rissler and Mellon, 1996).

A critical determinant of the future use of agricultural biotechnology products lies in the attributes consumers will demand of products and to what extent they will pay for these. At present this response is unclear and will be driven by conflicting concerns on environmental and food safety and the perception of biotechnology's impact on these, as opposed to the desire for quality characteristics biotechnology can deliver, such as improved taste and nutrition, enhanced shelf life, and also improved environmental performance associated with a reduction in pesticide use.

Considerable variations in consumer attitudes towards agricultural biotechnology products, particularly GMOs, are found in the potential markets for the products. Attitudes are often linked to income, with people from poorer countries having more positive attitudes than those from richer countries, although there are exceptions to the pattern (FAO, 2004). A survey conducted by Environics International in 34 countries revealed that, in general, people in developing countries are more likely to value the benefits of biotechnology over the potential risks, as compared with those in developed countries, particularly Europe. Consumer attitudes were also found to vary depending on the type of benefits biotechnology conveyed: Applications that address human health or environmental concerns were viewed more favorably than those that increase agricultural productivity.

Consumer rejection of GMOs has two major implications for the dissemination and adoption of agricultural biotechnology. Threat of loss of market share has caused exporting countries to ban the use of biotechnology in production, and this factor is now included in the risk-assessment procedures of some countries. For example, one of the largest soya-producing regions of Brazil banned the planting of GM soya and India stopped trials of BT cotton (FAO, 2004). Consumer demand for differentiated products has implications for the structure of the food processing industry as well. We have already seen the emergence of differentiated products in poultry and fresh fruits and vegetables in developed countries. Producers of these differentiated products are frequently either vertically integrated firms or a chain of firms that is linked through contracts. It is likely that some dominant firms in these industries (Proctor and Gamble, Gerber, etc.) will become actively involved in utilizing biotechnology to produce differentiated products. Both the marketing techniques and production structures that are associated with these industries are likely to transform the agricultural sectors that adopt biotechnology to meet differentiated consumer preferences. Increases in vertical integration and contracting in agriculture are likely to accompany the development of biotechnology to respond to these consumer demands.

4. DEMOGRAPHICS

Population growth and mortality rates will be key determinants of the composition and size of demand for agricultural production, and also the technology under which it is supplied. Increased populations generate increased demand for agricultural products, which must be supplied through an expansion or intensification of agricultural production. Demographic change is the key determinant of population pressures on the land, and thus important determinants of the rate and nature of agricultural intensification, with major implications for both biodiversity and biotechnology.

We are living in times of rapid and radical changes in population size and distributions. Global population growth rates are declining swiftly—from a peak of 2.04% per annum in the late 1960s to 1.35% per annum by the late 1990s (Bruinsma, 2003). It is projected to fall even further, to 1.1% per annum by 2015. Although rates are dropping, the absolute numbers of people added to the world's population each year are still quite large, particularly in developing countries. South Asia, East Asia, and Sub-Saharan Africa are the three areas where annual incremental population increases have been the highest over the past two decades and, thus, where a rapid growth in the working-age population is now occurring. Continuing large annual increases are projected to occur in South Asia and East Asia up to 2015. For Sub-Saharan Africa, however, the pandemic of HIV AIDS has resulted in a major shift in population projections and annual incremental increases, due to its impact on mortality rates among working age populations. In most of eastern and southern Africa the prevalence of HIV is over 10%. For some countries, negative population growth rates are projected by 2010 as the mortality from HIV outstrips new births (Jayne, Villareal, and, Pingali, 2004). Overall, the absolute numbers of adults projected to be alive in countries of Sub-Saharan Africa with HIV prevalence rates over 10% is roughly similar to what it is today. According to the projections, between 2000 and 2025 there will be a slight increase in the number of men between 20 and 59 years of age, but no change in the number of women (Jayne, Villareal, and, Pingali, 2004). However, HIV will also likely affect the productivity of the labor force, due to increased incidences of illness and lower capacity to perform work among afflicted laborers, as well as the need to divert labor to child care, funerals, and tending the sick among the population in general.

The impact of demographic change on agricultural technology choice and ultimately on biotechnology and agricultural biodiversity depends on the supply of factors of production aside from labor, such as land, capital, and technology. The distribution of these factors and policies that affect their relative prices will determine the degree to which an expansion in agricultural output will be met through increases in the extensive or intensive margin of agricultural production. FAO projects, which approximately 80% of the required growth in crop production will come from, increase in the intensive margin (i.e., increases in yields per hectare per year) (Bruinsma, 2003). Arable land expansion as a source of growth (the extensive margin) will be important in some Sub-Saharan and Latin American countries, although much less so than in the past.

In the past, and with the green revolution in particular, the intensification of agriculture and yield increases were accomplished partially through the adoption of improved varieties, which has also been associated with changes in crop genetic diversity, although there is some controversy over whether the direction has been negative or positive (see Chapter 3 for a detailed discussion; also Brush, 1999). The impacts of intensification on increasing crop genetic erosion and vulnerability are a serious concern (FAO, 1998; Matson et al., 1997). However, much of the areas where agricultural intensification through the adoption of monocultural systems has not yet taken place are characterized by a high degree of agroecological heterogeneity and poorly functioning markets, resulting in a higher value to maintaining diversity in the farming system (see Chapter 5). Intensification in these areas may require higher reliance on agricultural biodiversity due to the barriers to adoption of monocultural agricultural production systems (see Chapter 19; also Matson et al., 1997).

5. PRIVATIZATION AND DEVOLUTION

Many of the powers that governments wielded in the past have been transferred to the private sector or local governments in recent years. These processes of privatization and devolution are occurring parallel to the process of globalization; thus, we see a shift of power from national governments towards bigger international organizations as well as smaller, local governments and private firms. The logic of this devolution is an assignment of responsibilities that are scale appropriate and correspond to core competencies of organizations. There are several dimensions of privatization and devolution with implications for biotechnology and biodiversity, which will be discussed below.

5.1 The privatization of agricultural and life science research

One of the most striking areas where privatization has occurred is in the agricultural R&D industry, particularly those related to biotechnology. In the 1970s and 1980s developed countries experienced a major reduction in the amount of public funds devoted to agricultural research, accompanied by significant increases in private-sector spending (Pray and Umali-Deininger, 1998; Alston, Pardey, and Smith, 2000). In developing countries private sector-funded research is still a much smaller share of total research, but increases are occurring there as well. Declining public budgets, poor performance record of publicly funded research, increased appropriability of the returns to privately funded research due to IPRs, and the increased use of purchased inputs in agriculture as a result of increasing competition all contribute towards an increased role of private-sector funding in agricultural research.

Private firms in developed countries largely dominate the R&D of agricultural biotechnology with an estimated $2.6 billion invested in 1998 (Byerlee and Echeverria, 2000). Only a small share of this investment is directed towards developing countries. There are significant market failures in harnessing the benefits of biotechnology for the benefits of poor producers and consumers in developing countries, which are discussed in detail in Chapter 3. Several chapters (3, 4, 13, 14, 17, and 18) note that a key determinant of the degree to which biotechnology R&D can be harnessed for addressing the needs of developing countries is the capacity of the public sector research institutions to access the technologies generated in the private sector of developed countries. There is tremendous variation in this capacity among developing countries, both in terms of handling the science and the institutional issues involved (Byerlee and Echeverria, 2000). Forging innovative links between private and public R&D systems is an important way to create better access to biotechnology in developing countries and one which is taken up in other chapters of this book.

5.2 Privatization of natural resource property rights and expansion of trading schemes

Land reform and the decollectivization of commonly held properties have been major trends in transition economies and developing countries in recent years. Lands that previously belonged to the state or other forms of communal ownership have been allocated to individual owners who obtain property rights for utilization of the land and its resources. These measures are intended to eliminate inefficiencies that existed under centrally planned economies and inequities in distribution in others. At the same time a move to privatize natural resources and environmental services together with the introduction of market trading to improve environmental management has arisen—although on a much smaller scale. For example, individual rights to water and water trading are being introduced in countries such as Chile and the United States and are being considered in several countries in South America and South Asia. Carbon emission reduction credits is another area where trading regimes have been established and which have the potential for considerable expansion depending on the nature of future international agreements to control climate change. In the area of agricultural biodiversity, international agreements on the potential for establishing transfer mechanisms to pay for the conservation of resources, such as the International Treaty on the Conservation and Utilization of Plant Genetic Resources and the CBD have been established, although considerable work still needs to be done on the design and implementation (see Chapters 8 and 9).

The privatization of land and land reform could provide producer incentives for the adoption of biotechnology—to the extent that it contributes towards productivity gains, but impacts on biodiversity are less clear. Where land reform programs involve use of forested lands or previously uncultivated lands for agriculture, then impacts on wild biodiversity are likely to be negative.

The privatization and commoditization of other natural resources and environmental services provide farmers and natural resource owners with more flexibility and may provide them with incentives to provide environmental goods and services, such as biodiversity.

5.3 Privatization of extension and emergence of private agricultural consultants

Many countries are experimenting in privatizing some of the services that public sector agricultural extension has provided.[5] These reforms reflect both increased scarcity in public funds and the new reality where agriculture becomes more knowledge intensive, and farmers operating in the commercial sector are looking for more detailed and specialized knowledge and are ready to pay for it. In the United States and other industrialized countries (Wolf, 1998), dealers of input manufacturers (irrigation equipment and seed and chemical companies) have increased the amount of management information that they provide to farmers. The complexity of pest control decisions and the need to comply with environmental regulations have led to the emergence of independent pest control consultants. In specialty crops where contracting is prevalent, the buyers may dictate some production practices and provide technical assistance to the contractors. As farms grow in size, they may hire their own specialists in pest control and other aspects of production and design their own production systems.

In many regions, state extension specialists now provide advice and training to independent consultants, provide general retraining to farmers and farm workers, address some of the needs of smaller farms, specialize in treating regional problems (conflict resolution among farmers, environmentalists, and the urban sector), and provide information on the requirements and means to meet environmental regulations. Extension centers are also used to adapt and test new technologies under local conditions. With a decrease in the role of the public sector in providing information to farmers, a need for an overall increase in the resources allocated to education and the transfer of information has arisen. In developed countries there has been some response to this need, but in many developing countries there is still a considerable lack of resources devoted to education and information transfer with a consequent negative impact on the ability of farmers to assess and adopt new technologies.

5.4 Reduction in size and increased specialization of central governments

The reduction in the responsibilities of state governments is also associated with a reduction in taxation to support state governments (or at least a reduction in the rate of growth of taxation). Moreover, a larger share of the tax revenues of the central governments is returned to local governments that actually provide services. There are several

[5] See Wolf (1998) for evidence for England, Australia, and New Zealand. Some countries in Latin American, notably Nicaragua, are going through similar transformations.

government agencies now attempting to subcontract provision of key services (waste management and education) to private companies, thus significantly reducing the size of the public sector. Governments are attempting to concentrate on the areas that they do best, such as provision of public goods such as national defense, support for basic research, and monitoring and enforcement of environmental protection and economic competitiveness.

The declining role of central governments and the transfer of responsibilities to the private and nongovernmental sectors may lead to increased efficiency but may also lead to gaps in unsatisfied needs, and new arrangements need to be established to fill these gaps. In some cases, the reduced role of the central governments may negatively affect the poor, at least in the short run. On the other hand, the realignment of responsibilities will provide more resource mobility and flexible institutional infrastructure that will enable faster adoption of biotechnology innovations and better conservation of biodiversity.

Devolution changes the scale at which transfers are made and may create conflicts between local needs and the provision of goods and services that are national or global in scope. Biodiversity conservation clearly benefits a wide group but requires cost bearing at a local level, so there is a need for some sort of mechanism to address this. In terms of agricultural biodiversity, the relevant scale for management is often broader than the local level, which also creates some coordination problems. Thus, devolution may have opposing effects on the management of biodiversity, and increasing flexibility in management at the local level may be positive but can be offset by a decrease in the potential for coordination at higher levels.

6. THE EMERGENCE OF INFORMATION AND KNOWLEDGE ECONOMICS

Arguably, the dominant form of technological change in the last 25 years has been in the area of information, communications, and data processing. Over the last 25 years, we have witnessed drastic reductions in the cost of data processing and the proliferation of computer use among families and small firms, emergence of global communications networks that enable instantaneous financial transactions and fast, massive transfer of data across locations, and establishment of a network of satellites that facilitate monitoring of resource management with a high degree of accuracy. The emergence of the information economy has important implications for both biotechnology and biodiversity in terms of its impact on the capacity to develop new technologies and the institutions that are needed to promote such development, the introduction of modern production methods which are responsive to environmental heterogeneity, the analysis and monitoring of agricultural production impacts on environmental conditions, and the ability to inform and mobilize large groups of people over large geographic distributions.

The development of biotechnology has benefited largely from the increase in computational abilities. Biotechnology is data intensive, and mapping of genes would not have been feasible without advanced computer technologies. With information-intensive technologies such as biotechnology, most of the economic value is not attributed to equipment (hardware) but, rather, to management knowledge and information (which are in many cases embodied in software). Thus, with the evolution of information technologies, we have seen much more emphasis on establishing definitions and enforcement criteria for IPRs. Without the ability to capture accrued rents using software or new knowledge of information, private parties would not have the incentive to develop these items. Therefore, both patent and copyright laws have been modified to protect IPRs, and the extent of their coverage is being expanded through international trade agreements such as the TRIPS agreement under the WTO.

Establishing and protecting international IPRs for biotechnology innovations is a major challenge. A narrow definition of IPRs for biotechnology innovations may not provide sufficient incentives to cover R&D costs. On the other hand, a definition that is too broad may give owners of these rights excessive monopolistic power and deter access and further innovations by others. IPRs for the knowledge embodied in biotechnology need to take into account the contribution that indigenous knowledge has played in the development of an innovation and assign value to these rights accordingly. However, assigning property rights to goods that were previously freely available and exchanged among farmers could also reduce the accessibility to those resources and actually reduce diversity (see Chapter 9).

6.1 Adoption of precision agriculture

Precision agriculture can be defined generically as a bundle of technologies that adjusts input use to variations in environmental and climatic situations over space and time and reduces residues associated with input use. Many of these technologies rely on space age communication technologies and incorporate the use of geographic positioning systems (GPS). Modern irrigation technologies that adjust input use according to variability in soil and weather conditions relying on weather stations and moisture-monitoring equipment are also examples of precision technologies. Precision technologies have the potential to increase input-use efficiency, increase yields, and reduce residues of chemicals that may contaminate the environment. In many cases it may lead to input saving, but in others the yield effect may also entail increased input use (National Research Council, 1997).

Thus far, there have been significant variations in adoption rates of technology that can be generically defined as precision technologies. Some modern irrigation technologies have high rates of adoption in high value crops. Some components of what is promoted as "precision agriculture" such as yield monitors are gaining significant acceptance. But, overall, adoption rates of many components of precision agriculture have not been very high even in developed countries (National Research Council, 1997). Adoption of precision farming technologies may be hampered by the cost of investment. Furthermore, the management software needed to take advantage of the information has not been fully developed. Adoption of precision farming technologies will likely increase in the future as their cost declines, as productivity increases, and as new management software becomes available.

Precision technologies may both complement and substitute for biotechnologies. Precision technologies that enable the planting of a field with several varieties of seeds will increase the demand for diversified genetic stock that can be adjusted to slight variations in soil conditions. Precision agriculture may also improve sorting and harvesting methods, making the production of high-quality produce more economical and improve incentives to develop higher quality varieties. On the other hand, precision farming may reduce significantly the environmental side effects of pesticides and provide more refined mechanical ways to address weed problems, thus, reducing the demand for some of the pest control applications of biotechnologies and reducing the loss of biodiversity stemming from inadvertent contamination.

6.2 Introduction of precision and information technologies for the management of biodiversity

Some of the major problems with biodiversity conservation and management may be better addressed with applications of precision technologies. By-catch, the destruction of nontarget species by fishermen, is a major environmental side effect in fisheries. Similarly, forest clearcutting is a major cause of biodiversity loss. Adoption of more refined harvesting technologies may reduce these side effects and, thus, result in higher levels of biodiversity preservation. However, both the development and adoption of such technologies may not occur, at least in a socially optimal manner, unless financial incentives are introduced. These may include subsidization of research and technology adoption as well as penalties and regulations on harvesting technologies that damage the environment. Monitoring and enforcement of such incentives provide a significant challenge, but taking advantage of emerging technologies in remote sensing can solve some of the technological aspects.

6.3 Improved marketing and product flows

Computer technologies enable the documentation and monitoring of sales in real time and instantly provide useful information on inventory conditions and producers' preferences. Thus, marketers and distributors can obtain a faster response and reduce inventory costs. Also, marketers may be able to better identify quality preferences at specific locations and respond to them more promptly. Indeed, some of the recent product diversification in agriculture, especially in the poultry and produce sectors, took advantage of new information technologies, resulting in a higher quality and more diversified product. The efficiency gains that modern information technologies provide in marketing quality-differentiated products is likely to enhance the introduction and adoption of biotechnology.

Information technologies reduce the cost of product differentiation in agriculture, but also increase the relative advantage of contracting and vertical integration. The introduction of a new brand of differentiated products requires precise coordination among retailers (who provide the shelf space), distributors, and producers. It is subject to a strict timetable. The organization responsible for providing a new differentiated product to a retail chain will prefer to contract with farmers to produce a new product or control the production itself. Thus, the introduction of differentiated biotechnology products will be associated with "industrialization" of agriculture, including increased contracting and vertical integration.

7. CONCLUSIONS

There are several forceful and rapidly moving processes occurring globally that will affect the management, and ultimately the status, of biotechnology and biodiversity. In this chapter we have given an overview of some of the major social, political, and economic forces that we believe will shape the way in with the two "bios" will co-evolve with humankind. Of course, the processes we have focused upon here are not the only ones which will affect how biotechnology and biodiversity issues are resolved, and their relative importance will vary among countries. Climate change could have a major impact on the demand for agricultural technology, as well as international agricultural supply and production, and thus affect both biodiversity and biotechnology.

REFERENCES

Alston, J. M., Pardey, P. G., Smith, V. H., 2000, Financing agricultural R&D in rich countries: What's happening and why, in: *Public-Private Collaboration in Agricultural Research: New Institutional Arrangements and Economic Implications,* K. O. Fuglie and E. Schimmelpfennig, eds., Iowa State University Press, Ames, pp. 25-54.

Anderson, K., and Nielsen, C. P., 2001, GMOs, food safety and the environment: What role for trade policy and the WTO, in: *Tomorrow's Agriculture: Incentives, Institutions, Infrastructure and Innovations,* G. H. Peters and P. Pingali, eds., Ashgate Publishing, Ltd, Aldershot, Hampshire, pp. 61-85.

Becker, G. S., 1965, A theory of allocation of time, *Econ. J.,* **75**:493-517

Berdegué, J. A., Balsevich, F., Flores, L., and Reardon, T., 2003, *The Rise of Supermarkets in Central America: Implications for Private Standards for Quality and Safety of Fresh Fruits and Vegetables.* Final report for USAID-RAISE/SPS project.

Bruinsma, J., ed., 2003, *World Agriculture: Towards 2015/2030.* FAO and Earthscan Publications Ltd., Rome and London.

Brush, S., 1999, Genetic erosion of crop populations in centers of diversity: A revision, paper presented at the Technical Meeting on the Methodology of the FAO World Information and Early Warning System on Plant Genetic Resources held at the Research Institute of Crop Production, Prague, Czech Republic (June 21-23, 1999); http://apps3.fao.org/wiews/Prague/

Byerlee, D., and Echeverria, R. G., 2002, *Agricultural Research Policy in an Era of Privatization*, CABI Publishing, Cambridge, Mass., 320p.

Correa, C., 1999, Access to plant genetic resources and intellectual property rights background, study paper number 8. Commission on Genetic Resources for Food and Agriculture, FAO Rome.

Cropper, M. L., Evans, W. N., Berardi, S. J., Ducla-Soares, M. M., and Portney, P. R., 1992, The determinants of pesticide regulation: A statistical analysis of EPA decisionmaking, *J. Polit. Econ.* **100**:175-198.

Crucible Group, 1994, *People, Plants, and Patents: The Impact of Intellectual Property on Trade, Plant Biodiversity, and Rural Society*, IDRC Online Book http://network.idrc.ca/

Damania, R., Fredriksson, P. G., and List, J. A., 2003, Trade liberalization, corruption, and environmental policy formation: Theory and evidence, *J. Environ. Econ. and Manage.* **46**(3):490-512.

Deacon, R., 1999, Political economy of environment-development relationships: A preliminary framework, Department of Economics, University of California at Santa Barbara, Economics Working Paper Series 1089.

FAO, 1998, *The State of the World's Plant Genetic Resources for Food and Agriculture,* FAO, Rome.

FAO, 2004, *The State of Food and Agriculture: 2003-04*, FAO, Rome.

Gollin, D., 1998, Valuing farmers' rights, in: *Agricultural Values of Plant Genetic Resources*, R. E. Evenson, D. Gollin, and V. Santaniello, eds., CABI Publishing, Wallingford, U. K.

Helfer, L., 2002, *Intellectual Property Rights in Plant Varieties: An Overview with Options for National Governments,* FAO Legal Papers Online #31 http://www.fao.org/Legal/prs-ol/lpo31-2.pdf, FAO, Rome.

Jayne, T. S., Villareal, M., and Pingali, P., 2004, Interactions between the agricultural sector and the HIV/AIDS pandemic: Implications for agricultural policy, ESA working paper, FAO, Rome.

Josling, T., 1999, The WTO and its potential role in GMO regulations, in: *The Economics and Politics of Genetically Modified Organisms in Agriculture: Implications for*

WTO 2000, G. Nelson, ed., Bulletin 809, University of Illinois Board of Trustees, Urbana-Champaign.

Laxminarayan, R., ed., 2002, *Battling Resistance to Antibiotics and Pesticides: An Economic Approach*, Resources for the Future, Washington, D. C.

Matson, P. A., Parton, W. J., Power, A. G., and Swift, M. J., 1997, Agricultural intensification and ecosystem properties, *Science* **25**(277):504-509.

National Research Council, 1997, *Precision Agriculture in the 21st Century: Geospatial and Information Technologies in Crop Management,* National Academy Press, Washington, D. C.

Oates, W. E., and Baumol, W. J., 1996, The instruments for environmental policy, in: *The Economics of Environmental Regulation,* W. E. Oates, ed., Elgar, Cheltenham, U.K. , pp. 91-124.

Parker, D., and Zilberman, D., 1993, Hedonic estimation of quality factors affecting the farm-retail margin, *Amer. J. Agri. Econ.* **75**:458-466.

Pray, C. E., and Umali-Deininger, D., 1998, The private sector in agricultural research systems: Will it fill the gap? *World Dev.* **26**(6):1127-1148.

Reardon, T., Timmer, P., Barrett, C., and Berdegué, J., 2003, The rise of supermarkets in Africa, Asia and Latin America, *Amer. J. Agri. Econ.* **85**(5):1140-1146.

Rissler, J., and Mellon, M , 1996, *The Ecological Risks of Engineered Crops*, MIT Press, Cambridge, Mass.

Welch, R. M., and Graham, R. D., 2002, Breeding crops for enhanced micronutrient content, *Plant and Soil* **245**(1):205-214.

Wolf, S., 1998, Privatization of crop production information service markets: Spatial variation and policy implications, unpublished Ph.D. dissertation, University of Wisconsin-Madison.

Wright, B. D., 1998, Intellectual property and farmers' rights, in: *Agricultural Values of Plant Genetic Resources*, R. E. Evenson, D. Gollin, and V. Santaniello, eds., CABI Publishing, Wallingford, U. K., pp. 219-232.

Chapter 3

PRIVATE RESEARCH AND PUBLIC GOODS: *IMPLICATIONS OF BIOTECHNOLOGY FOR BIODIVERSITY*

Terri Raney[1] and Prabhu Pingali[2]
[1]Senior Economist, Agricultural and Development Economics Division, Food and Agriculture Organization of the United Nations, Viale delle Terme di Caracalla, 00100 Rome, Italy; [2]Director, Agricultural and Development Economics Division, Food and Agriculture Organization of the United Nations, Viale delle Terme di Caracalla, 00100 Rome, Italy

Abstract: The pattern of crop genetic diversity in the developing world has changed over the past two centuries with the modernization of agriculture, accelerating with the advent of the green revolution. Since the green revolution, the locus of agricultural research has shifted from the public to the private sector. The growing importance of the private sector in agricultural R&D is changing the types of crop technologies that are developed and the ways they are delivered to farmers, perhaps best illustrated by transgenic crops which are being developed and commercialized almost exclusively by private multinational companies. The spread of transgenic crops will influence crop genetic diversity, but their implications for the availability of plant genetic resources and the resilience of agricultural ecosystems are not entirely clear. Transgenic crops, per se, may increase or decrease crop genetic diversity, depending on how they are regulated and deployed. This paper explores a range of policy options to increase the likelihood that private sector R&D, particularly in the form of transgenic crops, enhances rather than erodes crop genetic diversity.

Key words: agricultural research; agricultural transformation; biodiversity; biotechnology; gene revolution; green revolution; technological change; transgenic crops.

1. INTRODUCTION

The pattern of crop genetic diversity in the fields of the developing world has changed fundamentally over the past 200 years with the intensification and commercialization of agriculture. This process accelerated with the advent of the green revolution in the 1960s when public sector researchers and donors explicitly promoted the international transfer of improved seed varieties to farmers in developing countries. Since the green revolution, the locus of agricultural research and development (R&D) has shifted from the public sector to the private multinational sector, driven by the commercialization of agriculture, the scientific discoveries underpinning the "gene" revolution, stronger intellectual property rights protections, and more open international markets. The growing importance of the private sector in agricultural R&D is changing the types of crop technologies that are developed and the ways they are delivered to farmers. Transgenic crops—which have been developed and disseminated almost exclusively by the private sector—provide perhaps the clearest illustration of the changes arising from the growth of private sector agricultural R&D. These crops will influence crop genetic diversity, but their implications for the availability of plant genetic resources and the resilience of agricultural ecosystems are not entirely clear.

The germplasm that dominates the area planted to the major cereals has shifted over time from the locally adapted populations that farmers historically selected from the seed they saved—often called "landraces"—to the more widely adapted seed types produced by scientific plant-breeding programs and purchased by farmers—often called "modern varieties." The genetic content and the geographical distribution of landrace populations are influenced by natural selection pressures and the seed and crop management practices of traditional farming communities. In contrast, the spatial and temporal diversity among modern varieties in farmers' fields is determined more by the economic factors affecting their profitability and by the performance of agricultural research institutions and seed industries (Pingali and Smale, 2001). The spread of transgenic crop varieties will also be influenced by farm level profitability and the performance of agricultural research institutions and seed sectors, but institutional and regulatory issues (private sector dominance, intellectual property rights, and regulatory concerns and procedures) will have a greater influence over the spread of transgenic varieties than for conventional modern varieties. Finally, transgenic technology itself may influence biodiversity by enabling the more targeted exchange of genetic materials in breeding programs and through the inadvertent spread of transgenes to related modern varieties and landraces.

Private firms are responsible for most transgenic crop R&D and almost all of the commercialization of transgenic crop varieties being undertaken today. This is in sharp contrast with the development and diffusion of modern green revolution varieties for which the public sector—national and international—played a strong role. Four interrelated forces are transforming the system for providing improved agricultural technologies to the world's farmers. The first is the ongoing process of agricultural modernization, i.e., the intensification and commercialization of agriculture. The second is the strengthening environment for protecting intellectual property in plant innovations. The third is the rapid pace of discovery and the growing importance of molecular biology and genetic engineering. Finally, agricultural input and output trade is becoming more open in nearly all countries, enlarging the potential market for new technologies and older related technologies. These developments have created powerful new incentives for private research, and are altering the structure of the public/private agricultural research endeavor, particularly with respect to crop improvement (Pingali and Traxler, 2002).

This chapter explores the linkages among the modernization of agriculture, the changing locus of agricultural research and technology transfer and the resulting patterns of crop genetic diversity in the developing world. Section 2 describes the modernization of agriculture and the evolution of plant improvement research from prehistory through the era of conventional scientific plant breeding to the current gene revolution. Section 3 discusses the changing locus of agricultural research from the public to the private sector and the implications for crop variety development and technology transfer. Section 4 explores the implications of these changes—particularly the spread of transgenic crops—for varietal use patterns and crop biodiversity. Section 5 concludes with some recommendations for the promotion of crop genetic diversity within the existing environment for agricultural research and technology transfer.

2. THE TRANSFORMATION OF AGRICULTURE AND THE EVOLUTION OF PLANT IMPROVEMENT RESEARCH

Modern cereal cultivars have developed through four main phases of selection: (i) subconscious selection by earlier food growers in the process of harvesting and planting, (ii) deliberate selection among variable materials by farmers living in settlements and communities, (iii) purposeful selection by professional breeders using scientific principles of inheritance and observable physical traits, and (iv) selection based on genomic characteristics and the application of molecular markers and transgenic techniques to crop improvement. The latter two phases have emerged as a result of the intensification and commercialization of agriculture.

2.1 The transformation of agriculture

The transformation of agriculture over the past 200 years has involved the interrelated processes of intensification and commercialization. The intensification of agriculture refers to the increase in output per unit of land used in production, or land productivity. Population densities explain much about where and under which conditions this process has occurred (Boserup, 1981). The transition from low-yield, land-extensive cultivation systems to land-intensive, double- and triple-crop systems is only profitable in societies in which the supply of uncultivated land has been exhausted. It is no accident that the modern seed-fertilizer revolution has been most successful in densely populated areas of the world whether traditional mechanisms for enhancing yields have been exhausted (Hayami and Ruttan, 1985).

Intensification could also occur in the less densely populated areas for two reasons: (i) in areas that are well-connected to markets, higher prices and elastic demand for output imply that the marginal utility of effort increases, hence farmers in the region will begin cultivating larger areas, and (ii) higher returns to labor encourage migration into well-connected areas from neighboring regions with higher transport costs. Intensification of land use and the adoption of yield-enhancing technologies have occurred in both traditional and modern agricultural systems.

Economic growth, urbanization, and the withdrawal of labor from the agricultural sector lead to the increasing commercialization of agricultural systems. Commercialization, in turn, leads to greater market orientation of farm production, progressive substitution of nontraded inputs in favor of purchased inputs, and the gradual decline of integrated farming systems and their replacement by specialized enterprises for crop, livestock, poultry, and aquaculture products (Pingali, 1997). Agricultural output and input use decisions are increasingly guided by the market and are based on the principles of profit maximization. This, in turn, influences patterns of crop genetic diversity through changes in land-use patterns and through crop choice changes.

2.2 The evolution of plant improvement

2.2.1 Domestication of wild species

Humans have manipulated the genetic makeup of plants since agriculture began more than 10,000 years ago (Table 3-1). Primitive societies of hunters and gatherers recognized wild species of cereals and harvested them for food. Societies of shifting cultivators gradually domesticated these wild species, creating the basis for sedentary or permanent agricultural systems. These early farmers unconsciously managed the process of domestication over several millennia, selecting and planting the best seeds through many growing cycles. The main attainment of this first phase of crop improvement was to develop domesticated crops more suitable for human cultivation—planting, harvesting, threshing, or shelling—and consumption. Higher germination rates, more uniform growing periods, resistance to shattering, and improved palatability were some of the achievements of this effort. The human selection pressures that accompanied domestication narrowed the genetic base for these crops as farmers selected among the full range of plant types for those that produced more desirable traits (Smale, 1997).

2.2.2 Development of landraces

In the second phase of crop improvement, farmers deliberately selected plant materials suited to local preferences and growing conditions. Many farmers in many locations exerted pressures continuously in numerous directions, resulting in variable crop populations that were adapted to local growing conditions and consumption preferences. These populations, broadly known as landraces, often differ radically from their early ancestors. Although more genetically uniform than these early relatives, landraces are nonetheless characterized by a high degree of genetic diversity within a particular field.

2.2.3 Conventional breeding of modern varieties

The third phase of crop improvement through scientific plant-breeding programs relied on the application of classical Mendelian genetic principles based on the phenotype or physical characteristics of the organism concerned. Conventional breeding, which began about 100 years ago, has been very successful in introducing desirable traits into crop cultivars from domesticated or wild relatives or mutants. The first high-yielding hybrid maize varieties were produced about 50 years ago and the high-yielding, semidwarf varieties of wheat and rice that gave rise to the green revolution were developed less than 50 years ago. The products of this third phase—often called modern varieties—have been widely adopted in intensive agricultural production systems.

As a result of the spread of modern varieties, fields of cereals have become more uniform in plant types with less spontaneous gene exchange. Planned gene migration increased, however, with the worldwide exchange of germplasm among research institutions that was an integral part of the green revolution research paradigm (Pingali and Smale, 2001). Although the nature of crop genetic diversity has changed as a result of the spread of modern varieties, it is neither straightforward nor particularly meaningful to discuss whether genetic diversity has increased or decreased, because a simple count of the varieties in a particular area or measures of genetic distance among varieties may not tell us much about the resilience of crop ecosystems or the availability of crop genetic resources for breeding program (see section 4).

Table 3-1. An agricultural technology timeline

Technology	Era	Genetic interventions
Traditional	About 10,000 BC	Civilizations harvested from natural biological diversity, domesticated crops and animals, began to select plant materials for propagation and animals for breeding.
	About 3,000 BC	Beer brewing, cheese making, and wine fermentation.

Conventional	Late 19th Century	Identification of principles of inheritance by Gregor Mendel in 1865, laying the foundation for classical breeding methods.
	1930s	Development of commercial hybrid crops.
	1940s to 1960s	Use of mutagenesis, tissue culture, plant regeneration. Discovery of transformation and transduction, discovery by Watson and Crick of the structure of DNA in 1953, identification of genes that detach and move (transposons).
Modern	1970s	Advent of gene transfer through recombinant DNA techniques. Use of embryo rescue and protoplast fusion in plant breeding and artificial insemination in animal reproduction.
	1980s	Insulin as first commercial product from gene transfer. Tissue culture for mass propagation in plants and embryo transfer in animal production.
	1990s	Extensive genetic fingerprinting of a wide range of organisms, first field trials of genetically engineered plant varieties in 1990 followed by the first commercial release in 1992. Genetically engineered vaccines and hormones and cloning of animals.
	2000s	Bioinformatics, genomics, proteomics, metabolomics

Source: FAO (2004).

2.2.4 Genomic selection in plant breeding

The latest phase of crop improvement research is based on the identity, location, and function of genes affecting economically important traits and the direct transfer of these genes through transgenesis. Transgenesis permits the introduction of genetic materials from sexually incompatible organisms, greatly expanding the range of genetic variations that can be used in breeding programs. Unlike conventional breeding, transgenesis allows the targeted transfer of the genes responsible for a particular trait, without otherwise changing the genetic makeup of the host plant. This means that a single transgenic innovation can be incorporated into many varieties of a crop, including perhaps even landraces (see Chapter 14). Compared with conventional breeding in which an innovation comes "bundled" within a new variety that typically displaces older varieties, transgenesis allows an innovation to be disseminated through many varieties, preserving desirable qualities from existing varieties and maintaining or, potentially increasing, crop genetic diversity.

On the other hand, the widespread incorporation of a single innovation, such as the *Bacillus thuringiensis* (Bt) genes that confer insect resistance, into many crops/varieties may constitute a type of genetic narrowing for that particular trait. Furthermore, transgenic crops that confer a distinct advantage over landraces may accelerate the pace at which these traditional crops are abandoned or augmented with the transgenic trait. Regulatory regimes are concerned with the potentially harmful consequences of gene flow from transgenic crops to conventional varieties or landraces. In this context, it is important to recognize that gene flow from conventional varieties to landraces frequently occurs (especially for open-pollinated crops such as maize) and is often consciously exploited by farmers. In the same way, it is likely that farmers would consciously select for transgenic traits that confer an advantage (de Groote et al., 2004) unless biological or legal methods are used to prevent them from doing so. How these offsetting forces will ultimately affect crop genetic diversity depends on the incentives and constraints facing researchers, plant breeders, and farmers. The changing locus of agricultural research from the public to the private sector is a key element in this regard.

3. THE CHANGING LOCUS OF AGRICULTURAL RESEARCH

3.1 The green revolution research paradigm

Most of the conventional breeding research that launched the green revolution was conducted by the public sector with the explicit goal of creating technologies that could be transferred internationally. International and national public sector researchers bred dwarfing genes into elite wheat and rice cultivars, causing them to produce more grain and shorter stems and enabling them to respond to higher levels of fertilizer and water. These semidwarf cultivars were made freely available to plant breeders from developing countries who further adapted them to meet local production conditions. Private firms were involved in the development and commercialization of locally adapted varieties in some countries, but the improved germplasm was provided by the public sector and disseminated freely as a public good.

The initial focus of the green revolution research was on raising yield potential for the major cereal crops. During the early decades of the green revolution, the crops grown by poor farmers in less favorable agro-ecological zones (such as sorghum, millet, barley, cassava and pulses) were neglected, but since the 1980s modern varieties have been developed for these crops and their yield potential has risen (Evenson and Gollin, 2003). In addition to their work on shifting the yield frontier of cereal crops, public sector plant breeders continue to have successes in other important areas of applied research. These include development of plants with durable resistance to a wide spectrum of insects and diseases, plants that are better able to tolerate a variety of physical stresses, crops that require significantly lower number of days of cultivation, and cereal grain with enhanced taste and nutritional qualities.

3.1.1 The public sector and international technology transfer

Prior to 1960, there was no formal system in place that provided plant breeders access to germplasm available beyond their borders. Since then, the international public sector (the Consultative Group on International Agricultural Research (CGIAR) system) has been the predominant source of supply of improved germplasm developed from conventional breeding approaches, especially for self-pollinating crops such as rice and wheat and for open-pollinated maize. These CGIAR-managed networks evolved in the 1970s and 1980s, when financial resources for public agricultural research were expanding and plant intellectual property laws were weak or nonexistent. The exchange of germplasm is based on a system of informal exchange among plant breeders, which is generally open, and without charge. Breeders can contribute any of their material to the nursery and take pride in its adoption elsewhere in the world, while at the same time they are free to pick material from the trials for their own use.

The international flow of germplasm has had a large impact on the speed and the cost of crop development programs of national agricultural research systems (NARS), thereby generating enormous efficiency gains (Evenson and Gollin, 2003). Evenson and Gollin (2003) report that even in the 1990s, the CGIAR content of modern varieties was high for most food crops; 35% of all varietal releases were based on CGIAR crosses, and an additional 22% had a CGIAR-crossed parent or other ancestor. Thus, while the green revolution promoted the spread of genetically uniform modern varieties in the developing world, the genetic pedigrees of these modern varieties were more complex than the landraces they replaced.

3.2 The emergence of private sector agricultural research

In the decades of the 1960s through the 1980s, private sector investment in plant improvement research was limited, particularly in the developing world, due to the lack of effective mechanisms for proprietary protection on the improved products. This situation changed in the 1990s with the emergence of hybrids for cross-pollinated crops such as maize. The ability of developers to capture economic rents from hybrids led to a budding seed industry in the developing world, started by multinational companies from the developed world and followed by the development of national companies (Morris, 1998). Despite the rapid growth of the seed industry in some developing countries, its activity has been limited to date, leaving many markets underserved.

The incentives for private sector agricultural research increased further when the United States and other industrialized countries permitted the patenting of artificially constructed genes and genetically modified plants. These national protections were strengthened by the 1995 Agreement on Trade-Related Aspects of Intellectual Property Rights (TRIPs) of the World Trade Organization (WTO), which obliges WTO members to provide patent protection for biotechnology inventions (products or processes) and protection for plant varieties either through patents or a *sui generis* system. These proprietary protections provided the incentives for private sector entry in agricultural biotechnology research.

The relative importance of the private sector in agricultural research, particularly in transgenic crop biotechnology, is shown in Table 3-2. While these estimates are imperfect, they reveal a sharp dichotomy between public and private research expenditures and between industrialized and developing countries. Industrialized countries spend 10 times as much on crop biotechnology research as developing countries, and this constitutes a higher percentage of their total agricultural research budget. While total research expenditures in the industrialized countries are almost evenly split between the public and private sectors, the latter concentrates a higher share of it total expenditures on transgenic crop biotechnology. In the developing countries, in contrast, the public sector spends a smaller total amount on agricultural research and devotes a smaller share of its total research budget to transgenic crop biotechnology. The CGIAR Centers (where much of the green revolution research was conducted) have a combined annual budget for crop biotechnology research of less than $50 million, less than 5% of the private multinational budget. Comprehensive data on private sector crop biotechnology research in developing countries are not available, although most of this research appears to be carried out by multinationals conducting trials of their transgenic varieties (Byerlee and Fischer, 2002).

Table 3-2. Crop biotechnology research expenditures

	Biotech R&D (million US$/year)	Biotech as share of sector R&D
Industrialized countries	1900-2500	
• Private sector[a]	1000-1500	40
• Public sector	900-1000	16
Developing countries	165-250	
• Public (own resources)	100-150	5-10
• Public (foreign aid)	40-50	n.a.[b]
• CGIAR Centers	25-50	8
• Private sector	n.a.	n.a.
World total	2065-2730	

[a] Includes an unknown amount of R&D for developing countries.
[b] Data not available.
Source: Byerlee and Fischer (2002).

The large multinational agrochemical companies invested early in the development of transgenic crops, although much of the basic scientific research that paved the way was conducted by the public sector and made available to private companies through exclusive licenses. The agrochemical companies entered the plant improvement business by purchasing existing seed companies, first in industrialized countries and then in the developing world (Pray and Naseem, 2003). These arrangements among the public sector, large multinational corporations, and national seed companies are economically rational because the three specialize in different aspects of the seed variety development and delivery process (Pingali and Traxler, 2002). This process is a continuum that starts upstream with basic scientific research (largely in the public sector), moves on to generating knowledge about economically valuable genes and engineering transgenic plants (public sector and large multinationals), and moves downstream to the more adaptive process of backcrossing the transgenes into commercial lines and delivering the seed to farmers (mostly private sector at the national or subnational level).

The products from upstream activities have worldwide applicability across several crops and agroecological environments. On the other hand, genetically modified crops and varieties are typically applicable to specific agroecological niches. In other words, spillover benefits and scale economies decline in the move to the more adaptive end of the continuum. Similarly, research costs and research sophistication decline in the progression towards downstream activities. Thus, a clear division of responsibilities in the development and delivery of biotechnology products has emerged, with the multinational firm providing the upstream biotechnology research and the local firm providing crop varieties with commercially desirable agronomic backgrounds (Pingali and Traxler, 2002).

As discussed above, private sector research focuses on the more applied end of the research spectrum. Indeed, the private sector has developed all of the genetically transformed crops that have been commercialized in the world so far with the exception of insect-resistant cotton in China and virus-resistant papaya in Hawaii, USA. The dominance of the private sector suggests that most transgenic crop development will focus on crops and traits that are aimed at commercially viable markets, to the neglect of smallholders in marginal production environments. Evidence on field trials and commercialization of transgenic crops supports this thesis. More than 11,000 field trials have been performed for 81 different transgenic crops in at least 58 countries since 1987; however, most R&D efforts focus on a few crops and traits of interest to temperate-zone commercial farmers. Data on commercialization are even more concentrated: six countries, four crops, and two traits accounted for 99% of all transgenic crops planted commercially in 2003 (James, 2003). In contrast, the crops and agronomic traits of particular importance to developing countries and marginal production areas are the subject of very few field trials and no commercialization thus far. This neglect is due to the limited commercial potential of these so-called "orphan" crops and to the technical difficulty of finding transgenic solutions for complex traits such as potential yields and abiotic stress tolerance (e.g., drought and salinity).

3.2.1 The private sector and international technology transfer

One of the lessons of the green revolution was that agricultural technology could be transferred internationally. This was especially true for countries that had sufficient national agricultural research capacity to adapt the high-yielding cultivars developed by the international public sector to suit local production environments. Unlike the high-yielding varieties disseminated in the green revolution, the products of the gene revolution are encountering significant regulatory and market barriers. Companies are unwilling to develop and commercialize transgenic crops for countries that lack transparent, science-based regulatory procedures. Furthermore, many of the technical innovations of the gene revolution are held under patents or exclusive licenses. The improved germplasm and varieties that were responsible for the green revolution, in contrast, were disseminated freely as international public goods. While stronger intellectual property protections have greatly stimulated private sector research in developed countries, they can restrict access to new technologies where countries lack appropriate regulatory structures or where farmers lack the financial means to pay for proprietary technologies. Public sector breeders in developing countries may not have access to proprietary genes and enabling technologies, and their farmers may be unable to afford the technology fees charged by private technology developers.

Unlike the green revolution technologies, transgenic technologies are transferred internationally primarily through market mechanisms. The commercial relationship between the multinational bio-science firms and national seed companies was described above. This system of technology transfer works well for commercially viable innovations in well-developed markets, but perhaps not for the types of innovations needed in developing countries: crops and traits aimed at poor farmers in marginal production environments. These "orphan" technologies have traditionally been the province of public sector research. Given the dominance of private sector research in transgenic crop research and meager resources being devoted to public sector research in most developing countries, it is unlikely that public sector research can play this role for transgenic crops.

The options available for public research systems in developing countries to capture the spillovers from global corporations are limited. Public sector research programs are generally established to conform to state or national political boundaries, and direct country-to-country transfer of technologies has been limited (Pingali and Traxler, 2002). Strict adherence to political domains severely curtails spillover benefits of technological innovations across similar agroclimatic zones. The operation of the CGIAR germplasm exchange system has mitigated the problem for several important crops, but it is not clear whether the system will work for biotechnology products and transgenic crops, given the proprietary nature of the technology.

Pingali and Traxler (2002) suggest three possible avenues for public sector institutions in developing countries to gain access to transgenic technologies: (i) directly import private- or public-sector transgenic varieties developed elsewhere, (ii) develop an independent capacity to develop and/or adapt transgenic varieties, and (iii) collaborate on a regional basis to develop and/or adapt transgenic varieties. The second option is the most costly and requires the highest degree of national research capacity, while the first option depends on the availability of suitable varieties developed elsewhere. The third option would require a higher degree of cooperation across national boundaries than has typically characterized public sector research. Pingali and Traxler (2002) ask whether incentives exist or can be created for public/private partnerships that allow the public sector to use and adapt technologies developed by the private sector. The implications of these options for crop genetic diversity are discussed below.

4. AGRICULTURAL MODERNIZATION, VARIETAL ADOPTION, AND CROP BIODIVERSITY

Crop genetic diversity has changed over time along with the modernization of agriculture and the evolution of plant improvement and the changing locus of agricultural research. Teasing out the effects of private sector research from those caused by the structural transformation of agriculture is not a simple task. This section examines the forces that have influenced the spatial and temporal spread of modern cereal varieties, including transgenic varieties, and their implications for crop genetic diversity.

4.1 Modern cereal varietal adoption patterns

Modern cereal varieties, developed by scientific breeding programs, began to spread through many of the countries now considered "industrialized" in the late 19th century. The green revolution accelerated this process and extended it into much of the developing world. The adoption of modern cereal varieties has been most widespread in land-scarce environments and/or in areas well connected to domestic and international markets, where the intensification of agriculture first began. Even in these areas, the profitability of modern variety adoption has been conditioned by the potential productivity of the land under cultivation. For instance, while modern rice and wheat varieties spread rapidly through the irrigated environments, their adoption has been slower in the less favorable environments—the drought-prone and high-temperature environments for wheat and the drought- and flood-prone environments for rice. Maize has an even spottier record in terms of farmer adoption of modern varieties and hybrids. For all three cereals, traditional landraces continue to be cultivated in the less favorable production environments throughout the developing world (Pingali and Heisey, 2001).

Evenson and Gollin (2003) provide information on the extent of adoption and impact of modern variety use for all the major food crops. The adoption of modern varieties (for 11 major food crops averaged across all crops) increased rapidly during the two decades of the green revolution, and even more rapidly in the following decades, from 9% in 1970 to 29% in 1980, 46% in 1990 and 63% by 1998. Moreover, in many areas and in many crops, first-generation modern varieties have been replaced by second- and third-generation modern varieties (Evenson and Gollin, 2003).

According to Smale (1997), the adoption of modern cereal varieties has been characterized first by a concentration on a few varieties followed by diversification as more varieties became available. In the 1920s, for example, a single variety accounted for more than 60% of the wheat crop in the northern and central parts of Italy. Single cultivars became similarly dominant in many countries in Europe and North America, as mechanization created a need for uniform plant types and uniform grain quality. As the process of modernization proceeded and the offerings of scientific breeding programs expanded, the pattern of concentration declined in many European and North American countries (Lupton, 1992; and Dalrymple, 1988, cited in Smale, 1997). Similarly, in the early years of the green revolution, the dominant cultivar occupied over 80% of the wheat area in the Indian Punjab, but this share fell below 50% by 1985. By 1990, the top five bread wheat cultivars covered approximately 36% of the global wheat area planted to modern varieties (Smale, 1997).

4.1.1 Implications of modern varietal distribution for crop genetic diversity

Whether the changes in crop varietal adoption described above have resulted in a narrowing of genetic diversity remains largely unresolved due to conceptual and practical difficulties[3] (Pingali and Smale, 2001). Scientists

---

[3] Crop genetic diversity broadly defined refers to the genetic variation embodied in seed and expressed when challenged by natural and human selection pressure. In applied genetics, diversity refers to the variance among alternative forms of a gene (alleles) at individual gene positions on a chromosome (loci), among several loci, among individual plants in a population, or among populations (Brown et al., 1990). Diversity can be measured by accessions of seed held in gene banks, lines or populations utilized in crop-breeding programs, or varieties cultivated by farmers (cultivars). But crop genetic diversity cannot be

disagree about what constitutes genetic narrowing or when such narrowing may have occurred. Several dimensions of diversity must be considered in this regard, including both the spatial and temporal variations between landraces and modern elite cultivars and the variation within modern cultivars. Hawkes (1983, cited in Smale, 1997) argued that the genetic diversity of landraces and modern varieties is incomparable by definition because landraces, which are mixtures of genotypes, "could not even be called varieties," and he called the range of genetically different varieties available to breeders the "other kind of diversity" (pp. 100-101). Smale (1997) argued that the range of genetic material available to breeders is not directly correlated with the number of varieties in use because a single modern variety may contain a more diverse range of genetic material than numerous landraces.

Scientists also disagree about what constitutes genetic narrowing within modern varieties. For example, Hawkes (1983) cites the introduction of the *Rht1* and *Rht2* dwarfing genes into wheat breeding lines as an example of how diversity has been broadened by scientific plant breeders, while Porceddu et al. (1988) argue that the spread of semidwarf wheat varieties during the green revolution led to a narrowing of the genetic base for that crop (Pingali and Smale, 2001).

These points imply that comparing counts of landraces and modern varieties or changes in the number of modern varieties over time may not provide a meaningful index of genetic narrowing. They also imply that even if reliable samples of the landraces originally cultivated in an area could be obtained, analyses comparing their genetic diversity might provide only part of the answer regarding genetic narrowing. Although the landrace in the farmers' field is a heterogeneous population of plants, it is derived from generations of selection by local farmers and is, therefore, likely to be local in adaptation. In contrast, the plants of a modern variety are uniform but the diverse germplasm in their genetic background may enable them to adapt more widely. The diversity in a modern variety may not be expressed until challenged by the environment. On the other hand, the landrace may carry an allele that occurs rarely among modern varieties and is a potentially valuable source of genetic material not only for the farmer who grows it today but also for future generations of producers and consumers. (Pingali and Smale, 2001).

4.2 Transgenic crop adoption

Like modern varieties, the adoption of transgenic crops depends in the first instance on economic factors. In addition to their purely agronomic

---

literally or entirely observed at any point in time; it can only be indicated with reference to a specific crop population and analytical perspective (Smale, 1997).

characteristics, a number of institutional factors will affect the farm-level profitability of transgenic crops, particularly in developing countries. Economic research is beginning to show that transgenic crops can generate farm-level benefits where they address serious production problems and where farmers have access to the new technologies. So far, however, these conditions are only being met in a handful of developing countries. These countries have been able to make use of the private sector innovations developed for temperate crops in the North. Furthermore, they all have relatively well-developed national agricultural research systems, intellectual property rights regimes, regulatory systems, and local input markets.

Qaim, Yarkin, and Zilberman (Chapter 14) summarize the available evidence on the varietal adoption of transgenic crops. The most widely adopted transgenic crops are available in a large number of varieties in the major markets (e.g., there are more than 1,100 varieties of RR soybean and more than 700 varieties of Bt maize in the United States). Traxler (2004) reports that more than 35 different Bt and Bt/HT cotton varieties are on the market in the United States.

The Chinese Academy of Agricultural Sciences (CAAS) has developed the only source of transgenic insect resistance independent of the Bt genes patented by Monsanto. Pray et al. (2002) reports that CAAS has developed more than 22 locally adapted transgenic cotton varieties for distribution in each of the Chinese provinces. The Monsanto *Cry1Ac* gene is also available in China through at least five varieties developed by D&PL (Pray et al., 2002). In contrast, in Argentina, Mexico, South Africa and elsewhere, only a few Bt cotton varieties are available, all containing the Monsanto *Cry1Ac* gene, and often imported directly from the United States without local adaptive breeding (FAO, 2004).

4.2.1 Implications of transgenic crops for genetic diversity

The impact of transgenic crops on crop genetic diversity, like that of conventional crops, is a complex concept. Multiple dimensions of diversity must be considered, including the diversity of plant types in farmers' fields and the genetic pedigrees of those plant types. Whether the introduction of transgenes, per se, will increase or decrease crop genetic diversity is a matter of debate. Transgenesis, by definition, broadens the genomic content of plants by introducing genetic material from organisms that would not naturally breed with the host plant. Furthermore, since transgenic techniques are more targeted than classical breeding approaches, it is technically feasible for many individual varieties or landraces to be transformed with selected transgenes, retaining a wider range of genetic diversity in the background material. However, widespread gene flow from transgenic crops to other modern varieties or landraces could eliminate nontransgenic options, arguably reducing crop genetic diversity.[4]

How transgenic crops will influence the diversity of plant types in farmers' fields depends largely on the forces shaping agricultural research, variety development, and adoption. If only a few transgenic traits or crop varieties are available and they are widely adopted, the spatial genetic diversity within agricultural fields could be reduced. The proprietary nature of private sector transgenic crop research means that germplasm is less readily shared between plant breeding programs than it was during the green revolution. The reliance on a narrower range of germplasm may lead to genetic narrowing beyond any effect associated with the transgenic trait, per se. On the other hand, if many genetically diverse locally adapted varieties become available at affordable prices, spatial diversity could increase. Temporal diversity could increase if the introduction of transgenic crops results in higher seed replacement rates among farmers, but unless a continual supply of new transgenic varieties is available, temporal diversity could subsequently decline.

Thus far, little evidence is available on the impacts of transgenic varieties on crop genetic diversity (Ammann, 2004). Sneller (2003) used coefficient of parentage analysis to determine whether the introduction of transgenic herbicide resistance in soybeans and the associated proprietary restrictions on germplasm exchange between breeding programs have resulted in a narrowing of genetic diversity within elite North American soybean cultivars. He concluded that the advent of transgenic herbicide-tolerant cultivars has had little impact on the diversity of soybean cultivars because of the wide use of this technology by many programs and its incorporation in many lines. In contrast, he found that restricted germplasm exchange among breeders has reduced the diversity among the elite lines available from some companies and cautioned that the elite soybean population was becoming subdivided by the source of germplasm.

Bowman et al. (2003, cited in Ammann, 2004) examined genetic uniformity among cotton varieties in the United States. They found that genetic uniformity has not changed significantly with the introduction of transgenic cotton cultivars. On the contrary, the dominance of both the single most popular cotton cultivar and the five most popular cultivars has declined compared with the years immediately prior to the introduction of transgenic varieties, suggesting that spatial diversity may have increased. These examples suggest that the impacts of transgenic varieties on crop genetic diversity may depend more on the economic and institutional setting in which they are deployed than on the technology itself.

---

[4] Scientists agree that gene flow is possible, although they differ on whether it matters in and of itself. Technical methods and crop management strategies can reduce the risk of gene flow (International Council for Science, 2003).

4.2.2 Scenarios for transgenic crop deployment and implications for crop biodiversity

Three scenarios for making transgenic technologies available in developing countries were mentioned above: (i) direct import, (ii) local development/adaptation, and (iii) regional cooperation. Each has different implications for transgenic crop adoption and crop genetic diversity.

In the first scenario, transgenic crop varieties developed elsewhere are imported directly on a commercial basis. In this scenario, farmers pay for the technology through the seed price and technology fees. Although some countries are currently planting imported transgenic varieties, it is unlikely that imported varieties provide optimal performance outside their original agroecological zone. Furthermore, commercial transgenic innovations are unlikely to be available for crops grown by small farmers in marginal areas, who are unlikely to have the financial means to afford them. Qaim and Traxler (2004) argue that the transgenic Bt cotton varieties available in Argentina were originally developed for the U. S. market and have lower agronomic potential yields than locally adapted conventional varieties. They identified this as one reason for the relatively slow adoption of Bt cotton in Argentina. The second (and main) factor cited by Qaim and Traxler (2004) were the high seed costs and technology fees for Bt cotton in Argentina, for which strict IPR protections are enforced. Due to the lack of local adaptation and their potentially high cost, imported transgenic varieties probably would not be widely adopted and the range of available varieties for a particular trait would be narrower. In these circumstances, the impact of transgenic technology on spatial genetic diversity would be small. In the areas where imported varieties are adopted, the narrow range of available varieties could contribute to genetic narrowing.

In the second scenario, each country would develop its own transgenic innovations or adapt imported technologies for local use. This scenario would depend crucially on the national research and regulatory capacity and the availability of transgenic constructs, either from the public sector or the private sector. Thus far, only China has brought independently developed transgenic constructs to the market. A few other countries may have the capacity to do so, but they are exceptional, and thus most countries will have to rely on imported constructs. In these countries, adaptive research could be conducted by the public sector, perhaps in cooperation with local seed companies that, in turn, are linked with a multinational firm through a joint venture or a licensing arrangement. Under these arrangements, licensing fees would be paid to the multinational company, but farmers would receive locally adapted varieties that potentially would be more profitable than imported varieties. The availability of a wider array of transgenic crops in locally adapted varieties would be expected to increase adoption rates, but the impact on crop genetic diversity is complex. The availability of many locally adapted transgenic varieties would promote both spatial and temporal diversity within the transgenic area. While the area planted to landraces and conventional varieties would probably be reduced, it is unlikely that they would disappear completely, as landraces have survived through the green revolution period. It is possible that transgenes could flow to landraces or conventional crops—inadvertently or by design— especially for open-pollinated crops, but the effect of this gene flow on biodiversity is a matter of debate. Gene flow could create legal and economic problems relating to the coexistence of transgenic varieties and other types of agriculture—landraces, conventional varieties, or organic—but the biodiversity implications are not clear.

The third option identified above would involve regional cooperation among public sector institutions in developing countries to develop and/or adapt transgenic innovations for local conditions. In this scenario, several small institutions could work together, or institutions in small countries could work with their counterparts in the International Agricultural Research Centers (IARCs) or in large neighboring countries. China, for example, could develop transgenic crops for its own tropical regions and share these with smaller neighboring countries where similar agroecological conditions apply. Regional cooperation would permit greater economies of scale in research, and could place small national research institutes in a stronger position to negotiate licensing fees with the multinational companies.

Regional cooperation would assist countries that have weaker research capacity, and could make a wider range of transgenic crops and varieties available than would occur in either of the previous scenarios. This would tend to promote adoption of a larger number of transgenic crops and a wider array of varieties. Area planted to landraces could be reduced but, as with the green revolution modern varieties, this would not necessarily constitute genetic narrowing. The availability of a wider range of varieties could contribute to genetic diversity.

While regional cooperation could be beneficial to the smaller countries, it is unclear whether larger countries would have the necessary incentives to participate. Public sector research institutes have generally conformed to national boundaries, often with the explicit goal of promoting the economic competitiveness of the national agricultural sector. Incentives that promote cooperation would need to be put in place for such a scenario to materialize. The IARCs could play a stronger role in promoting regional cooperation, as they did during the green revolution, but given their declining resources, it is unclear whether they will be able to do so.

5. CONCLUSIONS

The changing locus of agricultural research from the public to the private sector is influencing the kinds of crop technologies that are being developed and the ways they are being disseminated. This in turn will influence both spatial and temporal patterns of crop genetic resources. Transgenic crops, per se, may increase or decrease crop genetic diversity, depending on how they are regulated and deployed. For example, regulatory regimes that focus on the transgenic innovation rather on than the individual variety would tend to promote the development of a larger number of transgenic varieties.

The green revolution modern varieties were developed and disseminated largely by the public sector. The IARCs developed the improved germplasm and made it freely available to researchers in national institutions. The countries that most widely benefited from the green revolution were those that had or quickly developed strong national capacity in agricultural research. Researchers in these countries were able to make the necessary local adaptations to ensure that the improved varieties suited the needs of their farmers and consumers. However, since transgenics are often proprietary, they are more expensive and less accessible than green revolution technologies were. This means that national researchers may not have access, on affordable terms, to appropriate transgenic technologies and a diverse range of germplasm for breeding purposes. Thus there is a much stronger imperative for regionalized R&D to capture economies of scale and enhance the bargaining power of public research institutions relative to the technology suppliers.

The capacity to develop locally adapted transgenics is likely to lead to a wider range of relevant transgenic products (so more diversity of transgenics) and thus higher adoption and higher benefits to farmers. It is also more likely to lead to losses of areas planted to landraces and conventional varieties. Whether this would lead to genetic narrowing, however, is not entirely clear because varietal adoption cannot be directly associated with genetic diversity. It is unlikely that transgenics would entirely replace the landraces that have survived through the last 200 years of agricultural intensification and commercialization. Furthermore, transgenic varieties could be more genetically diverse than the landraces and conventional varieties they replace. The experience with varietal adoption in the early phases of agricultural modernization suggests that the rapid spread of a few transgenic varieties could be followed by a diversification as more varieties become available.

The international community, specifically the IARCs, could facilitate access to biotechnology for developing countries through sharing and coordinating research. Given their declining resources, however, the IARCs may not be able to play as strong a role in this area as they did during the green revolution. The IARCs and other international institutions can facilitate developing countries' access to biotechnology through other means, such as capacity-building and networks for research, regulation and IPR management. These institutional contributions may be as important, as the scientific research in making a wider range of transgenic crops and varieties available in developing countries and ensuring that transgenic technology promotes rather than detracts from crop genetic diversity.

REFERENCES

Ammann, K., 2004, Biodiversity and agricultural biotechnology: A review of the impact of agricultural biotechnology on biodiversity, Botanischer Garten Bern. (March 2004); http://www.botanischergarten.ch/EFB/Report-Biodiv-Biotech6.pdf.

Boserup, E., 1981, *Population and Technological Change: A Study of Long Term Change,* University of Chicago Press, Chicago.

Bowman, D., May, O., and Creech, J., 2003, Genetic uniformity of the US upland cotton crop since the introduction of transgenic cottons, *Crop Science*, **43**:515-518.

Brown, A., Clegg, M., Kahler, A., and Weir, B., eds., 1990, *Plant Population Genetics: Breeding and Genetic Resources,* Sinauer Associates, Sunderland, Massachusetts.

Byerlee, D., and Fischer, K., 2002, Accessing modern science: Policy and institutional options for agricultural biotechnology in developing countries, *World Dev.*, **30**(6):931-948.

Dalrymple, D. G., 1988, Changes in wheat varieties and yields in the United States, 1919-1984, *Agri. History* **62**(4):20-36.

de Groote, H., Mugo, S., Bergvinson, D., and Owuor, G., 2004, Debunking the myths of GM crops for Africa: The case of Bt maize in Kenya, Paper presented to FAO (February 2004).

Evenson, R. E., and Gollin, D., 2003, Assessing the impact of the green revolution: 1960-1980, *Science*, **300**:758-762.

FAO, 2004, Agricultural biotechnology: Meeting the needs of the poor, *The State of Food and Agriculture: 2003-04*, FAO, Rome.

Hawkes, J. G., 1983, *The Diversity of Crop Plants*, Harvard Univ. Press, Cambridge.

Hayami, Y., and Ruttan, V. W., 1985, *Agricultural Development: An International Perspective.* Johns Hopkins University Press, Baltimore, Maryland.

International Council for Science (ICSU), 2003, *New Genetics, Food and Agriculture: Scientific Discoveries – Societal Dilemmas* (March 2004); http://www.icsu.org.

James, C., 2003, *Preview: Global Status of Commercialized Transgenic Crops: 2003*, ISAAA Briefs No. 30, ISAAA: Ithaca, NY (March 2004); www.isaaa.org/kc/CBTNews/press_release/brief30/es_b30/pdf.

Lupton, F. G. H., 1992, Wheat varieties cultivated in Europe, in: *Changes in Varietal Distribution of Cereals in Central and Western Europe: Agroecological Atlas of Cereal Growing in Europe*, Vol. 4, Wageningen University, Wageningen, the Netherlands.

Morris, M., 1998, *Maize Seed Industries in Developing Countries.* Lynne Rienner Publishers: Boulder, Colorado.

Pingali, P., 1997, From subsistence to commercial production systems: The transformation of Asian agriculture, *Amer. J. Agri. Econ.* **79**:628-634.

Pingali, P., and Heisey, P. W., 2001, Cereal-crop productivity in developing countries: Past trends and future prospects, in: *Agricultural Science Policy*, J. M. Alston, P. G. Pardey, and M. Taylor, eds., IFPRI & Johns Hopkins University Press, Baltimore, Maryland.

Pingali, P., and Smale, M., 2001, Agriculture, industrialized. *Encyclopaedia of Biodiversity, Vol. 1.* Academic Press, pp. 85-97.

Pingali, P., and Traxler, G., 2002, Changing locus of agricultural research: Will the poor benefit from biotechnology and privatization trends? *Food Policy,* **27**:223-238.

Porceddu, E., Ceoloni, C., Lafiandra, D., Tanzarella, O. A., and Scarascia Mugnozza, G. T., 1988, Genetic resources and plant breeding: Problems and prospects, *The Plant Breeding International, Cambridge Special Lecture*, pp. 7-21. Institute of Plant Science Research, Cambridge, U. K.

Pray, C. E., Huang, J., Hu, R., and Rozelle, S., 2002, Five years of Bt cotton in China—The benefits continue, *The Plant Journal,* **31**(4):423-430.

Pray, C. E., and Naseem, A., 2003, The economics of agricultural biotechnology research, ESA Working Paper No. 03-07, FAO, Rome, Italy (March 2004); http://www.fao.org/es/esa.

Qaim, M., and Traxler, G., 2004, Roundup-ready soybeans in Argentina: Farm level, environmental and welfare effects, *Agr. Econ.*, forthcoming.
Smale, M., 1997, The Green Revolution and wheat genetic diversity: Some unfounded assumptions, *World Dev.*, **25**(8):1257-1269.
Sneller, C., 2003, Impact of transgenic genotypes and subdivision on diversity within elite North American soybean germplasm, *Crop Science*, **43**:409–414; http://crop.scijournals.org/cgi/reprint/43/1/409.pdf.
Traxler, G., 2004, The economic impacts of biotechnology-based technological innovations. ESA Working Paper No. 04-08, FAO, Rome, Italy (June, 2004); http://www.fao.org/es/ESA/pdf/wp/ESAWP04_08.pdf.

PART II.
Genetic Resources and Biodiversity: Economic Valuation and Conservation

Chapter 4

THE ECONOMIC VALUE OF GENETIC DIVERSITY FOR CROP IMPROVEMENT: *THEORY AND APPLICATION*

R. David Simpson
Resources for the Future, 1616 P Street NW, Washington, DC 20036

Abstract: It is often argued that maintaining genetic diversity is a valuable insurance policy against crop failure. In this paper the economic value of diversity is related to the scarcity of genetic resources. Different modes of analysis are proposed for qualitative and quantitative genetic attributes, although both approaches can be related to the theory of order statistics. Economic value is "value on the margin." While the total worth to society of agricultural genetic resources is immeasurable, the marginal value is typically much smaller. Additional biological diversity for use in agricultural improvement may be thought of as more "draws" to be taken from a random sample of potential outcomes in which only the best is chosen for cultivation. Value on the margin is then the expected improvement in the welfare realized from the best in the sample. When the numbers of potential progenitors available for agricultural improvement is large, the marginal value of additional biodiversity can become quite small. These principles are illustrated with an example from a teak forestry improvement program.

Key words: agriculture; biodiversity; economic valuation; genetic resources; qualitative characteristics; quantitative characteristics; social welfare; teak.

1. INTRODUCTION

One of the arguments often made for the maintenance of genetic diversity among agricultural crops and their wild relatives is that such diversity acts as a sort of "insurance policy" against the effects of unanticipated risks. New pest infestations or changes in climate, for example, can result in large crop losses or other reductions in yield unless resistance or adaptability can be bred in. In addition to this insurance function, there is also the more mundane matter of agricultural improvement. The broader is the set of materials from which breeders can draw to improve cultivated varieties, the more valuable those varieties will be.

The basic economic principle involved in the valuation of genetic material is straightforward, but reducing this principle to actual implementation is far from simple. Rather than launching immediately into a discussion as to why this is "far from simple," I will defer these matters to the final section of this chapter. I will begin instead by discussing some highly simplified, but illustrative models. I will present two basic models because, as I noted above, two considerations motivate concern for maintaining genetic diversity in agriculture. First, dramatic events could largely or perhaps totally wipe out a crop. Pest infestations or diseases are examples. Resistance to such threats is a qualitative characteristic. Qualitative characteristics often are related to the presence or absence of a single gene. Second, crops can be improved or adapted to small changes in circumstances by "optimizing" the combination of genes they contain. Characteristics involved in such incremental improvement or adaptation are generally linked to large numbers of genes. Such characteristics are categorized as quantitative characteristics, and involve attributes such as height and weight.[1]

In what follows I will develop two very simple models of the valuation of genetic resources. The first will treat considerations with respect to the potential to provide qualitative attributes, and the second, quantitative attributes. Some basic principles are the same between each. First, the social value of genetic improvement, either qualitative or quantitative, is related to the change in benefits arising from the particular improvement. Second, the contribution of a larger pool of genetic resources is, in expectation at least, to enhance the characteristics of those individual organisms chosen for cultivation. The principle of marginal analysis that underlies economic valuation, then, implies that the social value of genetic resources is related to the expected difference in attributes between the best individuals drawn from larger as opposed to smaller sets.

I will illustrate these points in the sections that follow. My intention here is more illustrative than descriptive. Thus, while I will provide an example drawn from some empirical work with which I have been involved, I will not attempt to provide an actual estimate of value. Again, I will defer a discussion of "how things really work" to a later section, and confine myself to a stylized example. That example is as follows. Consider a situation in which a pool of individual organisms of size N is being evaluated for their potential to develop a superior variety for commercial cultivation. I will suppose that:

---

[1] The actual observed values of quantitative attributes also depend on environmental circumstances. I will abstract from these for present purposes.

1. A single parent is identified, and any number of offsprings can be developed from this single parent organism. Moreover, the attributes of the parent are replicated exactly in the commercial offspring. Note that I am abstracting from both sexual reproduction and environmental variation in making these assumptions.
2. The parent is selected for a single attribute (although this might be a complex index incorporating a number of dimensions). In other words, I will not be considering a situation in which different organisms are selected for different purposes or growers care about diversifying their risks by planting different varieties.
3. I will suppose that the selection is made for purposes of planting a single generation of offspring. As I will argue later, one of the most complicated aspects of valuing genetic resources concerns the weight to be assigned to the contribution of one generation to the propagation of others. For the purposes of expositional clarity, I am going to abstract from this consideration for now.

Much of what I write here may be interpreted as applying to situations in which crops are propagated by conventional methods (although assumption 1 above might only literally be accomplished via large-scale cloning). With the advent of the gene-splicing methods of biotechnology, new methods of inserting valuable attributes into commercial varieties might be adopted. Rather than simply propagating the "best" of a set of parent plants, one might pick and choose among the genes of many members of the set, inserting only the "best of the best" in the commercial cultivars. This may presume, however, greater knowledge of genetic function than is now, or will likely soon be, available.

More generally, the effects of developments in biotechnology on the economic value of genetic diversity may be difficult to predict. On one hand, the potential scope over which the useful genes of one organism might be applied is greatly increased when genes can be inserted, rather than having to be bred into only those organisms that are sexually compatible. On the other hand, however, the same principle works in reverse: The number of potential sources of genetic material for use in the improvement in any one particular crop is greatly expanded.

In the next section I discuss the basic economic value of genetic resources in crop improvement. The second section that follows illustrates the application of this valuation framework to qualitative improvements. The third section illustrates its application to quantitative improvements. Following that, I present an application using data from "provenance trials" of teak trees in Thailand. The fifth section discusses the impediments to applying such approaches to "real world" valuation problems, and the sixth briefly concludes.

2. THE ECONOMIC VALUE OF GENETIC IMPROVEMENTS

Let us define an expected welfare function $W(N)$, i. e., the welfare, W, expected to be derived from conducting genetic improvements starting from a gene pool of N potential parent organisms. I will suppose that the welfare realized by consumers can be measured by the surplus they realize from the consumption of the commercial product, which is the area under the demand curve between zero and the quantity of actual consumption.[2] Let inverse demand be $p(q)$ where q is quantity consumed, and suppose that actual consumption is q_1. Then consumer surplus is

$$CS = \int_0^{q_1} p(q)dq \tag{1}$$

The cost of production may depend on both the total quantity produced and the attribute(s) for which selection occurs. For example, the identification of genes that convey pest resistance will save on the costs of pesticides employed.[3] Perhaps the most straightforward example of cost-saving genetic improvements is an increase in yield per plant. If a genetically improved plant can produce more grain, for example, per unit area planted than would a traditional variety, less land area will be required to produce the same total volume of grain.

Let us suppose now that the measure of the attribute of interest in the variety chosen for commercial propagation is θ. Let the cost of growing a quantity q_1 of a "type θ_1 variety" be $C(q_1, \theta_1)$. Then the welfare to be realized from growing improved crops would be

$$W = \int_0^{q_1} p(q)dq - C(q_1, \theta_1) \tag{2}$$

Now how does welfare change as a result of finding a "better-θ" variety of the crop? Suppose that the value of the attribute for which selection is

---

[2] I am, then, implicitly supposing that income effects are not important. This assumption is probably not unreasonable when purchases of the commodity in question comprise a relatively small share of the consumer's budget. If we are talking about potentially catastrophic collapses in the yield of important crops, however, this assumption would be suspect.

[3] It may be worth noting here that such improvements would also save on the *social* costs of pesticides employed, such as contamination of groundwater and incidental poisoning of beneficial organisms.

undertaken absent any efforts at improvement would be θ_0. Then if a "type θ_1" variety is planted, the change in welfare will be

$$\Delta W = \int_0^{q(\theta_1)} p(q)dq - C(q(\theta_1), \theta_1) - \left[\int_0^{q_1(\theta_0)} [p(q)dq] - C(q(\theta_0), \theta_0) \right].$$

This relationship is illustrated in Fig. 4-1. The output of this agricultural crop is initially determined by the intersection of the demand curve, $D = p(q)$ with marginal cost, $MC_0 = \partial C/\partial q$. It seems reasonable to suppose that an "improvement" in a crop will reduce the cost, and, specifically, the marginal cost, of growing it. Thus, I show the change in θ from θ_0 to θ_1 shifting marginal cost down from MC_0 to MC_1. This new intersection of supply and demand results in quantity produced shifting from $q_0 = q(\theta_0)$ to $q_1 = q(\theta_1)$.

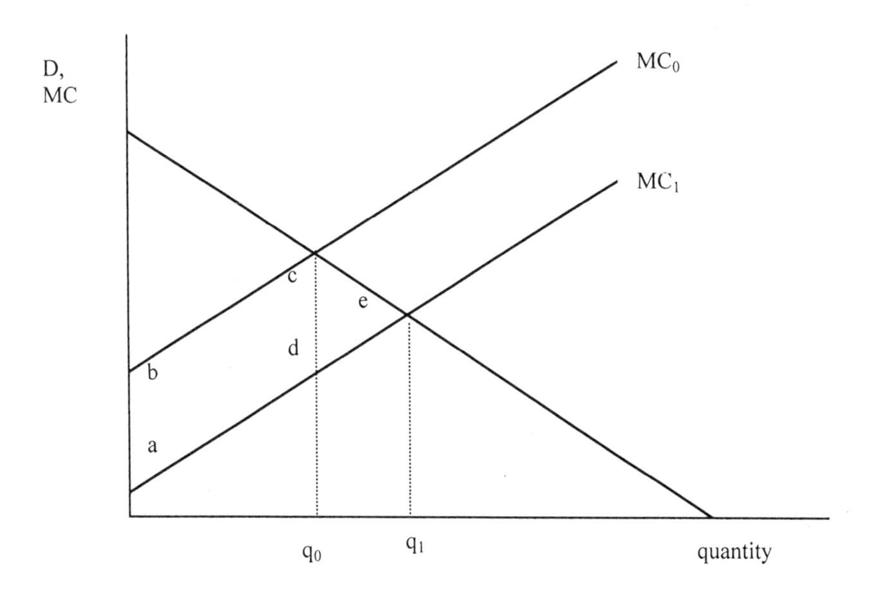

Figure 4-1. Welfare changes induced by genetic improvement

Now the nature of the change in welfare induced by the improvement in θ depends on the magnitude of that change. I have drawn the "before" and "after" marginal cost functions as straight lines, so part of the welfare change can be thought of as a "cost reduction" effect, and measured by the parallelogram abcd in the figure. The remainder of the welfare change can be thought of as an "output" effect, and measured by the triangle cde. In short, abcd measures the reduction in cost of producing q_0, while cde measures the benefits to consumers, net of incremental costs of production, of expanding output from q_0 to q_1.

Clearly, different aspects of the welfare change are going to be important depending on the nature of the genetic improvement. At one extreme we might think of the search for a genetic improvement that is absolutely essential if the crop is to be grown at all. For example, if a disease threatens to wipe out the entire crop, it may be essential to find a gene that confers resistance. In this case the welfare gain arising from the discovery of the gene would be the entire area between D and MC_1. On the other hand, if we're thinking of a situation in which plant breeders have already developed a relatively high-yielding variety, the welfare gain from finding a marginally better one will consist largely of the cost reduction effect, with the output effect being of the second order of importance.

In the examples considered in the next two sections, it seems that qualitative genetic attributes are likely to give rise to discrete improvements, and hence to situations in which welfare gains would combine cost reduction and output effects. Conversely, quantitative genetic improvements are more likely to give rise to incremental improvements, so welfare changes would be dominated by cost reduction effects. In short, then, we are likely to encounter a common but vexing problem in environmental valuation: comparing very unlikely, but potentially large effects with more likely but probably small effects.

2.1 Qualitative characteristics

The economic value of a genetic resource is related to the value of the expected outcome of a process of search for improved attributes with and without that particular genetic resource included in the set over which search is conducted. When we are considering single-gene qualitative characteristics, the probability distribution of outcomes is simple: a gene providing the required service is either available or it is not. More formally, the probability distribution function simplifies to a Bernoulli trial: The desired trait either is or is not present in the genome of a particular organism. Let us denote the probability with which the gene is found in a particular organism sampled as p, so it is not present with probability $1 - p$.

To keep the analysis tractable, suppose that each of N organisms in a population may contain a crucial genetic attribute with the same, independent, probability p (see Rausser and Small, 2000, for an example in which different organisms contain the desired gene with different probabilities). If an organism contains the crucial gene, a payoff of R is realized. This payoff, R, could be related to social welfare as in the previous section. For some crops, R might be astronomical: Consider, for example, the costs society would bear were it impossible to grow wheat, maize, rice, or potatoes. If the desired gene is not found, let us normalize the payoff to zero. Let us suppose that there is a cost c of evaluating any particular organism to determine if it exhibits genetic resistance to a particular pest. Combining the probability, payoff, and cost considerations, the value of the "marginal organism" with respect to its expected contribution to the development of genetic resistance to a particular pest is

$$v = (pR - c)(1 - p)^N \tag{3}$$

Heuristically, the value of the "marginal organism" is its expected value net of the cost of testing, $pR - c$, times the probability that the desired gene is not found among any of the N other organisms.

For fixed values of R and c, the maximum value v can take on is

$$v^* = \frac{R - c}{(N+1)e}\left(\frac{R - c}{R}\right)^N \tag{4}$$

(Simpson, Sedjo, and Reid, 1996).[4] Clearly, this becomes a small number as N gets large. Eq. (4) results when it is assumed that the probability with which any particular organism contains the necessary gene just happens to be that which maximizes the value of the marginal organism. For values of p other than those that approximately maximize the value of the marginal organism, v would be smaller yet. The issue then concerns the magnitudes of the payoff to successful discovery, R, the cost of testing, c, the probability, and uncertainty concerning the probability, p, with which success occurs, and the number of organisms over which testing can occur.

It is difficult to say exactly how these considerations interact with one another. Any conjectures are necessarily controversial given the magnitudes of the uncertainties involved. Perhaps if, as suggested above, humanity were truly confronted with the loss of a widely grown staple crop such as wheat, maize, rice, or potatoes, the prospect of so calamitous a loss would translate into a substantial value for the "marginal" element of genetic diversity. Given the existing quantity of potential "solutions" to such challenges, however, the value of diversity may still not be great with respect to the potential to provide qualitative genetic traits even in important crops.

2.2 Quantitative characteristics

---

[4] This expression is derived by differentiating Eq. (3) with respect to p, setting the result equal to zero, and using the approximation, $(n + 1)^n/n^n \approx e$, where e is the base of the natural logarithm, approximately 2.718.

Let us now consider quantitative characteristics. These characteristics, recall, depend on the combined effects of a number of genes. While the analysis could be presented at a higher level of generality by supposing arbitrary probability distribution functions defined over quantitative genetic attributes observed in populations, both theory and empirical evidence suggest that many quantitative attributes are distributed normally. Quantitative attributes can be regarded as arising from the approximately additive effects of a large number of individual genes. It is a fundamental principle of genetics—the "law of independent assortment"—that the random combination of genetic attributes is statistically independent (Falconer and Mackay, 1996). Hence, a large number of statistically independent random variables are added together, and the conditions for the central limit theorem are satisfied: The quantitative attribute is normally distributed.[5]

Returning to Eq. (2), the value of production of a variety of type θ is

$$W = \int_0^{q(\theta)} p(q)dq - C(q(\theta), \theta)$$

Assume that the values of θ encountered in N trials are independently and identically normally distributed with probability $\phi(\theta|\mu, \sigma^2)$. The distribution of the greatest value of θ encountered in these N trials is, then, the distribution of the greatest order statistic from N draws. Suppressing the mean and variance arguments of the normal distribution, then, the probability density of this greatest order statistic is

$$N\phi(\theta)\Phi(\theta)^{N-1}$$

where $\Phi(\theta)$ is the cumulative normal density (David, 1981). Thus, I can write the expectation of welfare resulting from choosing the best among N potential parents as

$$E(W) = \int_{-\infty}^{\infty}\left(\int_0^{q(\theta)} p(q)dq - C(q(\theta), \theta)\right)N\phi(\theta)\Phi(\theta)^N d\theta$$

[5] Some quantitative attributes arise from combinations of normally—and not necessarily independently—distributed variables. Consider, for example, the weight of an individual whose height is drawn from a normal distribution. It is likely related to some multiplicative combination of height and girth, which will not be normally distributed. While acknowledging such possibilities, the resulting variables may still be "closely enough" to normally distributed as to make the approximation reasonable.

Technically speaking, in order to find the effects of a change in N on expected welfare, one would need to consider the derivative of the above expression with respect to N. The problem can be made simpler, however, by taking a reasonably close approximation. This approximation is derived by ignoring Jensen's inequality[6] substituting the "function of the expectation" for the "expectation of the function," i. e.,

$$E(W \mid N) \approx \int_{0}^{q[E(\theta \mid N)]} p(q)dq - C\left(q[E(\theta \mid N)], E(\theta \mid N)\right) \tag{5}$$

Note that I have stated the expectation of welfare conditioned on the size of the set from which samples are drawn. This arises because the greatest order statistic is also stated as a function conditioned on N, $E(\theta|N)$. Differentiating with respect to N, I have

$$\frac{\partial E(W)}{\partial N} \approx \left(p - \frac{\partial C}{\partial q}\right)\frac{\partial q}{\partial \theta}\frac{\partial E(\theta \mid N)}{\partial N} - \frac{\partial C}{\partial \theta}\frac{\partial E(\theta \mid N)}{\partial N} = -\frac{\partial C}{\partial \theta}\frac{\partial E(\theta \mid N)}{\partial N}$$

where the second equality results because price is equal to marginal cost in competitive equilibrium.

Tractable results are further facilitated by taking advantage of another approximation. It is intuitively straightforward that the expectation of the greatest order statistic in a sample of size N is approximately that value of the random variable for which a fraction $N/(N + 1)$ of the sample is less than that value (a formal proof, including bounds on the approximation, is presented in David, 1981). Thus,

$$\Phi[E(\theta \mid N)] \approx \frac{N}{N+1}$$

or

$$1 - \Phi[E(\theta \mid N)] \approx \frac{1}{N+1} \tag{6}$$

Differentiating with respect to N and rearranging, we have

---

[6] Jensen's inequality is a theorem establishing that the expectation of a concave (convex) function is less (greater) than the value of the function evaluated at the expectation of its argument.

$$\frac{\partial E(\theta\,|\,N)}{\partial N} \approx \frac{1}{\phi(N+1)^2}$$

or using (6),

$$\frac{\partial E(\theta\,|\,N)}{\partial N} \approx \frac{1}{N+1}\frac{1-\Phi}{\phi} \tag{7}$$

where the cumulative and probability density functions are evaluated at $E(\theta|N)$. The expression $\phi/(1 - \Phi)$, whose inverse appears in Eq. (7), is encountered frequently in statistical and econometric applications. It is known as the hazard rate, defined as the probability with which some event occurs for values of θ in excess of $E(\theta|N)$, conditional on it not having occurred yet for any value of θ less than $E(\theta|N)$. The hazard rate is often abbreviated as λ. Using this shorthand, we can combine expressions to restate the change in welfare with respect to a change in the genetic diversity from which superior varieties can be drawn as

$$\frac{\partial E(W)}{\partial N} \approx -\frac{\partial C}{\partial \theta}\frac{1}{(N+1)\lambda} \tag{8}$$

The remaining conceptual task is to consider $\partial C/\partial\theta$. The form of this expression will depend in general on the nature of the genetic improvement introduced. While this could take many forms, let me offer one general consideration and then examine one special case. The general consideration is the following. Eq. (7) can be rearranged as

$$\frac{\partial E(W)/E(W)}{\partial N/N} \approx \frac{\dfrac{\partial C/C}{\partial\theta/\theta}}{E(\theta\,|\,N)\lambda}\frac{C}{E(W)} \tag{9}$$

That is, the elasticity of expected welfare with respect to N is approximately[7] equal to the elasticity of cost with respect to θ times the ratio of cost to expected welfare, all divided by the hazard rate times the expectation of θ conditioned on N. There are a number of instances—the special case I am about to discuss being a prominent example—in which one

---

[7] Another, albeit small for large N, source of error is introduced in the approximation by using $N + 1$ rather than N.

would expect the elasticity of cost with respect to θ to be relatively small. We might also expect the ratio of costs to welfare to be small in many instances of interest. This leaves the expression $E(\theta|N)\lambda$ in the denominator, which is the elasticity of the probability of finding an individual with a θ value greater than $E(\theta|N)$, $1 - \Phi[E(\theta|N)]$, with respect to $E(\theta|N)$. It can be shown that this expression grows without bound in the limit as $E(\theta|N)$ grows large, so, not surprisingly, the marginal value of additional resources must eventually be negligible.[8]

The special case arises when θ denotes yield per unit area planted and land is the only purchased input in production. If land devoted to a particular crop is T and yield per unit land is θ, total production is

$$q = \theta T \tag{10}$$

Let the rent on land be r, so the cost of production is

$$C(q, r, \theta) = \frac{rq}{\theta} \tag{11}$$

Thus,

$$\frac{\partial C}{\partial \theta} = -\frac{rq}{\theta^2} = -\frac{C}{\theta} \tag{12}$$

or

$$\frac{\partial C/C}{\partial \theta/\theta} = -1 \tag{13}$$

Returning to Eq. (9), let us think of how the fraction $C/E(W)$ might be evaluated. Suppose for the purposes of illustration that demand is of the form $p(q) = \beta q^{-\eta}$ for some constant β. Since $p = \partial C/\partial q$, and we are assuming marginal cost is constant given θ, we have[9]

---

[8] This result, while not implausible, is in some ways an artifact of the specification. Craft and Simpson (2001), for example, show that, with multiple interacting products, the value of "marginal leads" need not be negligible even in the limit.

[9] If η were greater than one the integral would not be bounded; the crop would be essential and willingness to pay would go to infinity as quantity shrank to zero.

$$\frac{C}{E(W)} \leq \frac{E(q_1)^{1-\eta}}{\int_0^{E(q_1)} q^{-\eta}dq - E(q_1)^{1-\eta}} = \frac{1-\eta}{\eta} \tag{14}$$

Finally, in order to evaluate $E(\theta|N)$, we can revert to Eq. (6). Inverting the cumulative density function—which is monotonic, of course, and thus invertible—we have

$$E(\theta \mid N) \approx \Phi^{-1}\left(\frac{N}{N+1}\right)$$

Thus, combining the results emerging from our specific assumptions, we have

$$\frac{\partial E(W)/E(W)}{\partial N/N} \approx \frac{1}{\Phi^{-1}\left(\frac{N}{N+1}\right)\lambda\left(\Phi^{-1}\left(\frac{N}{N+1}\right)\right)}\frac{1-\eta}{\eta} \tag{15}$$

3. AN APPLICATION TO TEAK IMPROVEMENT

Teak (*tectona grandis*) is native to Thailand, Burma (Myanmar), Laos, and India. Plantations were established in Indonesia as early as the 14th century, and more recently in Bangladesh, Sri Lanka, China, Vietnam, and New Guinea. In addition, the tree has been grown in Mexico, Puerto Rico, Cote d'Ivoire, Ghana, Togo, Tanzania, and Nigeria (Keiding, Wellendorf, and Lauridsen, 1986). There are today some 2.2 million hectares under cultivation in teak (Ball, Pandey, and Hirai, 1999).

Teak is the subject of selective breeding experiments in Thailand (Kaosaard, Suangtho, and Kjær, 1998). It has been estimated that selective breeding and the establishment of plantations incorporating genetically superior trees could result in improvements in yield of 17% or more. Areas of natural teak forest are being felled when land is converted to agriculture or other purposes. There is concern expressed then that genetic reservoirs that could be used to improve subsequent generations of commercial teak are being eliminated.

Teak breeders have found it useful to establish "provenance trials" in which they grow trees under controlled circumstances in order better to distinguish between genetic (and consequently, heritable) and environmental factors in performance. A provenance trial is an experiment in which teak trees (or other types of organisms) from different regions ("provenances") are grown under controlled circumstances in order to identify the genetic contribution to the appearance of attributes of commercial importance. A number of such trials have been conducted in Thailand and elsewhere. I will employ data from a long-running experiment on several international provenances conducted in Thailand (Keiding, Wellendorf, and Lauridsen, 1986; Kjaer, Lauridsen, and Wellendorf, 1995; Anderson, 1997).

Such trials collect data on a large number of quantitative and qualitative characteristics of the trees in the sample. As my purpose at present is merely illustrative, I will concentrate only on "commercial height." Commercial height is defined to be the maximum height at which diameter is at least 10 centimeters. More generally, of course, one would care not only about height, but also diameter and, more generally, wood volume, as well as quality of wood. In addition, owners of commercial plantations would want to plant trees that are known to be resistant to infestations and have other desirable survival and input cost minimization properties. Again, however, let us, for simplicity, simply concentrate on commercial height as a measure of yield-per-hectare.

The sample mean and standard deviation are sufficient statistics for normally distributed variables. The mean commercial height among 578 trees from the provenance trial we are considering was 11.6 meters. The standard deviation was 4.0 meters. From these sufficient statistics, we can derive the cumulative normal distribution, its inverse, and the corresponding hazard rate. These figures can then be employed in Eq. (15) above. The remaining variables are N, the number of potential parent organisms, and η, the elasticity of demand. The number, N, is varied in the left-most column of Table 4-1, and the elasticity, η, is varied in the top row of the table. Entries in each cell of the table show, for that number and elasticity, the elasticity of expected social welfare with respect to the size of the size of the genetic base from which selection can occur.

Table 4-1. Elasticity of expected welfare as a function of number of potential parents and elasticity of demand

Number of potential parents			Elasticity of demand		
	0.10	0.25	0.50	0.75	0.90
10	1.181	0.394	0.131	0.044	0.015
100	0.645	0.215	0.072	0.024	0.008
1,000	0.446	0.149	0.050	0.017	0.006
10,000	0.344	0.115	0.038	0.013	0.004
100,000	0.281	0.094	0.031	0.010	0.003
1,000,000	0.237	0.079	0.026	0.009	0.003

4. DISCUSSION

The calculations reported in Table 4-1 indicate that, as expected, the elasticity of expected welfare in the number of potential parent organisms declines in the number of potential parents, keeping the elasticity of demand fixed, and in the elasticity of demand, keeping the number of potential parents fixed. As the elasticity I have reported is the percentage change in welfare resulting from a 1% change in the number of potential parents, the change in welfare resulting from a small change in the absolute number of potential parents would be small indeed in most of the instances I have reported. While it would be difficult to estimate the total welfare derived from the consumption of something like teak wood, the results I have reported suggest that the incremental value of additional genetic resources would be modest.

This conclusion should be qualified in a number of ways. First, I have been assuming that consumer surplus is measured under uncompensated demand curves. As is well known, this approach is valid when income effects are not important. While this may not be too unreasonable of an assumption in discussing the demand for teak wood, it would be more questionable when applied to, for example, rice or wheat. Staple crops may claim large shares of income, at least when they become rare. Clearly, the stakes rise as the scope of use increases. Martin Weitzman (2000) has recently done interesting work in which he considers the tradeoffs between, on the one hand, maintaining a diversity of organisms to protect against the failure of each type of crop and, on the other, the opportunity costs of growing varieties anticipated to be less productive. The general issue he identifies—the desire to prevent very low probability but also very catastrophic outcomes—is important, but difficult to resolve.

Another major omission of the approach I have taken here concerns the treatment of dynamic considerations. One could make the approach I have illustrated dynamic by supposing that, in every period, breeders must identify a variety that best suits a set of conditions that have completely changed since the previous round of selection occurred. It may, in fact, be the case that this complete-change-in-circumstances scenario leads to the highest estimate of value for the "marginal potential parent": it would seem that, to the extent that selection continues to occur for the same traits over time, a finite upper bound would eventually be approached. More generally, however, each new generation of propagated organisms would comprise a new random draw from the gene pool. Thus, one might, in a more complex analysis, want to consider not only the direct benefits of genetic diversity in terms of producing superior varieties for immediate cultivation, but also for producing varieties that might in turn produce still other varieties.

This leads me to the final consideration that I will address here. I have supposed that the process of selection involves the identification of the single "best" individual, and that unlimited numbers of identical individuals can be replicated exactly from this source. In practice, selective breeding typically involves the identification of a group of individuals and the attributes of their offspring reflect only imperfectly those of the parents. The analysis I have presented here can be extended to consider selection of a set of parent organisms and to incorporate imperfect heritability of attributes.[10] Results are compromised somewhat by the practical necessity of abstracting from Jensen's inequality in order to maintain tractability (results are much more easily derived by supposing each offspring organism exhibits the attribute at the mean level). Given the other imprecisions inherent in the analysis, however, this does not seem to be a major impediment to deriving illustrative results of the type I have illustrated here.

I might note in closing that the model I have sketched is becoming dated by advances in biotechnology. I have supposed that the only way to generate improved commercial varieties is to identify promising "packages of genes" in the form of parent organisms. As the process of agricultural improvement comes to rely more and more on the insertion of favorable individual genes, the search for superior quantitative characteristics may come more and more to resemble that for favorable qualitative characteristics. Moreover, the ability to transplant genes even between different species (and sometimes higher phylogenetic taxa) may imply that genetic resources are becoming less and less scarce with respect to particular crop improvement applications and, hence, of lower economic value.

[10] A formula commonly encountered in quantitative genetics (see, e.g., Falconer and MacKay, 1996), holds that

$$\mu_1 - \mu_0 = h^2\left(\mu^s - \mu_0\right)$$

where μ_0 is the mean of the original population from which selection occurs, μ^s the mean of the selected population (where selection is accomplished by choosing all those individuals for which the observed value of some attribute exceeds a certain level; in other words, μ^s is the mean conditional on the observed attribute exceeding the selection criterion), μ^1 is the mean of the offspring of the selected parents, and h^2 is *heritability*, a measure of the correlation between parent and offspring attributes.

REFERENCES

Anderson, L., 1997, *Reassessment of an International Teak (Tectona Grandis Linn. f) Provenance Trial at Age 23*, Danida Forest Seed Centre, Humlebaek, Denmark.

Ball, J. B., Pandey, D., and Hirai, S., 1999, Global overview of teak plantations, paper presented to the Regional Seminar on Site, Technology and Productivity of Teak Plantations, Chiang Mai, Thailand (January 26-29, 1999).

Craft, A. B., and Simpson, R. D., 2001, The value of biodiversity in pharmaceutical research with differentiated products, *Environ. and Resource Econ.* **18**:1-17.

David, H. A., 1981, *Order Statistics,* Wiley, 2nd edition, New York.

Falconer, D. S., and Mackay, T. F. S., 1996, *Introduction to Quantitative Genetics*, Addison-Wesley, 4th edition, Boston.

Kaosaard, A., Suangtho, V., and Kjær, E. D., 1998, Genetic improvement of teak (*tectona grandis*) in Thailand, *Forest Genetic Resour.* **26**:21-29.

Keiding, H., Wellendorf, H., and Lauridsen, E. B., 1986, *Evaluation of an International Series of Teak Provenance Trials,* DANIDA Forest Seed Center, Humlebaek, Denmark.

Kjaer, E., Lauridsen, E. B., and Wellendorf, H., 1995, *Second Evaluation of an International Series of Teak Provenance Trials,* DANIDA Forest Seed Center, Humlebaek, Denmark.

Rausser, G., and Small, A., 2000, Valuing research leads: bioprospecting and the conservation of genetic resources, *J. Polit. Econ.* **108**:173-206.

Simpson, R. D., Sedjo, R. A., and Reid, J. W., 1996, Valuing biodiversity for use in pharmaceutical research, *J. Polit. Econ.* **104**:163-185.

Weitzman, M. L., 2000, Economic profitability vs. ecological entropy, *Quart. J. Econ.* **115**: 237-263.

Chapter 5

MANAGING CROP BIOLOGICAL DIVERSITY ON FARMS

Melinda Smale
International Plant Genetic Resources Institute and International Food Policy Research Institute, 2033 K Street NW, Washington, DC 20006-1002

Abstract: Managing the diversity of crop genetic resources on farms is of economic importance because it is a survival strategy for some of the world's rural poor, though conserving them on farms also reduces the loss of potentially valuable alleles in genetic stocks still held locally. International agreements encourage the design of benefit-sharing schemes to support conservation through rewarding farmers—but mechanisms for doing so are still unclear. "Win-win" policy solutions occur when crop biodiversity is maintained on farms for the benefit of future generations while farmers themselves benefit today from a wider set of crop variety attributes for consumption or sale. Empirical approaches, guided by theoretical principles in economics and genetics, can be used to identify locations with high benefit-cost ratios for on-farm conservation. Policy instruments to support conservation in those locations include supply-related mechanisms such as community genebanks, biodiversity registers, and the introduction of modern varieties that complement the range of traits found in local varieties. When markets are not well developed and transaction costs are high, farmers' supply of diverse crop varieties, and their derived demand for local landraces, can be enhanced by participatory plant breeding. As incomes rise and commercial markets develop, landraces will continue to be grown if there is consumer demand for a unique attribute that cannot be easily bred into or transferred to modern varieties. Yet, the necessary conditions for such market-based initiatives are not often met in the locations where landraces are cultivated. Protecting one landrace does not necessarily have desirable implications for either diversity conservation or social equity. The relative costs and benefits of such instruments have not yet been assessed rigorously.

Key words: crop biological diversity; crop genetic resources; economic development; on-farm conservation.

1. INTRODUCTION

Empirical research has documented that in harsh, isolated environments where climatic and soil conditions are variable, farmers may depend on the cultivation of multiple crops and varieties to meet their food and cash needs. For farm households to be food secure, they require stable supplies for consumption from either their own production or market purchases.

As markets develop, farm households generally specialize in fewer products oriented toward the demands of distant consumers, relying less on a portfolio of crop varieties and more on a portfolio of income sources to smooth their consumption. Yet, those in isolated areas continue to face heavy transactions costs because they have limited and uncertain options for buying and selling in markets. They have a "demand" for crop biological diversity[1] that is derived from the range of production traits and consumption attributes they require. Cultural autonomy may reinforce this demand through shaping the preferences of rural people for the food they consume, their perceptions of crop biological diversity, and its importance.

Managing the biological diversity of crop genetic resources on farms is of economic importance in part because it is a survival strategy for some of the world's rural poor. Conserving these resources on farms also reduces the loss of potentially valuable alleles in genetic stocks still held by farmers. Geneticists often hypothesize that rare, locally adapted genotypes may be found among the landraces[2] cultivated by farmers in such extreme or heterogeneous environments. Some genotypes are thought to contain tolerance or resistance traits that are not only valuable to the farmers who grow them but also to the global genetic resource endowment on which future crop improvement depends. Rare alleles are often discovered in centers of origin, though depending on the crop, valuable diversity can often be found elsewhere.

Genetic resources are renewable assets, but are renewed in farmers' fields only as long as farmers continue to sow the seed. When farmers replace "landraces" with "modern varieties" that may be more attractive to them as their economies industrialize, the gene combinations in landraces may be "lost" to future generations of farmers and consumers unless special efforts are made to collect them or encourage their continued cultivation.

Since the 1970s, large numbers of landraces and wild relatives of cultivated crops have been sampled and stored in *ex situ* gene banks. An alternative form of conservation *in situ* has also received some scientific attention (Brush, 2000; Maxted, Ford-Lloyd, and Hawkes, 1997). For cultivated crops, conservation of genetic resources *in situ* refers to the continued cultivation and management by farmers of crop populations in the

---

[1] The biological diversity of crops encompasses phenotypic as well as genotypic variation, including cultivars recognized as distinct by farmers and varieties recognized as genetically distinct by plant breeders.

[2] Generally, landraces are considered to be relatively more heterogeneous plant populations selected and adapted by farmers over generations, as compared to relatively more uniform, stable modern varieties bred by professional plant breeders.

agroecosystems where the crop has evolved (Bellon, Pham, and Jackson, 1997). Storing genetic resources in collections as backup seed stocks in *ex situ* collections therefore substitutes imperfectly for the evolution of crop plants in the fields of farmers.

Not only do genetic resources evolve differently when conserved *ex situ* and *in situ,* but the distributions of their economic benefits and costs also differ fundamentally. The costs of genetic resource conservation in gene banks are now borne largely by public investments, and consumers (as well as farmers who are consumers) benefit indirectly from the genetic resources incorporated into improved crop varieties when output expands and prices decline. In contrast, both the costs and benefits of conserving genetic resources *in situ* are felt directly (and in a very immediate sense) by the farmers who grow them.

To suggest that some of the poorest farmers of the world should shoulder the full burden for conservation of crop biodiversity seems inappropriate. For this reason, international agreements such as the Convention on Biological Diversity and the International Treaty on Plant Genetic Resources for Food and Agriculture encourage the design of benefit-sharing schemes to support conservation through rewarding farmers for their innova-tions—though mechanisms for doing so are still under discussion. National and local policy instruments that promote development and farmer management of crop biological diversity may also be feasible. A "win-win" policy solution occurs when crop biodiversity is maintained on farms for the benefit of future generations, while farmers themselves benefit today from a wider set of crop variety attributes for consumption or sale. Under those circumstances, private and social benefits coincide.

This chapter begins by reviewing how economic concepts can assist in identifying promising locations for on-farm management of crop biological diversity, supported by some findings from empirical analyses. Next, some of the policy mechanisms that have been invoked to support farmers in such locations are discussed and, where possible, evidence regarding their effectiveness is presented.

2. THE VALUE OF CROP BIODIVERSITY ON FARMS

As a production input and private good, seed is highly rival with low cost of exclusion (Fig. 5-1). The genetic resources embodied in seed are non-rival, however, and the costs of controlling their use on farms are relatively high. This means that two farmers cannot plant the same handful of seeds, but many farmers may grow the same variety simultaneously. Controlling the flow of genes among fields is difficult, especially with predominantly cross-pollinating crops as they are managed by farmers in less-commercialized agricultural systems.

The combinations of seed types grown by farmers produce a harvest that they consume and/or sell and from which they derive private value, but the pattern of genotypes across the landscape contributes to the biological diversity of the crop genetic resources from which people residing elsewhere and in the future may benefit. The public value of crop biological diversity includes insurance value for potential disasters and option value for any unforeseen events, such as changes in consumer tastes.

The private value includes utility or satisfaction from the agronomic traits and consumption attributes that these cultivars provide to farmers as producers and/or as consumers. In the special case of a commercial farmer producing for a well-defined market, that utility is related exclusively to profits derived from sales.

Since the biological diversity of crop genetic resources is never fully apparent to the farmers who provide and use it and is undervalued in markets, farmers are unable to consider the contributions of all other farmers to genetic diversity in their community or elsewhere when they make their decisions. Hence, biological diversity of crop plants has interregional and intergenerational dimensions (Fig. 5-1). Economic theory predicts that, as long as crop diversity is a (desirable) "good", farmers as a group will generate less diversity than is socially optimal (Cornes and Sandler, 1986; Heisey et al., 1997).

Institutional structures are needed to compensate for the inability of markets to provide sufficient incentives for farmers to allocate their resources in ways that are consistent with the needs of society. These structures will differ according to culture, as well as the temporal and spatial dimensions of the impure public good (Sandler, 1999). Some societies have much stronger collective behavior than others. All have norms related to management of genetic resources. At the community level, depending on the social and economic conditions farmers face, community awareness campaigns may be sufficient to ensure that certain materials of genetic importance continue to be grown. By contrast, complex structures are necessary to mediate conservation interests at the global level because actors may not perceive that they share common interests. The International Treaty on Plant Genetic Resources for Food and Agriculture and the Convention on Biological Diversity are elements of such structures, though these may not be consistent with local norms of use and access. Therefore, the extent of public investment and the policy mechanism needed to narrow the divergence between what individuals and societies perceive as optimal clearly depends on many factors.

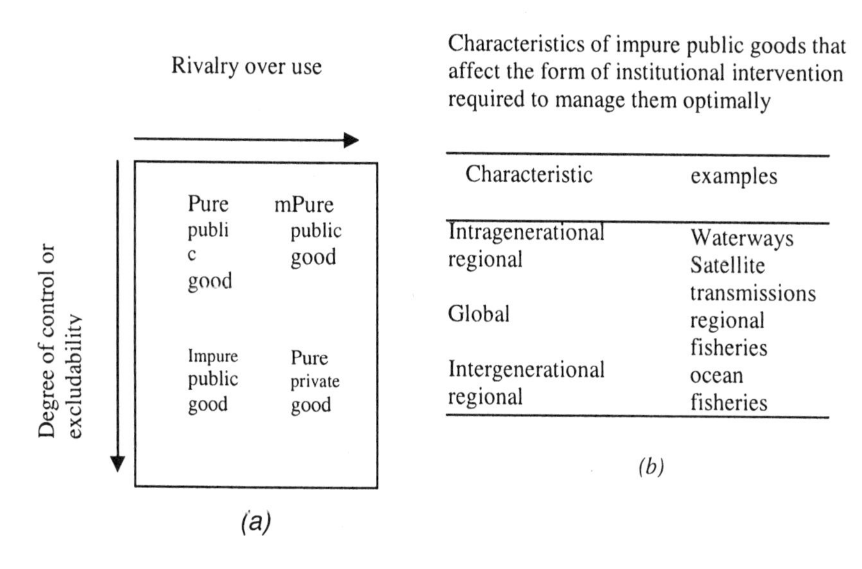

Characteristic	examples
Intragenerational regional	Waterways Satellite transmissions
Global	regional fisheries
Intergenerational regional	ocean fisheries

(b)

Figure 5-1. Simplified taxonomies of goods

Source: Adapted from Romer (1993, p. 72) and Sandler (1999, p. 24).

3. IDENTIFYING PROMISING LOCATIONS FOR MANAGING CROP BIODIVERSITY ON FARMS

3.1 High benefit-cost ratios

Not all global locations are equally promising candidates for managing crop biological diversity on farms. The highest benefit-cost ratios for on-farm conservation of diverse crop genetic resources occurs where the utility farmers derive from managing them as well as the public value associated with their biological diversity are high. This occurs conceptually in area II of Fig. 5-2. Since farmers are already bearing the costs of maintaining diversity de facto in those areas and they reveal a preference for doing so, the costs of public interventions to support conservation will also be least. Where genetic diversity is assessed as relatively low and farmers derive few benefits from it, there is no need to invest in any form of conservation (III, Fig. 5-2). Where the contribution to diversity is great but farmers derive little private value from it, *ex situ* conservation is the only option (I, Fig. 5-2). Where there is little diversity but farmers care a lot about it, there is no need for public investment at all since no value is associated with conservation (IV, Fig. 5-2).

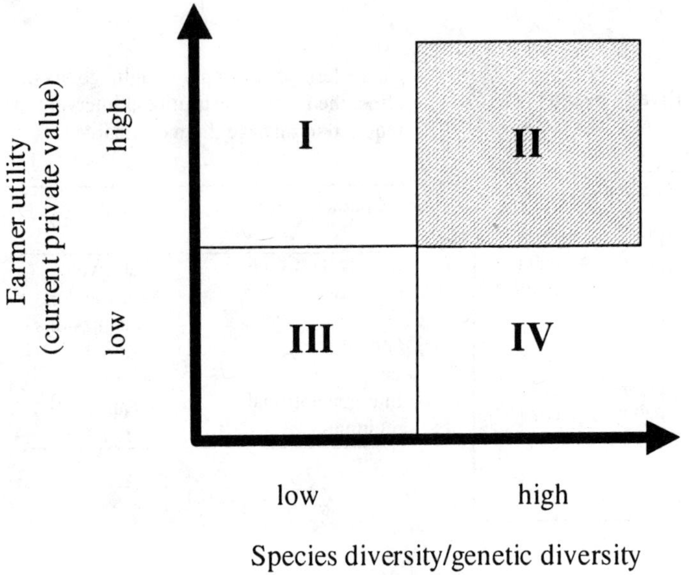

Figure 5-2. Sites with high benefit-cost ratios for on-farm conservation

Source: Adapted from Smale and Bellon (1999, p. 395).

An example of an "I" location was found in the uplands of Nepal for rice.[3] Farmers recognized many varieties but their genetic diversity when characterized was found to be relatively low. No modern varieties compete in that environment, so there were no opportunity costs associated with growing farmers' cultivars. An example of a "II" location was found in Mali.[4] There, despite 26 years of climatic change and drought, analysis of seed samples demonstrated that farmers maintained the same overall level of sorghum diversity, though its spatial distribution had shifted. This is strong empirical evidence that farmers depended on sorghum diversity to manage risk *ex ante* (as compared to those in more favored, less isolated environments who manage risk through markets, *ex post*). Geneticists also consider the range of traits found among those varieties to be important.

---

[3] Project entitled "*In Situ* Conservation of Agrobiodiversity On-Farm," implemented by the National Agricultural Research Council, Nepal, and the International Plant Genetic Resources Institute.

[4] Project entitled "Development of strategies for in situ conservation and utilization of plant genetic resources in desert-prone areas of Africa," funded by International Fund for Agricultural Development and implemented by the Institut d'Économie Rural, Unité des Ressources Génétiques and International Plant Genetic Resources Institute.

Based on empirical findings, we know that in other places in the world, rural people depend on the diversity of their crops and varieties to cope with climatic risk, match them to specific soil and water regimes, and meet a range of consumption needs when markets are unreliable. These locations are often characterized as "less favored," or "marginalized"; the people who live in them are often considered to be poor on a global scale.

With economic analysis, genetic analyses, and other scientific information, we can ascertain which locations are "promising candidates" for managing rare or diverse crop genetic resources on farms. Metrics can then be used to rank areas according to their expected social and private value.

To rank locations along the horizontal axis, diversity indices developed by scientists can serve as proxies for the public value of a set of crop varieties or populations. Indices are scalars constructed from any one of several types of data (Meng et al., 1998). For example, data may measure the physical characteristics of crop plants grown in controlled experiments. Alternatively, data may summarize the patterns in DNA taken from plant tissue and observed under a microscope. Option values other than those subsumed in scientists' assessments of information value are not likely to be estimable but would generally be positive (Brown, 1990).

Adaptations of econometric models of variety choice, and nonmarket valuation methods combined with random utility models, enable us to rank candidate locations along the vertical axis using probabilistic statements. Some empirical evidence from related analyses is summarized next.

3.2. Predicting locations where landraces will continue to be grown

Broadly speaking, three generic factors are hypothesized to determine the likelihood that modern varieties are attractive to farmers and, hence, the opportunity cost of growing landraces: population density, agroecology, and development of commercial markets.

The pattern of diffusion of modern varieties of wheat, rice, and maize illustrate this point. Population density, or the ratio of the supply of labor to the supply of land, explains much about where the transition from low-yield, land-extensive cultivation to land-intensive, double, and triple crop systems has occurred (Boserup, 1965; Hayami and Ruttan, 1985; Pingali, 1997). The genetic changes embodied in seed constitute one type of intensification, which refers more broadly to the increase in output per unit of land used in production (or yield). Predictably, the adoption of modern rice varieties in the less industrialized world has been most complete in densely populated areas of their cultivation where traditional mechanisms for enhancing yields per unit area have been exhausted (Pingali, 1997).

Population densities interact with agroecological conditions in explaining the adoption of modern varieties. Since the initial adoption and rapid diffusion of the first semidwarf varieties of wheat in the irrigated areas of the Asian subcontinent during the 1970s, more widely adapted descendants of these varieties spread gradually into environments less favored for wheat production and rain-fed areas. Today, wheat landraces are cultivated extensively only in portions of the drier production zones of the West Asia-North Africa region and highlands of Ethiopia. Similarly, the adoption of modern varieties of rice is virtually complete in irrigated areas and uneven in rainfed zones, while they are largely absent in the uplands and deepwater areas.

In contrast with wheat and rice, maize is grown over a greater range of latitudes, altitudes, temperatures, and moisture regimes. Maize also has the greatest proportion of area in the less-industrialized world that is still planted to landraces. For many of the environments in which maize is grown, suitable improved materials have not been developed by centralized breeding programs (Byerlee, 1996; Perales, Brush, and Qualset, 1998). The maize germplasm that performs well in temperate climates of industrialized countries cannot be introduced directly into the nontemperate regions of less-industrialized countries without considerable additional breeding for adaptation (Morris, 1998), nor are there always economic incentives for a commercial seed industry. Thus, even if adaptation problems could be overcome through breeding, farmer demand for improved seed may be small (Morris, 1998). In maize, agroecological factors have interacted with the development of commercial seed systems in slowing the expansion of area in modern varieties.

As the orientation of crop production shifts from subsistence toward commercial objectives, the locus of crop improvement and seed distribution also moves from individual farmers toward an organized seed industry composed of specialized private and public organizations (Morris, 1998). Maize has moved substantially faster than rice and wheat in terms of an increased reliance on commercially produced seed. In a stylized depiction of the maize seed industries in developing countries, subsistence production is characterized by open-pollinated varieties improved through farmer selection and on-farm seed production with local seed markets governed by custom. In a fully commercial system, the predominant seed type is a hybrid that is purchased annually. Seed is a globally traded product of specialized research that is both privately and publicly funded. The exchange of seed and the genetic resources used to improve it are enabled and protected by strict forms of intellectual property rights (Morris, Rusike, and Smale, 1998).

In rice and wheat, which are self-pollinating crops, the incentives for privatization of research have not been as strong as for maize, although this depends on the institutional and economic context. In the industrialized countries, profound changes in science and in intellectual property protection over the past 20-30 years have been associated with a higher rate of investment in agriculture by the private sector than the public sector and a shift in the composition of private investment from agricultural machinery and processing into chemical research and plant breeding (Alston, Pardey, and Smith, 1998; Fuglie et al., 1996). Privatization is greatest in the maize seed industry in industrialized countries and is increasing in the less-industrialized agricultural economies, but has only occurred to a limited extent for wheat—in Europe. Almost all of the seed research for rice has been and continues to be conducted by the public sector, and most research has occurred in Asia (Pray, 1998).

Historical experience in the diffusion of modern varieties of wheat, rice, and maize lead us to predict that, for highly bred, staple food crops, the opportunity costs of landrace cultivation will be lowest, and farmers' economic incentives to grow landraces higher, in less densely populated, commercially isolated areas for which either public or private breeding systems are unlikely to develop well-adapted materials. Where conditions are otherwise, the costs of designing institutions and mechanisms to encourage conservation are likely to be quite high—*unless* consumers demand specialized traits that are both unique to landraces and difficult to transfer genetically into modern varieties. On the other hand, the costs of designing institutions to support on-farm conservation is likely to be considerably lower for crops that are not so highly bred and have received less research attention but play significant roles in the strategies of small-scale farmers. These crops are sometimes known as "neglected" or "underutilized" species from the perspective of professional plant breeding and modern agricultural systems.

3.3 Predicting locations where crop biodiversity will continue to be managed by farmers

An extensive microeconomics literature on variety adoption and a growing set of case studies about on-farm management of crop biodiversity provide us with empirical evidence about the types of social and economic contexts that lead to a higher probability that farmers will continue to manage biologically diverse crop genetic resources. The first factor, which is important on a local as well as a regional scale, is agroecological heterogeneity. Environmental heterogeneity has also been advanced as an explanation of farmers' continued use of landraces in Turkey (Brush and Meng, 1998). Across a series of villages with differing agroclimatic conditions, heterogeneity in agroecological conditions increased the number of different crops and the varieties of maize, beans, and squash varieties grown by farmers in the state of Puebla, Mexico (Van Dusen, 2000). In the highlands of Ethiopia, land fragmentation, soil erosions, and the numbers of plots on farms had effects on the variety diversity that varied among crops (Benin et al., 2003).

The second factor that is important at both local and regional levels is the extent to which households trade on markets. The more removed a household is from a major market center, the higher the costs of buying and selling on the market and the more likely that the household relies primarily on its own production for subsistence. Van Dusen (2000) found that the more distant the market, the greater the number of maize, beans, and squash varieties grown by farmers. In the highlands of Ethiopia, findings regarding market distances depended on the cereal crop, how the variable was measured (distance from farm to household, distance from household to all weather road, or distance from community to market), and whether the diversity among crops or the diversity within crops was the dependent variable (Benin et al., 2003). In environmentally sensitive areas of Hungary, small-scale farmers of a more isolated site with poorer soils and fewer food markets valued ancestral varieties and higher levels of crop species richness more than farmers in an economically developed site with good road quality, fertile soils, and numerous food markets (Birol and Gyovai, 2003).

Small-scale farmers' choice to grow more than one variety simultaneously is likely to reflect their need to address numerous concerns that no single variety can satisfy (Bellon, 1996). Case studies demonstrate that in many of the regions of the developing world where landraces are still grown, either markets for commercially produced seed, markets for the crop output, or markets for the multiple attributes farmers demand from their varieties are incomplete (Brush and Meng, 1998; Brush, Taylor, and Bellon, 1992; Smale, Bellon, and Aguirre, 2001). This means that the traits demanded by farmers (grain quality, fodder, suitability for a certain soil type) cannot be obtained through the production of modern varieties or procured through impersonal market transactions, so that farmers must rely on their own or neighbors' production for their supply (Renkow and Traxler, 1994; Bellon and Taylor, 1993). Knowing the crop characteristics that most matter to farmers, how these are distributed across varieties and populations, and to what extent farmers can meet their needs through market transactions is, therefore, important. In Turkey, concern for bread quality in wheat, in addition to high household transaction costs such as transportation and uncertain prices, were associated with the choice to grow landraces rather than modern varieties (Brush and Meng, 1998).

"Promising candidates" may also be locations where both modern varieties and landraces are grown, if growing both types in that location represents an economic equilibrium. Though modern varieties have long been equated with a loss of diversity on farms (Frankel, 1970), like any new or exotic type that is introduced, a modern variety can add to the portfolio of distinct agro-morphological types grown in a community precisely because it has been bred with the ideal type of other farmer-breeders or professional breeders in mind. Modern varieties may possess a trait not found in the local varieties grown in a community (Louette, Charrier, and Berthaud, 1997), suiting particular production niches but not others (Gauchan and Smale, 2002). With cross-pollinating species, farmer seed management or deliberate introgression may mean that the introduction of modern varieties generates new types that exhibit traits from both (vom Brocke, 2001; Bellon and Risopoulos, 2001).

Farmers often choose to grow both landraces and modern varieties. Viewed in the conventional microeconomic literature as partial adoption, this observed pattern has been explained theoretically through attitudes toward risk and uncertainty, missing markets, and differential soil quality or nutrient response combined with fixity or rationing (reviewed in Meng, 1997; Smale, Just, and Leathers, 1994). Though treated as a transitional period to full adoption, the coexistence of modern varieties and landraces may represent an economic equilibrium if one or several of these aspects persist despite economic change. Meng, Taylor, and Brush (1998) concluded that multiple factors, including missing markets, yield risk, grain quality, and agroclimatic constraints, influence the probability that a Turkish household will grow a wheat landrace; a change in any single economic factor is unlikely to cause farmers to cease growing it.

Zimmerer (1996) found that the capacity of farmers to grow diverse food plants (including maize) in Peru and Bolivia depends on whether they can cultivate them in combination with commercially developed, high-yielding varieties that were easier to sell because they had more uniform grain quality. So far, introduction of modern varieties of wheat and maize has not meant that any single variety dominates or that modern varieties have replaced farmers' varieties in the Ethiopia highlands (Benin et al., 2003), most likely because they have limited adaptation and farmers face many economic constraints in this location. It is just as likely that small amounts of seed of modern varieties broadens the variety set of these farmers by meeting a particular purpose. Neither the physical terrain nor the market infrastructure network is favorable for specialized, commercial agriculture.

The relationship of household characteristics, such as wealth and income sources, to crop diversity depends on the measurement and empirical setting. Smale, Bellon, and Aguirre (2001) found that variety attributes such as suitability for food preparation (tortillas) far outweighed the importance of household characteristics in explaining the number of maize landraces grown by individual farmers and the average share of maize area planted to each. In three sites in Nepal, based on a composite variable for wealth rank, Rana et al. (2000) found that poor households cultivate more coarse-grained, drought-tolerant varieties of rice, while wealthier households grew high-quality varieties for premium market prices and special food preparations. In the state of Puebla, Mexico, Van Dusen (2000) found that the greater the wealth of the household, as measured by house construction and ownership of durable goods, the less likely the household is to plant a diversity set of maize, beans, and squash varieties. Benin et al. (2003) found that while larger farms were associated with more varieties of any major cereal crop grown in the highlands of Ethiopia, livestock assets had conflicting effects among cereals.

In many parts of the developing world, off-farm migration generates a growing proportion of the income of farm households. Brush, Taylor, and Bellon (1992) found that off-farm employment was negatively associated with maintenance of potato diversity in the Andes, indicating that the opportunity cost of cultivating many varieties—which requires labor-intensive seed selection and procurement tasks—is significantly higher where other employment possibilities exist. Van Dusen found that overall diversity in the *milpa* system decreased as local labor markets intensify, or as more migration to the United States occurs, though these effects were not as pronounced when each crop was considered singly. Yet off-farm income can also release the cash income constraint faced by some farmers, enabling them to shift their focus from growing varieties for sale to growing the varieties they may prefer to consume. In Chiapas, Mexico, Bellon and Taylor (1993) found that off-farm employment was associated with higher levels of maize diversity. Meng (1997) found the existence of off-farm labor opportunities to have no statistically significant effect on the likelihood of growing wheat landraces in Turkey.

4. POLICY MECHANISMS TO SUPPORT FARMER MANAGEMENT OF CROP BIODIVERSITY

Policy mechanisms to support farmer management of crop biodiversity in locations that are "promising candidates" may be classified as either related to (1) the demand for or (2) the supply of genetically diverse or distinct crop varieties (Bellon, 2000; Jarvis et al., 2000). Some illustrative examples are provided below, though many of these initiatives are new and their efficiency in meeting conservation goals given the level of investment required has not yet been assessed using cost-benefit analysis—at least in published literature.

4.1 Demand-related mechanisms

When markets are not well developed and transaction costs are high, the demand for distinct varieties reflects the tastes and preferences of local farmers who grow and consume them rather than those of distant urban consumers. Farmers' demand for local landraces can be enhanced by improving the traits they identify as important, including disease resistance, abiotic tolerance, and palatability as food or fodder. In a project in the

Philippines, molecular analysis revealed the genetic distinctiveness of the Wagwag group of traditional, nonglutinous varieties of rice that farmers also ranked highly according to consumption quality criteria. Researchers recommended a breeding intervention to address the long duration and low yield of the Wagwag group, reducing their disadvantages relative to modern rice varieties in the irrigated ecosystem (Bellon et al., 1998). Plant breeders can work with farmers to select superior local materials or transfer a preferred trait from exotic into local materials. In India, a drought-tolerant, locally adapted landrace was crossed with a modern variety with higher potential yield, and the offspring selected by farmers under their own conditions (Witcombe, Joshi, and Sthapit, 1996).

Known as "participatory" plant breeding, such approaches can require substantial time investments by farmers (Thiele et al., 1997; Rice, Smale, and Blanco, 1998). Oaxacan farmers as a group obtained nearly a 4:1 benefit-cost ratio from participating in a project to enhance local maize landraces and their diversity, although from the perspective of a private investor, benefits did not justify the cost (Smale et al., 2003). In Oaxaca, as in other maize-growing areas of the less-industrialized world, the high rate of cross-pollination in maize is reinforced by the simultaneous flowering of contiguously planted, fragmented fields. Under these circumstances, yield advantages of improved open-pollinated materials and landraces are difficult to maintain unless a cost-effective, local system of seed multiplication, diffusion, and replacement is established.

Smale et al. (2003) did not assess the public benefits of the project in Oaxaca, such as its contribution to maintaining the diversity of maize landraces, because of measurement difficulties. Not all participatory projects are undertaken with the goal of enhancing crop biological diversity, however. The release of well-adapted varieties, whether they are developed with or without the active participation of farmers, may cause a decline in the diversity of varieties grown because they are so popular (Friis-Hansen, 1996). Research has indicated that farmer involvement in the later stages of variety selection is associated with better local adaptation of materials (Sperling, Loevinsohn, and Ntambovura, 1993; Maurya, Bottrall, and Farrington, 1998; Sthapit, Joshi, and Witcombe, 1996). Witcombe and Joshi (1995) argue that while participatory variety selection may not increase diversity, participatory plant breeding will necessarily enhance both intra- and inter-variety diversity because of the methods employed. In 1995, Loevinsohn and Sperling found that the linkage between participatory crop improvement and diversity was not well documented, though more published evidence is now being generated (Friis-Hansen and Sthapit, 2000; Jarvis et al., 2000; Weltzien et al., 2000).

As incomes rise and commercial markets develop, landraces may continue to be grown when there is consumer demand for some unique attribute that cannot be easily bred into or transferred to improved varieties and if seed regulations permit. Advanced agricultural economies are characterized by growth in demand for an array of increasingly specialized goods and services (Antle, 1999). In general, though the income elasticity of demand for staple grains may be low or even negative, the income elasticity of demand for grain attributes is higher. For example, high income consumers spend more on rice by paying higher prices for varieties with preferred eating quality which they substitute for the lower-quality variety consumed when the income-level was lower (Unnevehr, Duff, and Juliano, 1992). Pingali, Hossain, and Gerpacio (1997) cite several examples of Asian landraces that are of higher quality and fetch premium prices in the market. In South Korea, the modern *tongil* variety was replaced by a relatively low-yielding, traditional *japonica* rice as consumers expressed a preference for *japonica* types by offering higher prices as their incomes rose.

Niche markets or branding enable farmers to recoup the premiums consumers are willing to pay for unique attributes or qualities. Farmers in niche markets do not control supply or prevent imitation. Branding enforces supply constraints through (1) some fixed and identifiable attribute such as a geographic origin, (2) membership in an exclusive producer group, (3) strict production standards or guarantee of process, or (4) control of an ingredient by an exclusive producer group through intellectual property rights (Hayes and Lence, 2002).

Restricted labeling systems have long been used to ensure consumer quality and authenticity for meat, cheese, and wine products in Europe. Falcinelli (1997) describes how labeling systems may promote conservation of *farro* (einkorn, emmer, spelt) and lentils in Italy. In a study of the Label Rouge system for quality poultry production in France, Westgren (1999) cautions that the investment in consumer education required to support significant, sustainable price premia is costly and may not be fully internalized in any single market supply chain. Though common in the European Union, farmer brands are relatively rare in the United States. Brush (2000) reports an example of a successful "green marketing" program for ancestral maize of Cherokee farmers in the United States. Vidalia onions of Georgia are another example based on geographic origin related to superior quality.

Labeling and marketing systems require public investments unless consumers are willing to pay price premiums large enough to cover the costs, and require fully commercialized, well-articulated markets for product attributes. Hayes and Lence (2002) report several necessary conditions for the successful differentiation of farmer-owned brands, including good transmission of price signals, a scale of production large enough to justify the costs of creating a differentiated image among consumers, and capacity to prevent imitation. These conditions are not likely to be met easily in most contexts where diverse landraces are grown in less-industrialized economies, especially for staple foodcrops that are grown by many atomistic farmers. For example, Gauchan, Smale, and Chaudhury (2003) found that except for traditional Basmati (aromatic high quality) rice, most rice landraces in upland Nepal are traded in small volumes through informal channels where price signals are weak. If desirable attributes can be clearly linked to specific landraces, the landraces themselves must be readily identifiable and their attributes maintained through careful seed multiplication and production. Even if branding is successful, protecting one landrace may not have desirable implications for diversity conser-vation—especially if seed regulations require that landraces be uniform and stable like modern varieties.

In either advanced or less-industrialized economies, public awareness initiatives can serve to increase farmer and consumer knowledge about the benefits generated by on-farm conservation, enhancing their demand for products and seed. In Nepal, Vietnam, and the Andes (Tapia and Rosa, 1993; Rijal et al., 2000; Jarvis et al., 2000, pp. 153-154), diversity fairs have been used to bring together farmers from one or more communities in order to exhibit the range of materials they use and raise awareness of the value of crop diversity. Rather than awarding prizes for the best individual variety (e.g., on the basis of yield or size), diversity fairs award farmers or cooperatives for the greatest crop diversity and related knowledge. In some communities, gatherings similar to diversity fairs are already customary events, so the incremental costs are low. Though such initiatives are gaining popularity in seed projects (Tripp, 2000, p. 116), there appears to be little evidence concerning their cost effectiveness or impact.

4.2 Supply-related mechanisms

Possible "supply" solutions encompass a range of seed market and genetic improvement options, as well as, potentially, protection of local varieties through Farmers' Rights legislation. India is one of the first countries in the world to have passed a legislation granting rights to both breeders and farmers under the Protection of Plant Varieties and Farmers' Rights Act in 2001. As of yet, however, the requirements for registering farmers' varieties have not been elucidated. Ramanna and Smale (2004) have argued that while this multiple rights system aims to equitably distribute rights, it could pose the threat of an "anticommons tragedy" where too many parties independently possess the right to exclude, resulting in the underutilization of crop genetic resources and discouraging the very innovation it was intended to promote.

In more isolated areas or more difficult growing environments of less-industrialized agricultural economies, agroecological and environmental factors exert a more decisive influence on crop biological diversity than commercial markets. From one year to the next, farmers may lose their seed stock due to disastrous harvests, or diminishing seed quantities may threaten the genetic viability of the variety. Initiatives aimed at supporting the range and total supply of local seed types have been suggested in response to such situations.

Community genebanks provide a mechanism for storing valuable landrace germplasm in a local *ex situ* form so that farmers' have more direct access to seed when they need it. Typically small in size, community genebanks can only maintain a limited number of accessions and replicates. The economic feasibility of community banks is also undermined by the high covariance of local crop yields, which means that many farmers in a community face similar seed deficits and surpluses. Cromwell (1996, p. 127) further cautions that access to seed through the community system is not always egalitarian, since seed is often hoarded or may be of poor quality.

If farmers prefer to store seed on an individual basis but desire access to knowledge about the location of other seed types in surrounding villages, biodiversity registers are one low-cost, modest alternative (Rijal et al., 2001). A community biodiversity register is a record of landraces cultivated by local farmers. In addition to the names of the farmers who grow them and place of origin, the register may include data about the agro-morphological and agronomic characteristics of landraces, agroecological adaptation, and special uses.

Registers can serve as an information tool that reduces the transactions costs of locating and exchanging diverse materials, but do not solve the problem of a shortage in seed supply relative to demand. National governments, nongovernmental organizations, and donors have invested in local-level seed projects as a means of delivering a better range and quantity of seed types where state seed enterprises have been ineffective and the commercial seed sector has been too slow to grow or too limited in focus. Reviewing these efforts to date, Tripp (2000, pp. 133-134) concluded that few of these projects have achieved the goal of establishing viable small-scale seed production enterprises. He cites as a principal obstacle the failure to recognize that seed provision requires more than multiplication. Projects internalize the costs of managing contacts involved in obtaining source seed and establishing quality control procedures, arranging for seed conditioning, and, in particular, marketing the seed. When projects cease to be funded, the effectiveness of the seed enterprise falters. He argues that more must be done to support institutional growth and strengthen farmers' links with markets and institutions that are already in place.

One prerequisite for building such institutions is an understanding of existing seed exchange networks, whether formal or informal. Another is an understanding of how public systems might be adapted in order to promote the use by farmers of a broader range of materials. In Nepal, informal research and development (IRD) has been used to test, select, and multiply seeds (Joshi and Sthapit, 1990). A small quantity of seed of recently released and/or nearly finished varieties is distributed to a few farmers in a community to grow under their own conditions with their own practices.

First practiced by Lumle Agricultural Research Centre (LARC), this approach has now been adopted by other organizations in Nepal and India for variety testing and dissemination (Joshi et al., 1997). Such approaches incur no substantive additional costs but speed the time to use of varieties since they shortcut release procedure. They are likely to enhance diversity since each farmer receiving seed adapts it through his or her own seed selection practices. Recent work by nongovernmental organizations has sought to identify the ways in which seed markets can be stimulated as a means of introducing and supporting local diversity (CRS, ICRISAT, and ODI, 2002). Rather than impose seed provisioning, attention has been shifted to reinforcing existing farmers' systems and seed system recovery (Jones et al., 2002).

5. CONCLUSIONS

It does not make economic sense to trade productivity for biological diversity when it means thwarting the opportunities of poorer farmers in less-industrialized economies. "Promising candidates" for on-farm management of crop biodiversity are sites where the local crop genetic resources are ranked highly with respect to both farmer utility and their biological diversity, and where empirical analysis predicts that farmers are likely to continue cultivating them.

When genetic analyses confirm that relatively high levels of diversity are found in a location where farmers are choosing to specialize in fewer modern varieties for commercial sale, programs to conserve landraces may be costly in terms of private opportunities foregone and public expenditures. Instead, careful introduction of larger numbers of modern types with distinct agronomic and consumption attributes, or participatory variety selection, may both benefit farmers and contribute to productivity enhancement over the longer term. Remaining landraces might be conserved *ex situ* following the criteria recommended for optimal sampling. In order to save misdirected public funds, it is equally important to know in which locations remaining landraces have little genetic interest. Not all landraces embody potentially valuable diversity. If farmers care about them, they will continue to grow them. If not, they will discard them, with few implications for society.

On a global scale, for highly bred, staple food crops (e.g., rice, wheat, and maize), historical factors such as labor to land ratios, agroecological features, and commercialization explain to a large extent in which regions landraces are still grown and will continue to be grown. At a more localized scale of analysis, studies suggest that, while the effects of incomplete local markets for crop products and market distance are fairly predictable, the effects of other economic variables, such as the extent of off-farm employment, income and wealth status, are not easy to predict a priori unless researchers already have extensive knowledge about farmer decision-making and local economies among candidate sites. Typically, it will therefore be necessary to undertake more empirical research at the household and community level once candidate sites have been identified, while controlling statistically for the regional-level factors described above. Only as more case studies accumulate can generalizations be drawn. Though some of the policy instruments mentioned here have been designed to support on-farm conservation through enhancing either the demand for or supply of diverse seed types, the balance sheet from the field has not yet been tallied.

REFERENCES

Alston, J. M., Pardey, P. G., and Smith, V. H., 1998, Financing agricultural R&D in rich countries, *Aust. J. Agr. Resource Econ.* **42** (1):51-82.

Antle, J. M., 1999, The new economics of agriculture, Presidential Address prepared for the Annual Meetings of the American Agricultural Economics Association, Nashville (August, 1999).

Bellon, M. R., 1996, The dynamics of crop infraspecific diversity: A conceptual framework at the farmer level, *Econ. Bot.* **50**:26-39.

Bellon, M. R., 2000, Demand and supply of crop infraspecific diversity on farms: Towards a policy framework for on-farm conservation, CIMMYT Economics Working Paper 01-01, International Maize and Wheat Improvement Center (CIMMYT), Mexico, D.F.

Bellon, M. R., Pham, J.-L., and Jackson, M. T., 1997, Genetic conservation: A role for rice farmers, in: *Plant Genetic Conservation: The In-Situ Approach*, N. Maxted, B. V. Ford-Lloyd, and J. G. Hawkes, ed., Chapman and Hall, London.

Bellon, M. R., Pham, J.-L., Sebastian, L. S., Francisco, S. R., Loresto, G. C., Erasga, D., Sanchez, P., Calibo, M., Abrigo, G., and Quilloy, S., 1998, Farmers' perceptions of varietal diversity: Implications for on-farm conservation of rice, in: *Farmers, Gene Banks, and Crop Breeding: Economic Analyses of Diversity in Wheat, Maize, and Rice*, M. Smale, ed., Kluwer Academic Publishers and International Maize and Wheat Improvement Center (CIMMYT), Dordrecht and Mexico, D.F., pp. 95-108.

Bellon, M. R., and Risopoulos J., 2001, Small-scale farmers expand the benefits of maize germplasm: A case study from Chiapas, Mexico, *World Devel.* **9**(5):799-812.

Bellon, M. R., and Taylor, J. E., 1993, 'Folk' soil taxonomy and the partial adoption of new seed varieties, *Econ. Devel. Cult. Change* **41**:763-786.

Benin, S., Smale, M., Gebremedhin, B., Pender, J., and Ehui, S., 2003, The economic determinants of cereal crop diversity on farms in the Ethiopian Highlands, Contributed paper, 25th International Conference of Agricultural Economists, Durban, South Africa (August 16-22, 2003).

Birol, E., and Gyovai, Á., 2003, The value of agricultural biodiversity in Hungarian home gardens: Agri-Environmental policies in a transitional economy, Contributed paper, Fourth BIOECON workshop on Economic Policies for Biodiversity Conservation, Venice International University, Venice (August 28-29, 2003).

Boserup, E., 1965, *Conditions of Agricultural Growth*, Aldine Publishing Company, Chicago, Illinois.

Brown, G. M., 1990, Valuation of genetic resources, in: *The Preservation and Valuation of Biological Resources*, G. H. Orians, G. M. Brown, Jr., W. E. Kunin, and J. E. Swierbinski, eds., University of Washington Press, Seattle, Washington.

Brush, S. B., ed., 2000, *Genes in the Field: On-Farm Conservation of Crop Diversity*, International Plant Genetic Resources Institute, International Development Research Centre, and Lewis Publishers, Rome, Ottawa, and Boca Raton.

Brush, S. B., and Meng, E., 1998, Farmers' valuation and conservation of crop genetic resources, *Genet. Resour. Crop Ev.* **45**:139-150.

Brush, S. B., Taylor, J. E., and Bellon, M. R., 1992, Biological diversity and technology adoption in Andean potato agriculture, *J. Dev. Econ.* **38**:365-387.

Byerlee, D., 1996, Modern varieties, productivity, and sustainability: Recent experience and emerging challenges, *World Dev.* **24**(40):697-718.

Cornes, R., and Sandler, T., 1986, *The Theory of Externalities, Public Goods, and Club Goods*, Cambridge University Press, Cambridge.

Cromwell, E., 1996, *Governments, Farmers, and Seeds in a Changing Africa*, CABI, Wallingford, U. K.

CRS, ICRISAT, and ODI, 2002, *Seed Vouchers and Fairs: A Manual for Seed-Based Agricultural Recovery after Disaster in Africa*, Catholic Relief Services, International

Crops Research Institute for the Semi-Arid Tropics, Nairobi, Kenya, and Overseas Development Agency, London.

Falcinelli, M., 1997, Il miglioramento genetico delle colture tipiche locali, *L'Informatore Agrario* **47**:73-75.

Frankel, O. H., 1970, Genetic dangers of the green revolution, *World Agr.* **19**:9-14.

Friis-Hansen, E., 1996, The role of local plant genetic resources management in participatory breeding, in: *Participatory Plant Breeding: Proceedings of a Workshop on Participatory Plant Breeding*, P. Eyzaguirre and M. Iwanaga, eds., Wageningen, Netherlands, and International Plant Genetic Resources Institute (IPGRI), Rome, pp. 66-76.

Friis-Hansen, E., and Sthapit, B., eds., 2000, *Participatory Approaches to the Conservation and Use of Plant Genetic Resources*, International Plant Genetic Resources Institute (IPGRI), Rome, Italy.

Fuglie, K., Ballenger, N., Day, K., Klotz, C., Ollinger, M., Reilly, J., Vasavada, U., and Yee, J., 1996, *Agricultural Research and Development: Public and Private Investments Under Alternative Markets and Institutions*, Economic Research Service, U. S. Department of Agriculture, AER No. 735.

Gauchan, D., and Smale, M., 2002, Choosing the "right" tools to assess the economic costs and benefits of growing landraces: An illustrative example from Bara District, Central Terai, Nepal, *Plant Genetic Resources Newsletter* **134**:41-44.

Gauchan, D., Smale, M., and Chaudhury, P., 2003, Market-based incentives for conserving diversity on farms: The case of rice landraces in Central Terai, Nepal, Contributed paper, Fourth BIOECON workshop on Economic Policies for Biodiversity Conservation, Venice International University, Venice (August 28-29, 2003).

Hayami, Y., and Ruttan, V. W., 1985, *Agricultural Development: An International Perspective,* Johns Hopkins University Press, Baltimore, Maryland.

Hayes, D. J., and Lence, S. H., 2002, A new brand of agriculture: Farmer-owned brands reward innovation, *Choices* (Fall):6-10.

Heisey, P., Smale, M., Byerlee, D., and Souza, E., 1997, Wheat rusts and the costs of genetic diversity in the Punjab of Pakistan, *Amer. J. Agri. Econ.* **79**:726-737.

Jarvis, D. I., Myer, L., Klemick, H., Guarino, L., Smale, M., Brown, A. H. D., Sadiki, M., Sthapit, B., and Hodgkin, T., 2000, *A Training Guide for In Situ Conservation On-Farm. Version 1*, International Plant Genetic Resources Institute (IPGRI), Rome, Italy.

Jones, R. B., Longley, C., Bramel, P., and Remington, T., 2002, The need to look beyond the production and provision of relief seed: Experiences from Southern Sudan, *Disasters* **26**(4):302-315.

Joshi, K. D., and Sthapit, B. R., 1990, *Informal Research and Development (IRD): A New Approach to Research and Extension*, Lumle Agricultural Research Centre Discussion Paper No. 90/4, Lumle Agricultural Research Centre, Pokhara, Nepal.

Joshi, K. D., Subedi, M., Kadayat, K. B., Rana, R. B., and Sthapit, B. R., 1997, Enhancing on-farm varietal diversity through participatory variety selection: A case study of *Chaite* rice in Nepal, *Exp. Agric.* **33**:1-10.

Loevinsohn, M., and Sperling, L., eds., 1995, Joining dynamic conservation to decentralized genetic enhancement: Prospects and issues, in: *Using Diversity: Enhancing and Maintaining Genetic Resources On-Farm*, Proceedings of a workshop International Development Research Centre, New Delhi, India, pp. 1-7 (June 19-21, 1995).

Louette, D., Charrier, A., and Berthaud, J., 1997, In situ conservation of maize in Mexico: Genetic diversity and maize seed management in a traditional community, *Econ. Bot.* **51**: 20-38.

Maurya, D. M., Bottrall, A., and Farrington, J., 1998, Improved livelihoods, genetic diversity and farmers' participation: A strategy for rice-breeding in rainfed areas of India, *Exp. Agric.* **24**: 311-320.

Maxted, N., Ford-Lloyd, B. V., and Hawkes, J. G., eds., 1997, *Plant Genetic Conservation: The In-Situ Approach*, Chapman and Hall, London.

Meng, E. C. H., 1997, Land allocation decisions and in situ conservation of crop genetic resources: The case of wheat landraces in Turkey, Ph.D. thesis, University of California, Davis.

Meng, E. C. H., Smale, M., Bellon, M. R., and Grimanelli, D., 1998, Definition and measurement of crop diversity for economic analysis, in: *Farmers, Gene Banks and Crop Breeding: Economic Analyses of Diversity in Maize, Wheat, and Rice*, M. Smale, ed., Kluwer Academic Press and International Maize and Wheat Improvement Center, Dordrecht.

Meng, E. C. H, Taylor, J. E., and Brush, S. B., 1998, Implications for the conservation of wheat landraces in Turkey from a household model of varietal choice, in: *Farmers, Gene Banks and Crop Breeding: Economic Analyses of Diversity in Wheat, Maize, and Rice*, M. Smale, ed., Kluwer Academic Press and CIMMYT, Dordrecht, pp. 127-143.

Morris, M. L., 1998, *Maize Seed Industries in Developing Countries*, Lynne Rienner, Boulder, and CIMMYT, Mexico.

Morris, M. L., Rusike, J., and Smale, M., 1998, Maize seed industries: A conceptual framework, in: *Maize Seed Industries in Developing Countries*, M. Morris, ed., Lynne Rienner, Boulder, and CIMMYT, Mexico.

Perales, R. H., Brush, S. B., and Qualset, C. O., 1998, Agronomic and economic competitiveness of maize landraces and *in situ* conservation in Mexico, in: *Farmers, Gene Banks and Crop Breeding: Economic Analyses of Diversity in Wheat, Maize, and Rice*, M. Smale, ed., Kluwer Academic Press and CIMMYT, Dordrecht, pp. 109-126.

Pingali, P., 1997, From subsistence to commercial production systems, *Amer. J. Agri. Econ.* **79**:628-634.

Pingali, P., Hossain, M., and Gerpacio, R. V., 1997, Asian rice market: Demand and supply prospects, in: *Asian Rice Bowls: The Returning Crisis?*, P. L. Pingali, M. Hossain, and R. V. Gerpacio, eds., CABI Publishing and International Rice Research Institute (IRRI), Wallingford, U. K., and Manila, Philippines, pp. 126-144.

Pray, C. E., 1998, Impact of biotechnology on the demand for rice biodiversity, in: *Agricultural Values of Plant Genetic Resources*, R. E. Evenson, D. Gollin, and V. Santaniello, eds., CABI Publishing, Wallingford, U. K., and FAO, Rome, pp. 249-259.

Ramanna, A., and Smale, M., 2004, Rights and access to plant genetic resources in India's new legislation, *Dev. Policy Rev.* **22**(4):423-442.

Rana, R. B., Gauchan, D., Rijal, D. K., Khatiwada, S. P., Paudel, C. L., Chaudhary, P., and Tiwari, P. R., 2000, Socioeconomic data collection and analysis: Nepal, in: *Conserving Agricultural Biodiversity In Situ: A Scientific Basis for Sustainable Agriculture*, D. Jarvis, B. Sthapit, and L. Sears, eds., International Plant Genetic Resources Institute (IPGRI), Rome, pp. 54-56.

Renkow, M., and Traxler, G., 1994, Incomplete adoption of modern cereal varieties: The role of grain-fodder tradeoffs, Selected paper presented at the annual meeting of the American Agricultural Economics Association, San Diego (August 7-10, 1994).

Rice, E., Smale, M., and Blanco, J.-L., 1998, Farmers' use of improved seed selection practices in Mexican maize: Evidence and issues from the Sierra de Santa Marta, *World Dev.* **26**(9):1625-1640.

Rijal, D. K., Rana, R., Subedi, A., and Sthapit, B. R., 2000, Adding value to landraces: Community-based approaches for in situ conservation of plant genetic resources in Nepal, in: *Participatory Approaches to the Conservation and Use of Plant Genetic Resources*, E. Friis-Hansen and B. Sthapit, eds., International Plant Genetic Resources Institute (IPGRI), Rome, pp. 166-172.

Rijal, D. K., Subedi, A., Upadhaya, M. P., Rana, R. B., Chaudhary, P., Tiwari, P. R., Tiwari, R. K., Sthapit, B. R., and Gauchan, D., 2001, Community biodiversity registers: Developing community-based databases for genetic resources and local knowledge in Nepal, paper prepared for the First National Workshop of the Project, Strengthening Scientific Basis of In situ Conservation of Agrobiodiversity On-farm: Nepal, Lumle, Nepal (April 24-26, 2001).

Romer, P. M., 1993, *Two Strategies for Economic Development: Using Ideas and Producing Ideas*. Proceedings of the World Bank Annual Conference on Development Economics 1992, The World Bank, Washington, D. C.

Sandler, T., 1999, Intergenerational public goods: Strategies, efficiency and institutions, in: *Global Public Goods,* I. Kaul, I. Grunberg, and M. A. Stein, eds., Oxford University Press and United Nations Development Programme, New York, pp. 20-50.

Smale, M., and Bellon, M., 1999, A conceptual framework for valuing on-farm genetic resources, in: *Agrobiodiversity: Characterization, Utilization, and Management*, D. Wood and J. Lenné, eds., CABI International, Wallingford, U. K., pp. 387-408.

Smale, M., Bellon, M., and Aguirre, A., 2001, Maize diversity, variety attributes, and farmers' choices in Southeastern Guanajuato, Mexico, *Econ. Devel. Cult. Change* **50**(1): 201-225.

Smale, M., Bellon, M., Aguirre, A., Manuel, I., Mendoza, J., Solano, A. M., Martínez, R., Ramírez, A., and Berthaud, J., 2003, The economic costs and benefits of a participatory project to conserve maize landraces on farms in Oaxaca, Mexico, *Agric. Econ.* **29**:265-275.

Smale, M., Just, R. E., and Leathers, H. D., 1994, Land allocation in HYV adoption models: An investigation of alternative explanations, *Amer. J. Agri. Econ.* **76**:535-546.

Sperling, L., Loevinsohn, M., and Ntambovura, B., 1993, Rethinking the farmer's role in plant breeding: Local bean experts and on-extension selection in Rwanda, *Exp. Agric.* **29**:509-519.

Sthapit, B. R., Joshi, K. D., and Witcombe, J. R., 1996, Farmer participatory cultivar improvement. III: Participatory plant breeding, A case of high altitude rice from Nepal, *Exp. Agric.* **32**:479-496.

Tapia, M. E., and Rosa, A., 1993, Seed fairs in the Andes: A strategy for local conservation of plant genetic resources, in: *Cultivating Knowledge*, W. de Boef, K. Amanor, and K. Wellard, with A. Bebbington, eds., Intermediate Technology Publications, London, pp. 111-118.

Thiele, G., Gardiner, G., Torrez, R., and Gabriel, J., 1997, Farmer involvement in selecting new varieties: Potatoes in Bolivia, *Exp. Agric.* **33**:275-290.

Tripp, R., 2000, *Seed Provision and Agricultural Development: The Institutions of Rural Change*, Overseas Development Institute, London; James Currey, Oxford; and Heinemann, Portsmouth, New Hampshire.

Unnevehr, L., Duff, B., and Juliano, B. O., 1992, Consumer Demand for Rice Grain Quality, Terminal Report of IDRC Projects, National Grain Quality (Asia) and International Grain Quality Economics (Asia), IDRC, Ottawa, and IRRI, Los Banos, Laguna, Philippines.

Van Dusen, E., 2000, In situ conservation of crop genetic resources in the Mexican *Milpa* System, Ph.D. Dissertation, University of California, Davis, California.

vom Brocke, K., 2001, Effects of farmers' seed management on performance, adaptation, and genetic diversity of pearl millet (*Pennisetum glaucum* [L.] R.Br.) populations in Rajasthan, India, Ph.D. Dissertation, University of Hohenheim, Germany.

Weltzien, E., Smith, M. E., Meitzner, L. S., and Sperling, S., 2000, *Technical and Institutional Issues in Participatory Plant Breeding—from the Perspective of Formal Plant Breeding: A Global Analysis of Issues, Results, and Current Experience.* CGIAR Systemwide Program on Participatory Research and Gender Analysis for Technology Development and Institutional Innovation.

Westgren, R. E., 1999, Delivering food safety, food quality, and sustainable production practices: The Label Rouge poultry system in France, *Amer. J. Agric. Econ.* **81**(5): 1107-1111.

Witcombe, J. R., and Joshi, A., 1995, The impact of farmer participatory research on the biodiversity of crops, in: *Using Diversity: Enhancing and: Maintaining Genetic Resources On-Farm,* L. Sperling and M. Loevinsohn, eds., Workshop Proceedings, International Development Research Center, New Delhi, India (June 19-21), pp. 87-101.

Witcombe, J. R., Joshi, K. D., and Sthapit, B. R., 1996, Farmer participatory cultivar improvement. I: Varietal selection and breeding methods and their impact on biodiversity, *Exp. Agric.* **32**:445-460.

Zimmerer, K. S., 1996, *Changing Fortunes: Biodiversity and Peasant Livelihood in the Peruvian Andes,* University of California Press, Berkeley, Los Angeles, and London.

Chapter 6

IN SITU CONSERVATION: *METHODS AND COSTS*

Detlef Virchow
InWEnt, Capacity Building International, Wielinger Str. 52, D-82340 Feldafing, Germany

Abstract: Conservation policies will be pursued with quite different sets of instruments and conservation methods depending on the objectives and the costs implied. In this chapter, the objectives of genetic resources conservation are discussed and *in situ* conservation methods are described, and the costs related to these conservation methods are analyzed. This chapter demonstrates that despite the intensive multilateral discussions regarding the potential and the political will of various countries to foster *in situ* conservation activities, the direct costs have not yet been assessed, much less the related indirect costs which will be even more difficult to assess. It is discussed that agrobiodiversity is largely produced by farmers as a positive externality without any conservation program costs. In the future, assuming a risk of an unplanned loss of traditional varieties, the question will be with what economic instruments and incentives can agrobiodiversity be kept at the social optimum, securing nonmarketable genetic resources? It is argued that efficient interventions as well as flexible and self-targeting incentive mechanisms are needed to enable the farmers to benefit from the product agrobiodiversity and at the same time to decrease the social opportunity costs of *in situ* conservation without losing varieties.

Key words: conservation costs; conservation methods; conservation policies; incentive mechanism; *in situ* conservation; plant genetic resources for food and agriculture.

1. INTRODUCTION

Plant collection and display, particularly in the form of botanical gardens, have a long history, dating back several hundred years. Without having the explicit objective of conserving plant species, botanical gardens became the first conservation sites for plants, and are still an important conservation institution. Since then, the range of conservation instruments for genetic resources has expanded. There is, however, a fundamental divergence of views on what objectives should be the focus of the conservation of genetic diversity. Ethical aims oriented at preserving all existing biodiversity stand in opposition to anthropocentric objectives which consider genetic diversity only worth maintaining to the extent that it serves human kind at present or in the future. Conservation policies will be pursued with quite different sets of instruments and conservation methods depending on the objectives and the costs implied. In this chapter, *in situ* conservation methods shall be described, and the costs related to these conservation methods shall be analyzed. This chapter seeks to demonstrate that despite the intensive multilateral discussions regarding the potential and the political will of various countries to foster *in situ* conservation activities (see Chapter 9), the direct costs have not yet been assessed, much less the related indirect costs which will be even more difficult to assess.

2. THE OBJECTIVES OF PGRFA CONSERVATION

Three fundamental objectives for the conservation of plant genetic resources for food and agriculture (PGRFA) can be identified in the ongoing discussion (Virchow, 1999; FAO, 1998):

- *Ensure future utilization* through long-term *ex situ* conservation, which conserves PGRFA in their present constellation for future generations to come.
- *Support adaptation of PGRFA to changing environmental conditions* through long-term *in situ* conservation, which exposes genetic resources to ecological pressure enforcing natural changes and permitting continued coevolutionary development, as well as adoption via selection activities by farmers.
- *Facilitate convenient access through storage activities* that support the supply of genetic resources as raw material. Stored collections need to be easy accessible for ongoing breeding programs by scientists, farmers, and interactive groups of farmers and scientists.

The term "*ex situ* conservation" is applied to all conservation methods in which the species or varieties are taken out of their traditional ecosystems and are kept in an environment managed by humans. Starting with the collecting activities of N. I. Vavilov, most conservation efforts for agricultural crops have until recently involved the use of *ex situ* conservation; particularly in the form of seed genebanks. Great emphasis was placed on germplasm collecting during the 1970s and 1980s. As a result, the conservation of agricultural plants is presently dominated by *ex situ* collections. Defined as the conservation of plants in their ecosystems, "*in situ* conservation," has been traditionally used for the conservation of forests and of sites valued for their wildlife or ecosystems (FAO, 1998). In recent years, however, the need for *in situ* conservation of PGRFA has been increasingly emphasized, above all at the United Nations Conference on

Environment and Development in 1992 (UNEP, 1994) and during the preparatory process for the International Technical Conference on PGRFA (FAO, 1998).

Both terms, *ex situ* as well as *in situ* conservation, originate from the conservation activities of wild plants and animals. These terms are now commonly used in the context of the conservation of PGRFA, but do not fit this application perfectly. By their very nature, PGRFA can only be maintained by human management, i.e., through a process of selection for cultivation in farmers' fields. Consequently, strictly following the terminology, all PGRFA are already in a state of *ex situ* management. On the other hand, *in situ* conservation for PGRFA is not possible, sticking to the terminology in the narrow sense, because domesticated plants do not have a natural habitat per se, and if left alone in the natural habitat of their wild relatives, they will have very little chance of survival. Therefore, the terminology adopted in this chapter is applied in a broader sense, i.e., *ex situ* conservation is defined as the management of domesticated plants or parts of them, outside of their common surroundings, mainly the farm as the agricultural production unit. Following the broader framework of definition, *in situ* conservation is defined as all activities to conserve PGRFA in their common surroundings, including the "on-farm management" of PGRFA. As defined in the Convention on Biological Diversity, *in situ* conservation of PGRFA *". . . means . . . the maintenance and recovery of . . . domesticated or cultivated species, in the surroundings where they have developed their distinctive properties"* (UNEP, 1994).

Experience from the existing *in situ* and *ex situ* conservation methods demonstrate that agrobiodiversity cannot be completely conserved by any single method. It should not be surprising that neither method is able to realize all the expected objectives of plant genetic resources conservation.

The two different concepts of conservation have been developed by the major actors involved in the various types of conservation activities: *ex situ* conservation—enforced and promoted by governmental and inter-governmental organizations as well as the private seed sector—has been managed for the conservation of those crops, which are mainly of interest at the global level. *In situ* conservation has been promoted mainly by NGOs, and handles important regional or local food crops, as well as crops which are not suitable for genebanks.

The limitations of *ex situ* conservation can be summarized as follows:

- Because of the current state of technology, many important species with unorthodox seeds cannot be stored in seed genebanks. Consequently, such species are underrepresented in germplasm collections.
- *Ex situ* storage methods do not guarantee long-term conservation without any negative impacts on the diversity of the plant genetic

resources. A genetic shift due to insufficient and inappropriate regeneration, storage, health care, and existing capacities results in a decline in the genetic variation that existed in the original collection sample.

- Because of the conservation method, genetic resources conserved *ex situ* are not exposed to natural and artificial pressure and cannot, therefore, be expected to evolve and adapt to environmental changes.

Although few studies examine the potential of on-farm /*in situ* conservation, the known limitations of this form of conservation are (Virchow, 1999):

- If PGRFA conservation through on-farm management is implemented by minimizing the area allocated to each variety to be conserved, there is a risk that some allelic diversity will be lost in the limited population.
- Cultural and socioeconomic factors are important for the development of diversity at the local level, but they are related to ecological, social, and technological development. Therefore, the future interest of local communities and individuals in conservation activities cannot be taken for granted.
- The promotion of *in situ* conservation of PGRFA is pointless without a very long time horizon of at least 50 to 100 years, which increases the risk for the unsuccessful implementation of conservation plans.

Given these limitations, the main advantages of *in situ* over *ex situ* conservation can be described as follows (Virchow, 1999):

- *In situ* conservation of PGRFA, as a dynamic form of plant genetic resources management, enables the processes of natural and artificial selection to continue. Consequently, despite the loss of some allelic diversity, on-farm management promotes the development of diversity (seen from a perspective of approximately 100 years).
- It allows the possibility of conserving a large range of potentially interesting alleles.
- It facilitates research on species in their natural habitats.
- It increases protection of associated species, which—in spite of their having no obvious economic value—may contribute to the functioning and long-term productivity of ecosystems.
- *In situ* conservation is especially desirable for crops that do not receive sufficient attention from the formal sector, i.e., government and intergovernmental organizations.
- *In situ* conservation may contribute to an agricultural development while conserving diversity. Linking the management of PGRFA to the improvement of landraces through breeding may be an appropriate strategy for improving farmers' livelihoods in marginal areas as long as there are no economic alternatives.

The links between *ex situ* and *in situ* conservation have, in the past, been generally limited to the transfer of germplasm samples from the farmers to the *ex situ* collections. This one-way traffic of information and goods clearly shows the present suboptimal utilization of the conservation methods. There is far more potential for interaction among the various actors to the mutual benefit of the whole conservation system. Hence, an efficient combination of *ex situ* and *in situ* conservation methods is important to improve overall conservation efficiency.

3. *IN SITU* CONSERVATION METHODS

Although *ex situ* conservation is the predominant method of conserving PGRFA, *in situ* conservation, which involves the conservation of ecosystems, habitats and the general interspecies diversity of wildlife therein, has recently become more focused on the conservation of intra-species diversity of PGRFA. Efforts have increased to maintain and use the diversity of PGRFA either in its natural habitat (especially wild relatives and forestry species) or in locations where the material has evolved, i.e., on farms or in home gardens. In these latter cases, farmers manage diversity and maintain it through use PGRFA in their production systems.

Because only a few countries have been involved in *in situ* conservation programs for PGRFA until now (see Table 6-1), a conceptual framework for *in situ* conservation of PGRFA is just emerging (see Fig. 6-1), Biosphere reserves,[1] protected areas, and natural reserves are all conservation methods for the *in situ* conservation of wild plant (and animal) diversity. In some cases, however, the conservation of wild plant genetic resources has a positive impact on the maintenance of plant genetic resources of the wild relatives of crops (see Fig. 6-1).

---

[1] Biosphere reserves are characterized as "*... areas of terrestrial and coastal/marine ecosystems, where, through appropriate zoning patterns and management mechanisms, the conservation of ecosystems and their biodiversity is combined with the sustainable use of natural resources for the benefit of local communities, including relevant research, monitoring, education and training activities*" (Robertson, 1992).

Table 6-1. Countries reporting *in situ* conservation programs involving PGRFA

	Countries	Dominating *in situ* conservation programs
Africa	Burkina Faso	Landraces of millet and sorghum
	Ethiopia	Wild relatives of coffee Landraces of teff, barley, chick pea, sorghum and faba bean
	Malawi	No detailed information available
Americas	Bolivia	No detailed information available
	Brazil	Wild relatives of cassava, peanut
	Colombia	No detailed information available
	Mexico	Wild relatives of maize
	Peru	No detailed information available
	USA	No detailed information available
	Azerbaijan	Wild fruit trees and shrubs
	Kyrgyzstan	Wild fruit trees and shrubs
Asia	Philippines	No detailed information available
	Sri Lanka	Wild relatives of rice, legumes, spices, wild fruit trees
	Thailand	No detailed information available
	Turkmenistan	Wild fruit trees and shrubs
	Bulgaria	Wild relatives of various crops
Europe	Czech Republic	Wild relatives of various crops
	France	No detailed information available
	Germany	Wild relatives of apples, pears
	Greece	No detailed information available
	Hungary	No detailed information available
	Turkey	Wild relatives of cereals, horticultural, and ornamental crops
Near East	Egypt	No detailed information available
	Israel	Wild relatives of wheat

Source: Virchow (1999) and FAO (1998).

One important example of the *in situ* conservation of wild crop relatives is the Global Environment Facility (GEF) initiated and funded project to conserve wild crop relatives of cereals, horticultural and ornamental flower crops, medicinal plants, and forest trees in Turkey (GEF, 1997). Another important example is the Sierra de Manantlan Biosphere Reserve, which was enlarged by the Mexican government specifically to protect maize and other crop wild relatives. It covers 139,000 hectares, and contains the site where a new species of perennial maize (*Zea diploperennis*) was first reported in 1979. In addition to the conservation of wild relatives of PGRFA, on-farm management and improvement in the conservation of landraces and old cultivars is another form of *in situ* conservation of PGRFA. Various *in situ* conservation methods can be classified according to the material conserved (see Fig. 6-1).

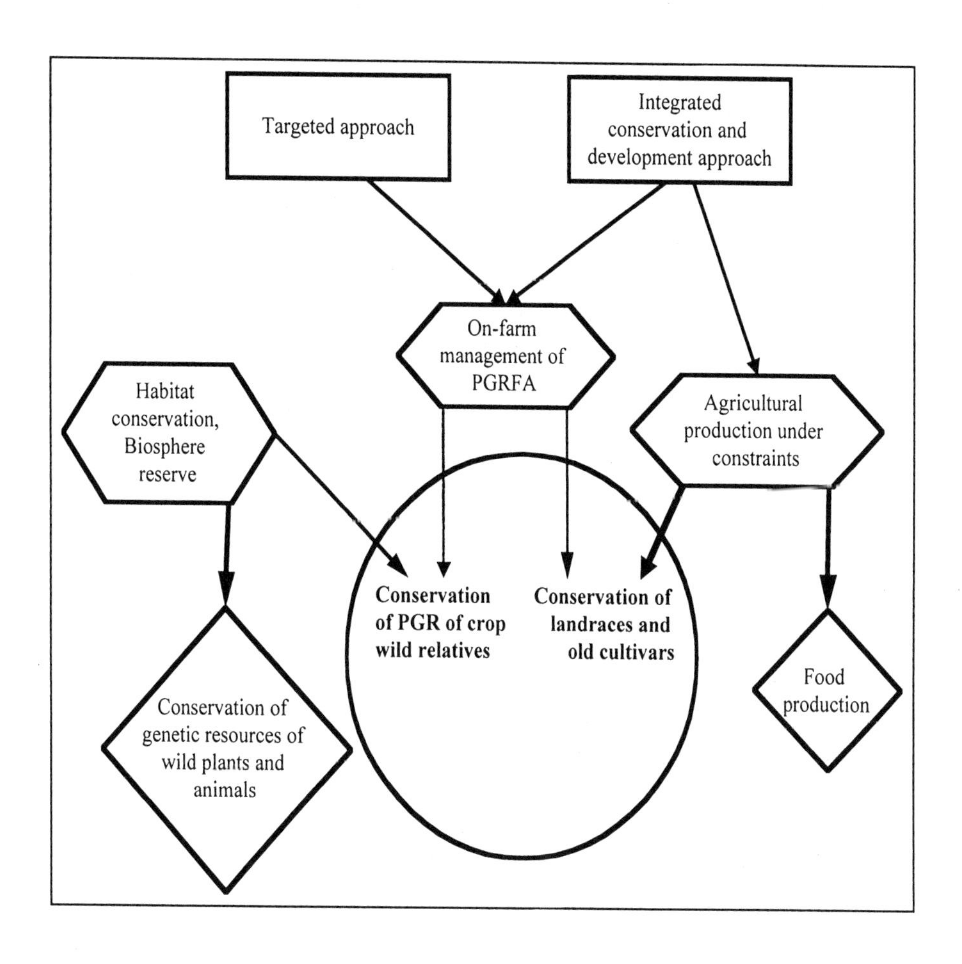

Figure 6-1. In situ conservation of PGRFA

Source: Virchow (1999).

On the one hand, there is the conservation of crop wild relatives, which is mainly a by-product of conservation of general natural areas. Only seldom, but more often in recent times, natural reserves are established explicitly for the conservation of wild crop relatives. The other major type of *in situ* conservation is the on-farm management of the conservation of landraces and old cultivars. Until recently, programs for the on-farm management of PGRFA have been rarely implemented or documented in the scientific literature.

In addition to these programs, the de facto on-farm conservation of crop genetic resources is carried out by numerous farmers all over the world. These farmers live in complex, diverse, risk-prone environments, where local livelihoods depend on subsistence farming. Their actions do not stem from an explicit objective of conserving landraces, but rather from the need to produce food and other agricultural products. The continued use and maintenance of landraces by farmers is carried out for a complex number of anthropological and socioeconomic reasons. One major reason is that these farmers are unable to utilize modern varieties for food production due to their agroecological or socioeconomic conditions (see section 2).

The concept of on-farm management and improvement "provides a mechanism by which the evolutionary systems that are responsible for the generation of variability are conserved" (Worede, 1992). The level of intra-species diversity in PGRFA is a result of a multitude of impacts and does not remain static, but rather continues to evolve. Therefore, in order to continue to influence genetic development, the concept of on-farm PGRFA management seeks to enable the processes and associated infrastructures that were responsible for creating the existing variability of landraces. The advocates of this concept stress the importance of local systems of knowledge and management, local institutions and social organization as well as several other cultural and socioeconomic factors, which determined the development of diversity in the past and which will continue to maintain and develop the diversity at present and in the future. Where they existed, such *in situ* activities tend to be more decentralized and more independently organized than is the case for *ex situ* conservation activities. Existing on-farm management programs can be categorized into:

- *Targeted approaches*, which prioritize the conservation of landraces with significant interest at local, national, regional, or global levels. Furthermore, the increased supply of enhanced seed for breeders and farmers resulting from such activities is also of relevance (Altieri et al., 1987). One of the first "targeted approaches" was initiated in 1988 in Ethiopia. In the drought-prone areas of Welo and Shewa provinces, 21 farms were selected for the project, covering sorghum, chickpeas, teff, field peas and maize[2] (Cooper and Cromwell, 1994).
- *Integrated conservation and development approaches*, which usually directly link the conservation of landraces and old cultivars with specific characteristics of value to the cultivation of these varieties. These valuable varieties are introduced or reintroduced into a certain agroecological region or production system. Additionally, these

[2] In collaboration with the national genebank, farmers select populations grown in their fields by phenotype. The populations are maintained as distinct from each other, although the system allows for pooling similar landraces and even the introgression of valuable genes from exotic sources (Cooper and Cromwell, 1994).

valuable varieties are promoted for breeding and adaptation purposes on farmers' sites ("participatory plant breeding" approaches). Breeders have increasingly turned to such sources and turned away from traditional collections, in which variability is stored in a static state (NRC, 1993). Programs with this approach often involve the participation of NGOs in "grass roots" PGRFA activities, e.g., the MASIPAG Program (Farmer-Scientist Partnership for the Advancement of Science and Agriculture) (Vicente, 1994).

As can be seen from the examples reported in Table 6-1, the *in situ* conservation of PGRFA is mainly directed towards the wild relatives of crops. This indicates that programs set up for the conservation of old cultivars or specific landraces are an exception so far. However, if all grassroots activities promoted by various NGOs are included, there are more conservation activities relating to old cultivars and landraces than the official reports of the national *in situ* conservation statistics suggest (see Chapter 5 for other examples)

As demonstrated by the development of the International Treaty on Plant Genetic Resources (Chapter 6), many countries are pressing for action in the on-farm conservation of PGRFA. Some promotion of activities contributing towards this goal exists, although there seems little experience concerning the successful management of this conservation approach (Virchow, 1999). The majority of existing projects adopts an integrated conservation and development approach and is not limited to purely *in situ* conservation of PGRFA. These programs are usually linked to the support of traditional agricultural systems, to crop improvement through participatory approaches to plant breeding, or are linked to community genebanks, which is a form of *ex situ* conservation.

4. COSTS OF *IN SITU* CONSERVATION

According to information provided by the national reports, only 24 of the 145 countries submitting a report to the International Technical Conference on PGRFA in 1996 have indicated some kind of *in situ* conservation of PGRFA (see Table 6-1). The information related to *in situ* conservation activities is, however, very limited. The officially reported activities refer mainly to the conservation of wild relatives of crops. Considering this limited information base, it is not surprising that the information on the costs of *in situ* conservation activities is negligible or nonexistent. In addition, taking into account that one major cost factor is the opportunity cost of foregoing use of high-yielding varieties, it is understandable that the cost accounting for *in situ* conservation activities is quite difficult. In the following section, a conceptualized framework for analyzing the costs of *in situ* conservation activities is presented, together with a specific example of cost assessment for an *in situ* conservation activity.

Besides some specific transaction and other overall costs, which do not relate directly or exclusively to the *in situ* conservation activities (see Chapter 8), three different types of costs can be identified for the different *in situ* conservation activities:

1. Direct costs for *in situ* conservation programs and projects, including:
   - Implementation costs for the programs and projects, e.g., establishment costs of infrastructure, materials, and personnel.
   - Management costs for the programs and projects, e.g., operating costs of infrastructure, materials, and personnel.
   - Pure production costs for the variety.
2. Costs for supporting activities, including payments made for compensation or incentives for maintaining PGRFA diversity in farmers' fields.
3. Opportunity costs for the use of land for the maintenance of farmers' varieties, including:
   - Private opportunity costs for the individual farmer.
   - Social opportunity costs for a country.
   - Opportunity costs will primarily be composed of the loss of income and food production through foregone utilization of high-yielding varieties on land enrolled in conservation programs and are discussed in greater detail below.

4.1 Direct costs for *in situ* conservation programs and projects

To date, cost estimates for *in situ* conservation are rare. An interesting study on the costs of *in situ* conservation is a survey on potatoes in Peru by Gehl (1997). It was estimated that the conservation of one traditional variety costs US$594 per year (see Table 6-2); 84% of these costs were fixed, determined by the *in situ* conservation project and its costs. Only 16% of the costs (US$98) were derived from the opportunity costs, defined as foregone benefits by cultivating the traditional variety, based on the assumption that farmers cultivated different traditional varieties because of conservation concerns.

Table 6-2. In situ conservation costs for potatoes in Peru

Conservation method	Costs / accession / in US $		
	Variable costs	Fixed costs	Total costs
In situ conservation / Peru[a]	98	496	594

[a] Including opportunity costs.
Source: Gehl (1997).

4.2 Compensation for *in situ* conservation activities

When technological and economic changes occur, institutional arrangements have to be implemented to foster PGRFA conservation with increased incentives. Given the appropriate economic incentives, farmers could continue to cultivate traditional varieties and do so for the sake of conservation. Few countries presently provide incentives to farmers to support *in situ* conservation of their landraces. Proposals for such incentives have, however, been put forward in countries like India, the Philippines, and Tanzania (Virchow, 1999). The costs of these incentives are essentially the opportunity costs of *in situ* conservation.

Attempts to influence the behavior of individuals (farmers) and of whole groups (countries) through the use of incentives may be divided into educational, institutional, and economic measures. Educational incentives, e.g., awareness promotion, may sensitize farmers to the social importance of the conservation of agrobiodiversity, but as Morris and Heisey (1997) emphasize, profit-motivated farmers will in general not be willing to renounce the additional benefit of a less agrobiodiverse production system to benefit society.

On the one hand, institutional incentives could internalize the positive external effects of PGRFA conservation by improving the property rights situation for farmers, allowing them to capture the value of their activities. On the other hand, institutional incentives could be imposed to force farmers to conserve PGRFA diversity. In addition to being undemocratic and inequitable, the efficiency of such a method may be questioned, and the enforcement of any sanction is difficult in countries with poor infrastructure.

The problem of implementing enforceable mechanisms to internalize the costs of diversity loss such as taxes or charges for maintaining agrobiodiversity at socially optimal levels in the field is largely insurmountable. The specific needs of agrobiodiversity conservation, namely, to reduce the area under different traditional varieties to a safe minimum, only permits the utilization of positive incentives.[3] Direct and indirect positive incentives may involve using market mechanisms to target

[3] The objective is to maintain the different crop varieties in farmers' fields, but to reduce the area so as to increase food production through cultivation of higher yielding varieties.

individual farmers and to influence their decision making. These incentives should be coordinated with market forces so as to develop a market for genetically coded information, but also to afford protection against market imperfections, i.e., recognition of the nonuse values of agrobiodiversity that do not enter into market transactions.

The best direct incentive measure for promoting the maintenance of agrobiodiversity among selected individual farmers would be a payment mechanism, through which farmers would be compensated for continuously cultivating a specific variety or maintaining a specific level of agrobiodiversity in his or her fields. The amount of compensation could be determined by the opportunity costs of foregoing alternative production systems, specifically a conversion to a system with modern varieties. Other direct incentives could be rewards for diverse production systems or other social-based incentives. However, the most important nonmonetary incentive would be improved cooperation between farmers and genebanks, especially enabling farmers to receive germplasm from the genebank for further utilization (Gupta, 1996).

4.3 Private and social opportunity costs of *in situ* conservation activities

To analyze the costs of *in situ* conservation, especially the opportunity costs, some conceptional considerations concerning the conservation of PGRFA should be raised. These considerations regard the determinants of the loss of PGRFA in farmers' fields as well as some basic considerations regarding the opportunity costs facing different groups of farmers in adopting conservation activities. The amount of PGRFA diversity maintained in farmers' fields is determined at three levels (see Fig. 6-2): the framework-level, the decision-making process on farm-level, and the level of land-use development.

The maintenance level of PGRFA in farmers' fields is determined by factors at the *framework level*, which directly or indirectly influence the present and future level of PGRFA in farmers' fields. These include socio-economic factors, the development of a relevant market, policies and institutions, as well as natural disasters, war, and civil strife.

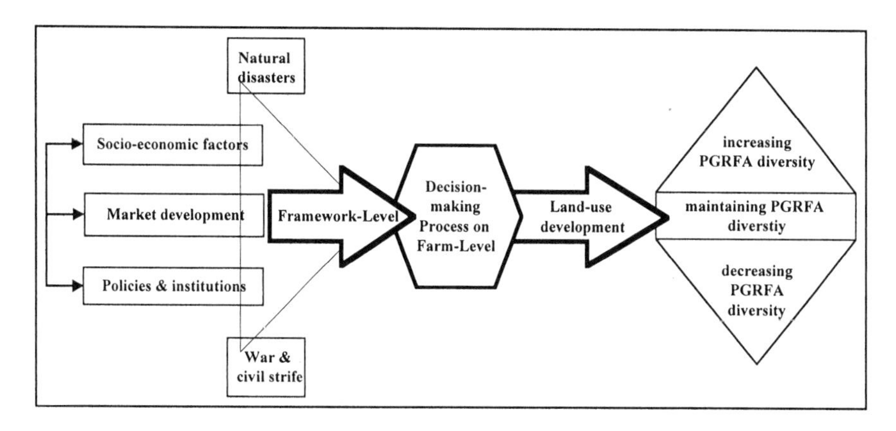

Figure 6-2. Factors determining the diversity-level of PGRFA in farmers' fields

Source: Virchow (1999).

The decisions made on the farm level depend on the individual or farm-specific objectives (see, e.g., Chapters 5 and 7). The practice of cultivating different varieties or crop species and thereby maintaining a specific level of PGRFA diversity is mostly a positive externality of the farm sector.

Farmers, being the main contributors to diversity of PGRFA in the past, are at present those who influence the state of diversity of PGRFA the most with their day-to-day activities. Decisions at the farm level may result in changes in agricultural production systems. These changes will have impacts on the *level of land-use development*. A major category of change in agricultural systems is its intensification through the application of new technology, which may lead to variety replacement, overexploitation of genetic resources, and habitat destruction. As a side effect, changes in agricultural systems may lead to the introduction of new pests, diseases, or weeds; or, in an extreme case, these changes may lead to the abandonment of agricultural production and the development of nonagricultural land-use systems. The site- and time-specific factors of the different levels and their inter- and intra-linkages determine whether the quantity of PGRFA diversity in farmers' fields is maintained, increased, or reduced.

The farm level is the key point for targeting the decision-making process to influence the diversity in *in situ* PGRFA. If the farmer expects greater benefits from the use of new varieties, the farmer will replace the old varieties with the new varieties. Brush and Meng (1996) found that many farmers, even though introducing improved varieties, are keeping farmers' varieties in their production systems. These farmers are doing this mainly because the new varieties do not have certain traits (i.e., taste) which are known in the traditional varieties. However, assuming that breeders are going to incorporate these specific traits into the improved varieties as well, the farmers will have no incentive for keeping the traditional varieties any longer.

According to the field experience of Pundis (1996), farmers adopt a new variety if the yield gains are over 15% greater than that of the traditional variety. The more the plant breeding improves the new varieties (i.e., incorporation of all needed or asked for traits), the more farmers' varieties will be replaced by the improved varieties. For each trait incorporated into an improved variety, the advantage of one or more farmers' varieties will be diminished or lost. As breeding is a long and difficult process, the replacement over all farms will take some time.

The farmers' decision to utilize modern varieties leads to the reduction of land cultivated with specific traditional varieties. If all farmers who cultivated a specific variety replace it with modern varieties, that traditional variety will disappear from farmers' fields. Each farmer will make his or her decision based on his or her private marginal benefits and marginal costs, which do not reflect the social costs of the variety loss. Depending on the population dynamics of each variety, less than 1 hectare of cultivated area may be sufficient to conserve one variety (Bücken, 1997). Technically speaking, the loss of traditional varieties through the transformation of land under traditional varieties to modern varieties, is a negative environmental externality with a very high buffer capacity determined by the specific ecological threshold effects (Perrings and Pearce, 1994). Consequently, the negative externality effect gains importance only with an increasing number of farmers making the same decision. In other words, the environmental damage curve representing the loss of a traditional variety due to land conversion from a traditional variety to modern varieties is increasing with increased land conversion (Swanson et al., 1994).

Farmers are often characterized in the literature as the main load-bearing actors in *in situ* conservation of PGRFA (e.g., Altieri and Montecinos, 1993). Except perhaps for hobby cultivation that is found mainly in industrialized countries, farmers do not maintain PGRFA diversity for its own sake and in accordance with the three objectives of PGRFA conservation (see section 2). Thus, *in situ* conservation of PGRFA diversity is a positive externality of the farm activities, based on the farmers' private benefit expectations for other reasons. Hence, resource-poor as well as resource-rich farmers may produce PGRFA diversity without additional costs.

The maintenance of PGRFA diversity in farmers' fields for other reasons than PGRFA conservation may be interpreted as "de facto conservation" of PGRFA diversity (Meng et al., 1997) noted earlier. However, as long as the diversity of PGRFA is not valued in the market, and therefore providing little incentive for its conservation, the maintenance of PGRFA diversity in farmers' fields through legislative means will be negatively correlated with overall agricultural development in a specific region.

The evolution of the marginal costs and benefits in the farmers' production systems determines the level of PGRFA diversity selected and thus the positive external effect generated for all the beneficiaries of PGRFA diversity without costs. Some farmers may conserve PGRFA diversity in a de facto manner by utilizing farmers' varieties on their whole farm (mainly resource-poor farmers) or only on parts of it, in spite of access to modern varieties and the agroecological possibilities (mainly resource-rich farmers). They do so because the private anticipated benefit to their production systems (including traditional varieties) in terms of insurance, breeding, taste, and so forth, is higher than that of an alternative system (including a higher or total share of land sown to modern varieties) (Virchow, 1999; Brush and Meng, 1996). This de facto conservation may imply that the marginal benefits of maintaining and utilizing a high quantity of traditional varieties is higher for resource-poor farmers than for resource-rich farmers (see Fig. 6-3).

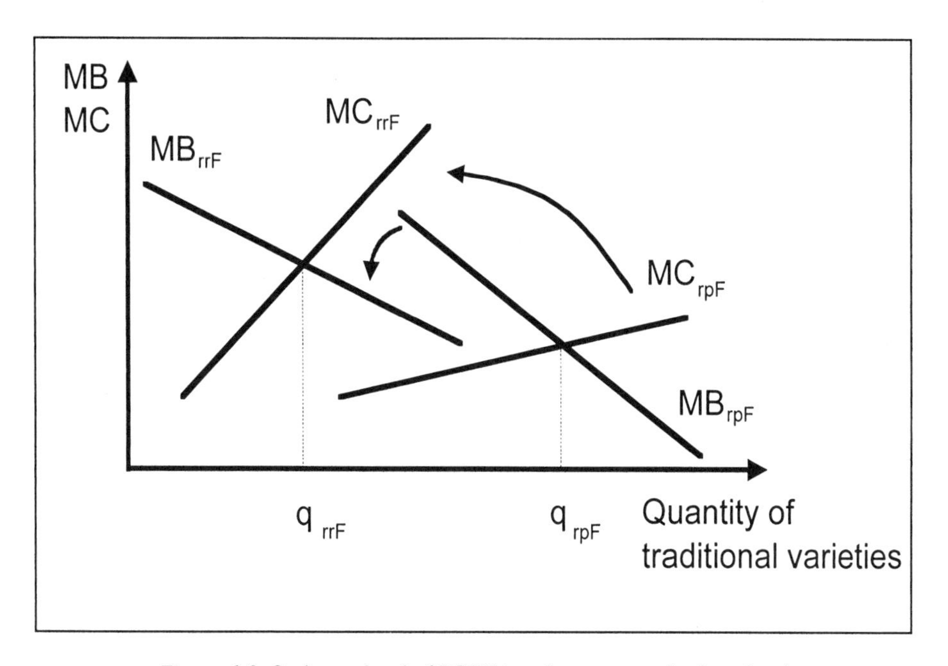

Figure 6-3. Optimum level of PGRFA maintenance at the farm level

Note: MC: marginal costs; MB: marginal benefit.
MCrrF / MBrrF: marginal costs / marginal benefit for resource-rich farmers.
MCrpF / MBrpF: marginal costs / marginal benefit for resource-poor farmers.
Source: Virchow (1999).

Farmers will go on maintaining their production systems, and consequently, a specific level of PGRFA diversity, as long as their private marginal benefit is higher than the private marginal costs they incur from the loss of higher yields or other benefits possible under a change in the production system. A change of the production system to a "modern" one would inevitably reduce the level of agrobiodiversity. Wherever farmers are able to change their production systems, the costs of maintaining the production system and therefore the de facto conservation of PGRFA diversity have to be reflected as opportunity costs of potential income lost by not utilizing modern varieties.

As described in Fig. 6-3, the marginal costs of maintaining PGRFA as traditional varieties are the lowest for the group of economic and ecological marginalized farmers without any other production possibilities, and hence, without any opportunity costs for maintaining PGRFA-diverse production systems (represented by the marginal cost curve of the resource-poor farmers MC_{rpF}). The only opportunity costs this group faces arise from either the abandonment of agricultural production due to out migration from the marginal areas or the change to nonagricultural occupation.

Technical improvement and development at the farm level creates the possibility of choice between traditional varieties and modern varieties and the introduction of modern varieties may therefore occur. Because of the change of their production system, there will be an inevitable decline in PGRFA diversity in their fields. This decline is characterized by the change of area under traditional varieties from q_{rpF} (resource-poor farmers) to q_{rrF} (resource-rich farmers). At the farm level, the decision is made whether to cultivate traditional varieties or not, and to what extent. This decision is solely determined by the individual or farm-level benefit-optimizing criteria, which are also influenced by the economic and ecological framework, i.e., the access to resources and technologies.

Countries involved in the conservation of PGRFA may expect direct market benefits for the breeding activities derived from the raw material as well as food security benefits through the insurance value of genetic resources (i.e., stabilized or increased food production) and the functional value (reduced degradation of natural resources) as part of the breeding value. However, while aiming to conserve their PGRFA *in situ,* these countries face one problem in determining their country-specific optima of PGRFA diversity level: as discussed above, farmers provide the country's current PGRFA diversity as a free good. At the country level, the social marginal costs are mainly derived from the opportunity costs for foregone increased food production through the renouncement of the utilization of modern varieties. To increase national food security, many countries have to continue and increase the integration of the resource-poor farmers, the "custodian" of PGRFA diversity, into the market (von Braun and Virchow, 1997). As a rule, one traditional variety may be maintained *in situ* on an area less than $100m^2$ for crops with orthodox seed and less than $250m^2$ for vegetatively propagated crops. For other crop species, e.g., with recalcitrant seeds, perennial species, and species with long life cycles, less than 1 hectare is necessary (Bücken, 1997).[4] To provide a margin of safety, it may be assumed that each traditional variety may be planted on average on $1{,}000m^2$ and on 10 to 50 farms so that crop failure on one farm does not risk loss of the variety. Therefore, the total area necessary to conserve one traditional variety of any crop species would be less than 5 hectares.

Because of this minimum safety standard (MSS), the country's conservation objective could be fulfilled with the minimum area requirement at q_n in Fig. 6-4. More area planted to the specific traditional variety will increase the social marginal costs of the *in situ* conservation in a specific country (MC_n). In contrast to the country's optimal area needed, farmers—according to their individual optima—utilize altogether far more land (characterized as q_p in Fig. 6-4) for the traditional variety than is socially optimal (q_n), reflecting the opportunity costs at the national level.

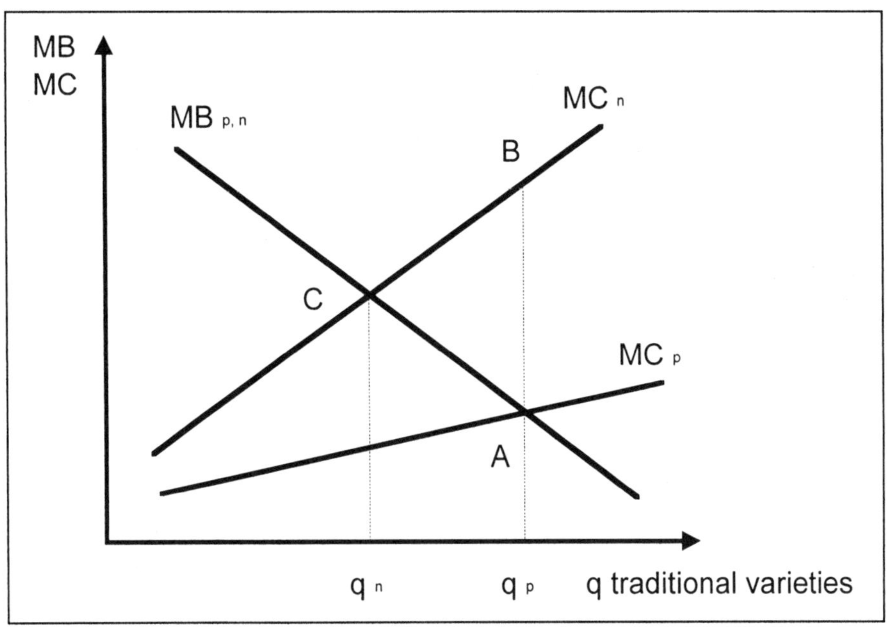

Figure 6-4. Scenario for which PGRFA conservation is greater than is socially optimal

Note: MC: marginal costs; MB: marginal benefit.
MC_n / MC_p: marginal costs on national / private level.
$MB_{p,n}$: marginal benefit on private and national level.
Source: Virchow (1999).

[4] For instance, the area requirement for maize is 60 m^2 and for wheat only 0.75 m^2.

At last count, only 74%, 70%, and 57% of the area planted to rice, wheat, and maize, respectively, in developing countries is planted with modern varieties[5] (Alexandratos, 1995). Pretty (1995) estimates that a total of almost 2 billion people are still not benefiting from modern agriculture. Due to the area still under traditional varieties, where farmers grow the same set or overlapping sets of varieties there is the negative external effect for society as a whole (marked in Fig. 6-4 as area *ABC*). The private decisions may provide optimal levels of on-farm production systems, thereby including a specific level of agrobiodiversity. But it may also provide sub-optimal levels of PGRFA diversity from a social perspective (e.g., an oversupply). This suboptimal level of PGRFA diversity from the national perspective is reflected in the opportunity costs at national level.

4. FINANCING *IN SITU* CONSERVATION

De facto on-farm conservation by farmers without any external incentives is the most important way of conserving PGRFA at present. For instance, the Arguarana Jivaro community in the Peruvian Amazon grows 61 distinct cultivars of cassava, while some small communities in the Andes grow 178 locally named potato varieties (Brush, 1991; Bolster, 1985). From the national perspective, the area cultivated with the traditional varieties can be minimized while maintaining diversity. The high opportunity costs for land with farmers' varieties calls for action. While such a level of agrobiodiversity currently *in situ* may represent an optimum level of *in situ* conservation, the large group of marginalized farmers is utilizing far too much land with farmers' varieties than is needed to maintain a national optimal level, given the few alternatives open to them.

Although the high opportunity costs are reason enough to reduce the area under traditional varieties as much as possible, the high fixed costs of *in situ* conservation programs and projects make these questionable. Hence, a system of *in situ* conservation has to be found which is flexible enough to react when needed and is less expensive.

A preliminary approach at estimating the costs of *in situ* conservation is based on the introduced "controlled *in situ* conservation" system.[6] Considering these details and taking into account that all agricultural crops consist of approximately 3 million distinct varieties (FAO, 1998), an MSS of roughly 15 million hectares of arable land must be utilized for a safe, but minimized *in situ* conservation; 1% of the existing 1.4 billion hectares arable

---

[5] As average figures they do not show the differences between countries or regions, so for instance, in sub-Saharan Africa, modern wheat varieties are planted only on 52% of the wheat area and in rice it may be even less.

[6] See the explanations of the MSS on page 23 and Virchow (1999) for detailed information.

land (Engelman and LeRoy, 1995) is necessary to conserve the estimated 3 million distinct varieties.[7]

As long as the economic and technological development has not yet transformed all marginalized agricultural areas into high potential agricultural areas, much more than 1% of the arable land is still utilized with traditional varieties and landraces. In India, more than 60% of arable land is still cultivated with traditional or farmers' varieties (CMIE, 1988). Hence, in most cases no intervention is required; no financial costs will be incurred. *In situ* conservation, however, is supposed to be sustainable even in the face of changes of comparative advantages of traditional varieties to modern varieties. Consequently, a rough estimate has to guide the decision whether *in situ* conservation will be justifiable in the light of scarce financial resources.

In the first place, financing the controlled *in situ* conservation may lay in the hand of each country, which has the sovereignty over its genetic resources, as was stressed by CBD (UNEP, 1994). There may, however, be countries, which need more than the calculated 1% of their arable land for the *in situ* conservation of all distinct varieties because of their richness in agrobiodiversity. On the other hand, there may be countries, which need less than the average 1%. Because of the interdependency in PGRFA already utilized and needed in the future, the countries having to spend less than average may compensate countries which need more than 1% of their arable land.

India and Germany are examples of agrobiodiversity-rich and -poor countries; it can be stated that while in India land under agricultural use is still expanding to meet the increasing food needs, in Germany 12% of arable land is left as fallow due to the EC's set-aside programs (GCR, 1995). Consequently, the social costs on a national level for controlled *in situ* conservation, in terms of opportunity costs for foregone benefit, will be much higher in India than in Germany. From a simplistic point of view, the resources would seem allocated best if Germany could take over India's controlled *in situ* conservation. The different ecological conditions, however, restrict a transfer of *in situ* conservation activities from India to Germany. Consequently, Germany and other agrobiodiversity-poor countries could compensate India for the conservation, which is in a global interest similar to the joint implementation programs, known from the UN Framework Convention on Biological Diversity.

---

[7] For the sake of comparison, India has set aside 4% of the country's surface for almost 500 wildlife protected areas (ICR, 1995).

6. CONCLUSIONS: THE COSTS OF *IN SITU* CONSERVATION ACTIVITIES AND ITS IMPACT ON PGRFA CONSERVATION POLICIES

Agrobiodiversity is largely produced by farmers as a positive externality without any conservation program costs. Farmers naturally maintain some agrobiodiversity through their farm-specific production systems according to the individual optima of the decision-making process at the farm level. Analysis of the *in situ* conservation levels suggests that the total number of traditional varieties is produced or maintained by farmers on more land than is necessary. This land has the potential of being used for the production of more food by utilizing "modern" production systems, including modern varieties.

Instruments like taxes or coercion will not be effective in reducing the land utilized for less productive production systems as farmers in marginal areas have no other choice than to utilize these systems including the traditional varieties. Therefore, economically and ecologically marginalized areas, in which most of agrobiodiversity is produced and conserved de facto to date, need external investment in infrastructure and technology to reduce production limitations and to increase national food production. In the long run, it should be in the interest of all countries to reduce the economic and ecological marginalization of areas through investments and at the same time to increase the marginal costs of the farmers utilizing traditional production systems by increasing the opportunity costs of maintaining traditional varieties. Consequently, the amount of area utilized by these systems would decrease. However, this development implies a risk of an unplanned loss of traditional varieties.

In the future, assuming an increase in the production in the present marginalized areas, the question will be with what economic instruments and incentives can agrobiodiversity be kept at the social optimum, securing nonmarketable genetic resources? Due to the high costs of *in situ* conservation projects and programs and as long as most of PGRFA still *in situ* are conserved by farmers without any conservation costs, such active *in situ* programs can be only cautiously promoted. As long as property rights for genetic resources are not well defined and as long as there do not exist any mechanisms to integrate the social value of agrobiodiversity into a market mechanism, there will be the need for government interventions to protect the existing diversity on a minimum of agricultural land. These interventions, however, have to be as efficient as possible. Therefore, a flexible and self-targeting incentive mechanism that comes only into effect when a traditional variety is endangered by extinction is needed for a controlled *in situ* conservation on the national level. For the time being, the following question will, however, remain: how can the marginalized farmer benefit from the producing "good" agrobiodiversity and how can the area utilized for traditional varieties be reduced to decrease the social opportunity costs of *in situ* conservation without losing varieties?

REFERENCES

Alexandratos, N., ed., 1995, *World Agriculture: Towards 2010: An FAO Study*, John Wiley and Sons, New York.

Altieri, M. A., Merick, L. C., and Anderson, M. K., 1987, Peasant agriculture and the conservation of crop and wild plant genetic resources, *Conservation Biology* **1**:49-58.

Altieri, M. A., and Montecinos, C., 1993, Conserving crop genetic resources in Latin America through farmers' participation, in: *Perspectives on Biodiversity: Case Studies of Genetic Resource Conservation and Development*, C. S. Potter, J. I. Cohen, and D. Janczewski, eds., AAAS Publication, Washington, D. C., pp. 45-64.

Bolster, J. S., 1985, Selection for perceptual distinctiveness: Evidence from Aguarana cultivars of Manihot esculenta, *Economic Botany* **39**:310-325.

Brush, S. B., 1991, Farmer conservation of new world crops: The case of Andean potatoes, *Diversity* **7**:75-79.

Brush, S. B., and Meng, E., 1996, Farmers' valuation and conservation of crop genetic resources, paper presented at the CEIS - Tor Vergata University, Symposium on the Economics of Valuation and Conservation of Genetic Resources for Agriculture (May 13-15, 1996).

Bücken, S., 1997, Personal communication at the genebank of the Federal Centre for Breeding Research on Cultivated Plants, Braunschweig.

CMIE (Centre for Monitoring Indian Economy), 1988, Basic Statistics Relating to the Indian Economy, Vol. I: All-India, CMIE Pvt. Ltd.

Cooper, H. D., and Cromwell, E., 1994, In situ conservation of crop genetic resources in developing countries: The influence of economic, policy and institutional factors, Draft Discussion Paper 1, Rural Poverty and Resources Research Programme, ODA.

Engelman, R., and LeRoy, P., 1995, *Conserving Land: Population and Sustainable Food Production*, Population Action International, Washington, D.C.

FAO, 1998, *The State of the World's Plant Genetic Resources for Food and Agriculture*, FAO, Rome.

GCR (German Country Report), 1995, *Country Report on Status of Plant Genetic Resources Germany*, Submitted to FAO in the preparatory process for the International Technical Conference on Plant Genetic Resources, 1995.

GEF (Global Environment Facility), 1997, *Quarterly Operational Report* (June, 1997), GEF, Washington, D. C.

Gehl, J., 1997, *Kosten-Wirksamkeits-Analyse zur Bewahrung genetischer Ressourcen, Fallbeispiele aus Deutschland und Peru*, Diplomarbeit, Rheinischen Friedrich-Wilhelms-Universität zu Bonn.

Gupta, A. K., 1996, Personal Communication, Indian Institute of Management, Ahmedabad.

ICR (Indian Country Report), 1995, *Country Report on Status of Plant Genetic Resources India*, Submitted to FAO in the preparatory process for the International Technical Conference on Plant Genetic Resources.

Meng, E., Taylor, J. E., and Brush, S. B., 1997, Household varietal choice decisions and policy implications for the conservation of wheat landraces in Turkey, Draft Paper presented at the Symposium: Building the Theoretical and Empirical Basis for the Economics of Genetic Diversity and Genetic Resources Conservation in Crop Plants, Held at Stanford University, Palo Alto, California (August 17-19, 1997).

Morris, M. L., and Heisey, P. W., 1997, Achieving desirable levels of crop genetic diversity in farmers' fields: Factors affecting the production and use of improved seed, paper presented at the Symposium: Building the Theoretical and Empirical Basis for the Economics of Genetic Diversity and Genetic Resources Conservation in Crop Plants, Held at Stanford University, Palo Alto, California (August 17-19, 1997).

NRC (National Research Council), 1993, *Managing Global Genetic Resources: Agricultural Crop Issues and Policies*, National Academy Press, Washington, D. C.

Perrings, C., and Pearce, D., 1994, Threshold effects and incentives for the conservation of biodiversity, *Environ. and Res. Econ.* **4**:13-28.

Pretty, J. N., 1995, *Regenerating Agriculture: Policies and Practice for Sustainability and Self-Reliance*, Earthscan Publications Ltd., London.

Pundis, R. P. S., 1996, Personal communication, ICRISAT.

Robertson, J., 1992, Biosphere reserves: Relations with natural world heritage sites, *Parks* **3**:29-34.

Swanson, T. M., Pearce, D. W., and Cervigni, R., 1994, *The Appropriation of the Benefits of Plant Genetic Resources for Agriculture: An Economic Analysis of the Alternative Mechanisms for Biodiversity Conservation*, FAO Commission on Plant Genetic Resources, Background Study Paper No. 1, FAO, Rome.

UNEP (United Nations Environment Program), 1994, *Convention on Biological Diversity*, Text and Annexes, Interim Secretariat for the Convention on Biological Diversity, Geneva Executive Center, Switzerland.

Vicente, P. R., 1994, The MASIPAG program: An integrated approach to genetic conservation and use, in: *Growing Diversity in Farmers Fields*, Proceedings of a Regional Seminar for Nordic Development Cooperation Agencies, Lidingo, Sweden (September 26-28, 1993).

Virchow, D., 1999, *Conservation of Genetic Resources: Costs and Implications for a Sustainable Utilization of Plant Genetic Resources for Food and Agriculture*, Springer, Heidelberg.

von Braun, J., and Virchow, D., 1997, Conflict-prone formation of markets for genetic resources: Institutional and economic implications for developing countries. *Quart. J. Inter. Agri.* **1**:6-38.

Worede, M., 1992, The role of Ethiopian farmers in the conservation and utilization of crop genetic resources, First Int. Crop Sci. Congress, Ames, Iowa.

Chapter 7

UNDERSTANDING THE FACTORS DRIVING ON FARM CROP GENETIC DIVERSITY: *EMPIRICAL EVIDENCE FROM MEXICO*

Eric Van Dusen
Wantrup Fellow, Agricultural and Resource Economics, University of California, Berkeley, CA 94720-3310

Abstract: This chapter outlines an empirical approach to understanding the determinants of farmers' access to, and use of, crop genetic resources (CGR) and the impacts of farmer behavior on crop populations. These are critical to modeling the current status of agricultural biodiversity on farm, as well as in designing and implementing effective programs for the *in situ* conservation of agricultural biodiversity. A case study from applied fieldwork in Mexico is also provided to illustrate the concepts presented. Empirical evidence is presented to demonstrate the type of data to be collected from farm level surveys that is necessary to measure population level processes from farmer behavior.

Key words: agricultural biodiversity; crop genetic resources; farmer seed systems; *in situ* conservation.

1. INTRODUCTION

This chapter outlines an empirical approach to understanding the determinants of farmers' access to, and use of, crop genetic resources (CGR) and the impacts of farmer behavior on crop populations. These are critical to modeling the current status of agricultural biodiversity on farm, as well as in designing and implementing effective programs for the *in situ* conservation of agricultural biodiversity. Recently, a series of applied studies of farmer seed systems has begun to generate data on local seed systems useful in seeking to understand farm level decisions on CGR management (Brush, Taylor, and Bellon, 1993; Perales Rivera, 1996; Meng, 1997; Louette and Smale, 1998; Rice et al., 1998; Smale et al., 1999; Aguirre Gómez, Bellon, and Smale, 2000). However, there is still further need to model, document, and understand the human-mediated impacts upon crop populations, utilizing the involvement of applied social science to complement the work of crop breeders, geneticists, biologists, and ecologists. Socioeconomic assessments documenting behavior towards seed selection or exchange, based upon sample surveys and statistical analysis, are one area where further research is required. More work is also needed to document the role of a given crop within a household's set of activities, and the interaction of that crop in the village economy across varying socio-economic and environmental conditions.

This chapter seeks to articulate some of the practical socioeconomic issues that have arisen in the attempt to understand farm level management of CGR and how these are necessary for the design and implementation of effective strategies for promoting *in situ* conservation. A case study from applied fieldwork in Mexico is also provided to illustrate the concepts presented. Empirical evidence is presented to demonstrate the type of data to be collected from farm level surveys that is necessary to measure population level processes from farmer behavior.

Deriving a deeper understanding of the factors that determine on-farm management of CGR is necessary for the design and implementation of policies to promote *in situ* conservation. For the purposes of the discussion here, *in situ* conservation will be defined as farmer-based maintenance of traditional crops in their fields (also referred to as on-farm conservation). This discussion will not cover wild species related to crops directly, but the maintenance of a traditional cropping system may contain wild relatives, favor geneflow between wild and cultivated species, and contain more complex ecological relationships such as with pests and symbiotic species. Traditional varieties are defined as crop varieties which are products of farmer selection processes and not the result of a scientific breeding program. The starting point for the empirical analysis of crop dynamics is to define the unit of analysis for conservation and for farmer management. This paper will use the definition of a seed lot (provided by Louette and Smale, 2000): the seed saved from a previous harvest for planting in the following year.

In identifying the empirical questions of farmer management of crop resources and the human-mediated impacts on crop genetics, it is important to keep in mind the following basic characteristics of *in situ* conservation:

(1) The maintenance of crop genetic diversity occurs in a dynamic, evolutionary context.

(2) Local farmer varieties or landraces can be maintained within larger crop populations with flows of genes into and out of such populations.

(3) *In situ* conservation is decentralized and disaggregated. The potential risks to long-term conservation of CGR are therefore diffuse, especially socioeconomic pressures such as economic development, market integration, and cultural change.

These characteristics provide a starting point for the design of *in situ* conservation programs, particularly in the ways in which they can complement existing *ex situ* conservation activities. Three related limitations of *ex situ* conservation are that (1) accessions are frozen in time, (2) accessions are kept in isolation from population characteristics and ecological contexts, and (3) *ex situ* collections are centralized in a single location and immune to economic conditions but are thus fragile to loss in fire, loss of power, natural disaster, etc. The dynamic component is a key difference, while an *ex situ* approach would seek to preserve materials and periodically regenerate them to maintain the original type, the *in situ* approach would seek to have farmers continuously selecting plants and populations in the face of, for example, pest pressure, climatic effects, or mutations.

The options for intervention and policy planning to conserve agricultural biodiversity can be described in three possible stages: monitoring, mitigation, and preservation. The most basic form of *in situ* conservation would be to monitor the viability of local crop populations. This starts from the documentation of the de facto conservation that takes place throughout the developing world, often in areas with marginal conditions where improved varieties may not offer superior performance (Brush and Meng, 1998). An intermediate step would be for the mitigation of the loss of local varieties, while seeking to enhance local crop production. These type of approaches have centered around improving local selection techniques or enhancing seed exchange and the seed distribution system, and also include attempts to encourage the integration of local crop populations into the breeding populations selected by formal breeders, often through collaboration with local farmers (MILPA Project, 1999). The most involved form of *in situ* conservation would entail the planned conservation of specific crop varieties, involving management planning and possibly some form of contract between farmers and conservation managers. A variety of issues remain unresolved for designing the institutions for this form of *in situ* conservation, including monitoring and compliance, moral hazard issues, and the opportunity costs of other activities.

For the above reasons, the focus of this chapter is not the costs of conservation, but practical issues that arise in the human management of crop populations. The policy relevance is not how to allocate scarce conservation funding between complementary forms of conservation, but specific characteristics of *in situ* conservation. However, an important contrast to *ex situ* conservation arises from dynamic nature of planning for future conservation. In *ex situ* conservation costs and activities can be reasonably forecast into the future given the current state of storage technology. In *in situ* conservation there is a need to predict from the current situation what risk or constraints will threaten the viability of a targeted crop population.

2. FACTORS DETERMINING ON-FARM CROP GENETIC DIVERSITY

2.1 Dynamics of seed selection

The dynamic nature of farmer seed management has been repeatedly highlighted as one of the most important determinants of the pattern of on-farm crop genetic diversity (Perales, 1996; Brush, 2000). Farmers and environmental selection are constantly affecting the population genetics of a given variety. This continual and repeated selection is fundamental to the evolutionary process that generates new diversity (Brown, 2000). Farmers also periodically renew seed lots, trying new varieties and losing varieties to changing weather or economic conditions (Rice, Smale, and Blanco, 1998; Louette and Smale, 2000).

Characterization of intertemporal behavior is thus important to understanding the sustainability of the conservation of CGR over a long time horizon. Shifts over time are fundamental to genetics of a crop population, whether the population is narrowed through erosion, changed through genetic drift, or broadened through introgression. Diversity can be continuously created over time, and the diversity that is observed in the present is the product of a long process of dynamic evolution in the past.

2.1.1 Selection pressures

The selection pressures applied by farmers when saving seed from harvest for planting in the following year are important determinants of the dynamic nature of on-farm management of CGR. These selection pressures can be understood as the ways (whether conscious or unconscious) that farmers determine how the seed selected from the total amount harvested differs genetically. It is necessary to document how, when, and by whom seed is selected in order to understand the intertemporal evolution of CGR. The criteria used by farmers for selection are important determinants of how a given variety will evolve over time. For example, in selection carried out by women, processing or consumption characteristics may be the most important criteria for selection, while selection carried out by men may be focused on production, storage, or animal feeding characteristics. The timing of selection can have important population consequences ranging from selection in the field with ability to relate seed to the plant's fitness, to selection in the storage bin, with the ability to compare the entire harvest. All of these interact with the household economic situation through the division of labor, household use of hired labor, and substitution between own production and market commodities.

For long-term genetic conservation, it is important to determine if there are risks to a crop population due to farmer selection practices. Two of the most relevant risks from *ex situ* management are genetic drift (a random loss of alleles due to using small sample sizes to renew a population) and genetic shift (a change in population due to regeneration under different environmental conditions than where the population was originally from). For *in situ* it will be necessary to see if selection practices will lead to changes or shifts in population structure, will be perceivable by the farmer or manager, and will lead to evolution of new varieties.

2.1.2 History of seed

Documentation of the history and recent management of seed can help to understand the life of a variety over time. One basic piece of information is the age of seed managed by farmers. The loss of the seed lot in any given period can happen for a variety of reasons—such as infestation, bad weather, or the need to consume all of the harvest due to poverty. Furthermore farmers may periodically "renew" seed lots, possibly due to decreased productivity from inbreeding depression, or in order to include new traits into the variety. In either case, data may be gathered on the age of seeds held by each farmer, or how frequently seed lots are lost or renewed in order to extrapolate the average age of seed lots held by each farmer. This information is highly relevant to the design of an *in situ* conservation program. If the program is seeking to maintain a certain variety in continuous planting and re-selection, these data on the age of seed lots may determine the number of farmers who need to be involved to ensure the survival of a given variety. A conservation program, which seeks to improve local varieties through farmer selection practices, will need to be built upon information on the frequency with which farmers change the seed they use.

2.1.3 Variety historical profiles

Historical data on each variety are needed to detect long-term trends in the agricultural system. It is necessary to determine whether the varieties found in the fields in the current period are typical historically or are transient. In a more general sense, historical questions on the extent of past agricultural practices relative to current ones can indicate whether the farming system is stable, or is in a process of transformation. This information on the prominence of current conditions in relation to past practices provides a basis for developing measures of the stability of *de facto in situ* conservation. Possible economic forces that may lead to decreased on-farm diversity of CGR include: the profitability of cash crops, yields of alternative staple crops, land rents, labor wages, and changes in cropping systems and intercropped varieties.

2.2 Population structure

Individual crop varieties may correspond to populations that share agroecological characteristics, or to certain criteria perceived by farmers that may not correlate to genetic criteria. A variety may be adapted to an ecological niche, such as soil type, soil moisture, exposure, cropping season, or intercropping system. The ecological or climate factors may affect the population structure by interacting with farmer efforts that cause mixing or inflow of genes. Mixing occurs between ecological regions, and re-segregation to local type occurs through human pressure (a preferred phenotype) and through environmental pressure (through selection of the best performing plants). In general, de facto conservation takes place under challenged conditions that may cause larger populations to segregate into local populations according to ecological niches. Understanding and measuring this may be difficult and costly, but are important in the design of a conservation program.

Any *in situ* conservation program will need to define its targets for conservation, and the population structure of each crop will inform a management strategy. A program for a self-pollinating crop like wheat may seek to preserve a set of distinct farmer varieties, or may seek to preserve fields that contain mixed populations that compete under local conditions. An out-crossing crop like maize may call for the maintenance of a crop population at a village or regional level spanning individual farmers (Louette and Smale, 2000). Another approach may be to work with a single or select group of farmers, who can later serve as a local seed source for their community.

2.2.1 Scale of analysis

Analysis of crop populations and management strategies may be influenced by the ecological model used. A model based on niche theory would seek to understand the diversity of varieties (and competition for resources) within a farmer's field or a village. A meta-population model may lay more stress on the flows between farms and between villages. Selecting the unit of analysis is also an important component in the design of *in situ* conservation projects, where the costs of expanding the scope of the project to a larger region have to be balanced against the inclusion of a more comprehensive level of crop diversity. The unit of analysis for an *in situ* conservation program could range from an individual cooperating farmer, a series of farmers within a village, or a series of villages within a region. In each case it will be important to document the flows of genes and seeds into, and within, each system.

2.2.2 Flows

The institutions that influence seed flows are basic elements of the farmer seed system. Characterization of these is needed both to understand the stability of a crop variety and the breadth of genetic sources used by farmers in an evolutionary process. Seed sources both within the village (or relevant unit of analysis) and outside the village should be identified. Other institutions may be regional markets, commodity traders, seed or input suppliers, or labor migration by local farmers. Furthermore, seed sources may be influenced by cultural factors such as common linguistic groups or kinship networks.

The flows of seeds or genetic materials can be documented at two principal levels. First are flows from the larger population (regional) to a smaller population (farmers seed lot) and the second are flows between subpopulations. Furthermore, the flows may be uni-directional, e.g., only from outside the farm into the population, or multidirectional, between many farmers simultaneously. Approximation of the rates of flow, and the human mediated transactions can be recorded through asking the sources of new seed lots, the customary sources for seed renewal, or potential sources in case seed is depleted. Finally, in out-crossing species, the level of cross-pollination in farms with small plot sizes is significant and could be important both within a given farmer's field and across fields in a village (Louette and Smale, 2000).

2.3 Socioeconomic factors influencing variety selection

At present, the primary means by which the *in situ* conservation of agricultural biodiversity resources is achieved is through de facto conservation achieved through farmers' choice of CGR. In order to understand the current status of agricultural biodiversity as well as where interventions will be necessary, it is important to document the socio-economic characteristics of the farm households involved in growing specific crops. A primary goal of such an effort will be to identify the possible economic reasons for de facto conservation, or why the farmers continue to cultivate the traditional variety despite competition from other varieties. This may include anything from the cultural attributes of a specific variety or trait, to economic constraints that prevent the farmer from changing crop technology. Another important result of such an analysis is an assessment of whether farmers will continue with current cropping patterns into the future and what implications this may have for the sustainability of crop genetic diversity levels. Finally, the analysis will be used to identify the opportunity costs farmers may face in participating in programs directed towards promoting the *in situ* conservation.

Economic models have advanced various hypotheses for farmers' choice of CGR. In particular, the motivations for partial adoption of modern varieties, e.g. where farmers maintain a share of traditional varieties in their fields, has been the subject of several studies. One finding is that traditional crop varieties may have significant taste attributes, as well as ties to cultural practices such as seasonal holiday dishes. Crop varieties may also be valued for multiple uses, such as straw for fodder or the taste as vegetables when harvested fresh. A risk or portfolio approach to analyzing the question indicates that risk-averse farmers seek to minimize food or income risk by planting different varieties with different yield variances.

Several empirical studies have indicated that households maintain traditional crop varieties because of imperfect markets. Markets may be characterized by high transactions costs for important factor inputs, such as agrochemical inputs, labor substitutability between family and hired labor, or specific consumption traits that are lost in a market for commodities. Labor market integration, including opportunities for off-farm work and migration, will affect conservation because of the intensity of family labor involved in household crop production, plus the labor involved in seed selection and maintenance that participation in a conservation project would involve. An increase in the imports of staple crops or the availability of commodity substitutes may diminish the market niches for particular consumption traits.

3. CASE STUDY FROM MEXICO

In the following section, empirical data illuminating the concepts outlined above are presented. The data for this research were gathered as a part of the McKnight Foundation Collaborative Crop Research MILPA project, composed of a joint Mexico-U. S. research team of botanists, biologists, crop breeders, and social scientists. Research teams were based around the principal crops of the milpa cropping system: maize, beans, squash, and quelites (a broad category of other edible plants found in the milpa). The fifth research group, the socioeconomic group, concentrated on local and regional analyses of the motivations behind farmer behavior.

The data presented below are from a survey of 281 households in 24 villages in the Sierra Norte de Puebla, a mountainous region roughly delimited (and isolated) by two major river valleys. The survey sample was structured to cover a representative sample of villages in the study area. The villages were chosen to incorporate a wide range of geographic, agro-ecological, agronomic, market, and cultural diversity. The variation in levels of economic development and market integration in the region can also be used to model the socioeconomic processes that may affect CGR conservation.

3.1 Maize seed dynamics

The seed histories were recorded to determine how old a farmer's seed lot was, and to extrapolate how frequently the seed lots change.

Table 7-1. Years with current maize seed, by color and total

Years	White	Percent	Yellow	Percent	Blue	Percent	Total	Percent
0-5	42	19	11	23	4	21	57	20
5-10	32	15	2	4	1	5	35	12
10-15	17	8	4	9	1	5	22	8
15-20	12	5	2	4	1	5	15	5
20-25	2	1	1	2	0	0	3	1
>25	115	52	27	57	12	63	154	54
Totals	220		47		19		286	

In Table 7-1, it is observed that across all maize colors we see that 20% of the farmers have not had their seed for more than 5 years and 32% have not had their seed for more than 10 years. On the other hand, 54% of farmers have had the seed for over 25 years, many for their entire lives. This bimodal structure is similar to findings by Perales (1996) and Louette and Smale (2000) that seed histories are either brief or long. This seems to be characteristic of landraces, many or most are held for an entire lifetime, but some farmers renew seed or try new types in the process of evolution and adaptation.

The question was later rephrased to get at farmers who may "renew" seed that they see as the same, but actually acquire new seed lots. When asked when was the last time they had to get seed from a neighbor, 58% reported within the last five years, and only 32% said they had never lost their own seed. Farmers were also asked if they had ever "changed" their maize seed. A large number, 82 farmers or 39%, said yes. Of those who changed, 87.5% reported using seed from the same village and 13.5% reported using seed from another village. This higher rate of looking outside of the village for seed illustrates farmer experimentation with new types. Farmers who reported changing were also asked why they changed seed**(s?),** and the responses are recorded in Table 7-2.

Table 7-2. Responses to why farmers changed maize seed

Reason	No.	Percent
Doesn't yield well	9	11
Changed parcels	8	10
Lost the seed	29	36
Tried other types	27	34
Other	7	6

The possibly unstable dynamic nature of farmer seed management is reflected by the 34% of farmers who changed seed lots in order to experiment with a new type. The farmers who answered that they changed when changing parcels reflect local opinions that seed could be adapted to the conditions of a specific parcel (correlated with slope, exposure, soil type, etc.). Finally the most common reason reported was the loss of seed, and this highlights the reasons why *in situ* conservation may need to focus on a community, rather than individual farm-household level.

3.2 History of the milpa system

Historic questions were asked to gather some background on the importance of the milpa system for each household. The questions were used to try to ask the households directly what the principal threats to the milpa system are. While most of the households in the survey sample planted milpa in the survey year (221 of 281, or 79%) many households had left the activity recently. The first question on the survey was whether the household had planted the milpa. Those households who reported not planting the milpa in the past year were asked why they chose not to.

Table 7-3. Stated reasons for not planting milpa

Limitation:	Land	Labor (migration/ sickness)	Capital	Coffee (land, labor)	Low yields/ bad weather	Not financially viable
No. of HHs	13	12	8	2	14	7

The stated reasons were grouped into the categories presented in Table 7-3. Many reported a shortage of land—either no available land or rent being too high for milpa production. Labor was reported as a constraint both in finding workers (hired labor) and because the head of household was too old or sick to continue farming (family labor). The most common answer, however, was that weather was unfavorable to production, or that yields were below acceptable levels. Finally, several households reported that the milpa was not viable because it ended up costing more than it benefited the household. Each of these answers reflects constraints and opportunity costs and highlights important interactions between economic decisions and seed management, e.g., fixed endowments of family time and land, and limited access to capital.

Many households reported that previously they had grown more maize than in the current period, and this was addressed by a survey question presented in Table 7-4. This linear trend of decreasing involvement in the maize sector may be as important to *in situ* conservation in the long term as seed management questions are in the short term.

Table 7-4. How long ago did you sow more maize?

Years	No. of HHs
0-5	32
5-10	28
10-15	12
15-20	10
20-25	1
gt 25	5

The dynamic process of the decreasing importance of maize was addressed in an additional question on the decreasing intensity of maize plantings which is reported in Table 7-5. The combination of different plot sizes and different times reported make it difficult to determine the decrease in hectares. Instead, all farmers were asked how many hectares they planted 10 years ago as a way to compare current activity levels to historical ones.

Table 7-5. Ratio of area planted 10 years ago to current

Ratio	No. of HH	Percent
0	16	8
0-1	17	9
1	59	31
1-2	12	6
>=2	87	46

While fewer than 20% reported growing less maize 10 years ago, over 50% reported growing more maize 10 years ago, and 46% at least twice as much. This implies that any sort of *de facto* equilibrium that describes farmers planting maize at this time is unstable as farmers are decreasing maize acreage, with corresponding consequences for number of varieties and effective population sizes. For many of these cases, farmers who previously sold some of their harvest are decreasing acreage to infra-subsistence levels. Over the last 10-15 years, the increase in coffee planted in lowland areas and the increase in migration across the region may be bidding up the wage rate and making maize production less economic. An important question for further (interdisciplinary) study is whether or when decreasing planting sizes affect the crop population genetics.

3.3 Maize population structure

The sources of farmer's maize seed are reported in Table 7-6. Most farmers had acquired their seed from their fathers, followed by others in the same village. The blue maize is a smaller population and more of the farmers have maintained it for their entire lives. This is another indication of the precarious status of the blue maize within the region. Farmers may rely on seeds from the same village because of the adaptation of seeds to local conditions. The steep and varied terrain may create very different climatic conditions in neighboring towns. Another reason for the predominance of same village seed is because of social networks that allow farmers to know who would be a good seed source.

Table 7-6. Source of maize seed

	White	Yellow	Blue
		percent	
Father	45	56	68
Same village	52	40	32
Other	3	4	0

Although the amount of seed coming from outside the village is low (3%-4%), the cumulative combination over a 10-20 year time horizon, combined with the periodic renewal of seed and trade within the village, can have a sizable impact on the flow of genes.

Finally, the seed questions were asked to the larger sample in order to see if the estimates of the flow of seeds into the village were robust. The question was asked where farmers would look for seed *if* they needed it.

Table 7-7. Potential seed source, by crop

	Maize	Beans	Squash
Father/same village	94%	82%	86%
Other village/store	5%	17%	14%
Number of observations	239	230	222

The rate of 5% seed flow of maize into a village is comparable to those reported previously. For the principal variety of the principal crop (maize) the seed networks are mostly closed, with a small but consistent inflow of seed from outside the community. The rate is again higher for beans at 17%, showing a higher level of inflow of germplasm, and possibly a different perception of adaptation. The level for squash is similar to that of beans, as again farmers may view squash as more widely adaptable than maize.

3.4 Dynamics and populations flows combined—bean seeds

The farmers were also asked about the history and sources of their bean seeds and a different pattern emerged. The results are reported in Table 7-8. As with maize, the age of *P. polyanthus* is basically bimodal with either a recent acquisition or a very long history and is principally locally sourced.

Table 7-8. Age and origin of bean seed

Years	*P. polyanthus*	*P. vulgaris* (bush)	*P. vulgaris* (vine)	*P. coccineus*	Other
0-5	25%	40%	10%	33%	36%
5 to 20	14%	19%	10%	33%	18%
>=20	61%	40%	81%	33%	45%
Source of seed					
Father	40%	33%	43%	33%	27%
Same village	47%	45%	48%	50%	36%
Other village	13%	21%	10%	17%	36%
Number of observations	104	42	21	6	11

However, the bush form of *P. vulgaris* appears less stable as a local landrace, and there appears to be a distinction between the vine form of *P. vulgaris* and the bush form. The vine form follows the maize landrace pattern where 80% of farmers' seed is greater than 20 years old, and only 13% of seed comes from outside of the village. The bush form, however, has a higher percentage of new seed lots, 40%, and 21% of the seed comes from outside of the village. The seed lots listed as "other" are mostly *P. Vulgaris* bush types as well and follow a similar pattern of recent acquisition and high levels of introduction from outside the village. This indicates that *P. Vulgaris* is less entrenched genetically in local sources, and farmers rely less on saved seed to maintain local populations.

Across all bean types, 27% of farmers reported having changed bean seeds at some time. Of those who reported changing, 50% reported using seed from local, village sources, 25% used seed from another village, and 25% used seed purchased in the market. Again, it is possible that the idea of seed adaptation to local conditions is much stronger for maize than for beans. Furthermore the large number of bean seed lots purchased as food seed in the market indicates a large inflow of germplasm and the more precarious nature of local *P. vulgaris* diversity. Furthermore, the market beans are principally imported from other states within Mexico and the United States and, thus, represent a flow of genetic material into local populations.

Farmers who reported changing their bean seed were asked why they had changed, and the results are reported in Table 7-9. The most common reason reported was that they had lost the seed from the previous season.

Table 7-9. Reasons for changing bean seed

	No.	Percent
Doesn't yield well	3	15
Changed parcels	1	5
Lost the seed	8	40
Tried other types	6	30
Other	2	10
Total	20	

The 30% of farmers who wanted to try another type were those seeking to experiment with new seed types. Furthermore, the fact that 40% of farmers reported changing because they lost seeds highlights the fact that many farmers have relatively small populations of beans, with an output equivalent to a one- or two-month food supply.

3.5 Socioeconomic characteristics—average number of varieties

As discussed above, there are a variety of possible socioeconomic factors that may cause the household to plant a greater number of milpa varieties. A preliminary approach is to divide the sample into relevant sub-samples. For a series of socioeconomic characteristics, the sample median was calculated and used to divide the sample into households above and below the median. Table 7-10 presents three household variables—age, family size, and wealth—which may affect the number of varieties planted by a household. In Table 7-10, the average number of maize varieties grown is slightly larger for those households with an older household head, with a larger number of adult family members. Both of these categories present results that may be expected, but for neither category are the means significantly different. The mean number of maize varieties is significantly lower for wealthier farmers. This agrees with the hypothesis generated by the household model that households with a higher level of wealth have less of a need to self-insure through a crop portfolio.

Table 7-10. Mean number of varieties for household subsamples

	Number of maize varieties		Total milpa varieties	
Age of HH head				
Below median	1.01		2.36	
Above median	1.05		2.46	
Adult family size				
Below median	0.97		2.25	
Above median	1.11		2.61	
Wealth				
Below median	1.13		2.58	
Above median	0.92	**	2.21	
Ecological zone				
Low elevation	0.93		1.96	
High elevation	1.17	**	3.03	**
Number of plots				
0-1	0.68		1.82	
>1	1.37	**	2.98	**
Owned land (ha.)				
Below median	0.98		2.29	
Above median	1.09		2.53	

Table 7-10. (continued).

Maize land (ha.)				
Below median	0.73		1.70	
Above median	1.45	**	3.38	**
Infrastructure level				
Small town	1.29		3.10	
Large town	0.83	**	1.86	**

** Indicates means are significantly different at 5% level (two-tailed, two-sample t-test).

The sample was also divided into subsamples in order to examine the average number of varieties planted by agroecological characteristics. The first category corresponds to the major ecological zones in the region, Tierra Caliente (Hot Lands - below 1200 masl) and Tierra Fria (Cold Lands - above 1200 masl). The average number of varieties grown is higher at the higher elevations, due to agroecological conditions. The second category is the number of plots farmed by the household, and this is used as proxy for whether the households are matching varieties to soil conditions. The average number of plots is significantly higher for households with multiple plots, indicating that the agroecological conditions also hold at the household level. The next two categories address the quantity of land, a key constraint to the number of varieties planted. Own hectares is the total hectares owned by the household, and maize hectares is the total hectares planted by a household to maize. For the own hectares, the means are not significantly different, while for hectares planted to maize, the average number of varieties planted is higher. This indicates that within the land planted to maize and milpa, land is a constraint to planting a greater number of varieties. Finally the market integration category is proxied by comparing small towns to larger towns. Large towns are a municipal capital, on a major paved road, or have a significant commercial sector and services. The average number of varieties planted by a household is significantly higher in the small towns, indicating that when the level of market integration increases, the number of varieties decreases.

However, each of these categories is showing a change in household levels of diversity in isolation from other factors. Furthermore, the effect of each condition could have a different effect, when all other effects are held constant, *ceteris paribus*. For instance, the age of the household head could increase or decrease diversity, if isolated from the effects of the number of plots farmed and the agroecological zone that the household is in. Therefore, the use of categories or correlation limits the ability to test for all of the effects that could be included in a household model. A general, nested model may be needed to test the effects of individual parameters and groups of parameters on the level of diversity maintained by households.

4. CONCLUSIONS

Several areas for the empirical measurement of farmer-based conservation of CGR were presented. These practical diagnostic tools can be used to target an *in situ* conservation program, as well as to understand the key constraints that a program would face over a longer planning horizon. Simple, analytic survey questions can be combined to yield a description of basic parameters of overall seed management. The history of seed showed that local landrace populations were characterized by a bimodal age distribution, divided between long periods of saving seed by the same farmer and recent renewal of seed. The documentation of seed sources shows the majority within the same family or the same village, with a small flow from outside the village. However, from a dynamic perspective, small amounts of seed renewal and inflow could accumulate to shape the evolution of local populations. This intertemporal accumulation of genetic flows would have different rates depending on the scale of analysis, whether it was for an individual farmer, a village, or a region. Furthermore, any conservation equilibrium appears unstable, as over time farmers are reducing system diversity and reducing area planted (and, therefore, effective population sizes). Finally, factors that seem unrelated to crop populations, such as labor markets, cash crop markets, transportation, or transactions costs, can have effects on the management of CGR and the stability of local populations.

The dynamic nature of crop populations creates difficulties for acquiring data over a long time horizon. In the empirical case presented here estimates were provided from current trends and from questions about recent history. The dimension of the crop population is important to conservation, and therefore the sample frame must be calibrated to take into account village and regional effects on populations. The intertemporal and interregional dimensions of seed systems combine to make the crop populations moving targets that may not have definite boundaries. The goal of the conservation program may be to focus on these aspects of the seed system related to flows of genetic material and continual selection by farmers under local conditions.

Finally, the issues of the long-term sustainability of *in situ* conservation remain unresolved because of contrary processes in market integration, agricultural specialization, and transactions costs for farm labor. In many areas of diversity of CGR, traditional varieties remain predominant or popular, but relying on this de facto conservation may require documentation of the forces affecting the decline in genetic diversity. The reduction of diversity may proceed on various fronts at the same, and it will be difficult to separate the simplification of agricultural production systems from the transfer of resources within the family to other income activities, such as cash crops or migration. However, a focus on the components of the evolutionary process, such as seed selection and exchange, may allow a conservation program to remain flexible in the face of such economic pressures.

REFERENCES

Aguirre Gómez, J. A., Bellon, M. R., and Smale, M., 2000, A regional analysis of maize biological diversity in Southeastern Guanajuato, Mexico, *Econ. Botany* **17**:60-72.

Bellon, M. R., and Taylor, J. E., 1993, "Folk" soil taxonomy and the partial adoption of new seed varieties, *Econ. Dev. and Cultural Change* **41**(4):763-786.

Brown, A., 2000, The genetic structure of crop landraces and the challenge to conserve them *in situ* on farms, in: *Genes in the Field: On-Farm Conservation of Crop Diversity*, S. B. Brush, ed., International Plant Genetic Resources Institute and International Development Research Centre, Rome.

Brush, S. B., 2000, *Genes in the Field: On-Farm Conservation of Crop Diversity*, International Plant Genetic Resources Institute and International Development Research Centre, Rome.

Brush, S. B., and E. Meng, 1998, Farmers' valuation and conservation of crop genetic resources, *Genetic Resources and Crop Evolution* **45**(2):139-150.

Brush, S. B., Taylor, J. E., and Bellon, M. B., 1992, Technology adoption and biological diversity in Andean potato agriculture, *J. Develop. Econ.* **39**(2):365-387.

Louette, D., and Smale, M., 2000, Farmers' seed selection practices and traditional maize varieties in Cuzalapa, Mexico, *Euphytica* **113**(1):25-41.

Louette, D., and Smale, M., 1998, *Farmers' Seed Selection Practices and Maize Variety Characteristics in a Traditionally Based Mexican Community.* CIMMYT, México, D.F., México.

Meng, E. C.-H., 1997, Land allocation decisions and *in situ* conservation of crop genetic resources: The case of wheat landraces in Turkey, Ph.D. dissertation, UC Davis.

MILPA Project, 1999, Annual Report of the McKnight Foundation MILPA Project, UC Davis - UNAM, Davis.

Perales Rivera, H. R., 1996, Conservation and evolution of maize in Amecameca and Cuautla valleys of Mexico, Ph.D. dissertation, UC Davis.

Rice, E., Smale, M., and Blanco, J. L., 1998, Farmers' use of improved seed selection practices in Mexican maize: Evidence and issues from the Sierra de Santa Marta, *World Dev.* **26**(9):1625-1640.

Smale, M., Bellon, M. R., Mendoza, J., and Manuel Rosas, I., 1999, *Farmer Management of Maize Diversity in the Central Valleys of Oaxaca, Mexico: CIMMYT INIFAP 1998 Baseline Socioeconomic Survey*, CIMMYT, México, D.F., México.

Chapter 8

COSTS OF CONSERVATION: *NATIONAL AND INTERNATIONAL ROLES*

Detlef Virchow
InWEnt, Capacity Building International, Wielinger Str. 52, D-82340 Feldafing, Germany

Abstract: With growing awareness of the irreversible loss of plant genetic resources for food and agriculture (PGRFA), there has been an immense effort in terms of human and financial resources devoted to the collection and conservation of plant genetic resources and the establishment of an institutional framework at international, national and local levels. Deficits of information and uncertainties are, however, hindering an economically efficient approach to optimizing agrobiodiversity conservation. Despite the existing uncertainties, the political will of countries stresses the importance of genetic resources conservation, even though long-term conservation activities face strong competition from other, often more short-term development activities for the allocation of financial resources. Considering these circumstances, there is a need for cost-effective and efficient strategies for PGRFA conservation. Cost-efficient conservation will reduce the risk of losing unique, genetically coded information and reduce the problem of allocating an excessive amount of financial resources to conservation activities. Therefore, this chapter analyzes the national and international actors' costs of PGRFA conservation activities and discusses existing and potential collaborations between the actors with the aim of increasing the efficiency of PGRFA conservation. Furthermore, this chapter highlights the opportunities and limitations of funding of conservation activities, especially applied to regional and international collaborations.

Key words: actors of PGRFA conservation; cost-effective and efficient-PGRFA conservation; funding of conservation activities; funding mechanisms; international and regional collaboration; national and international conservation costs; plant genetic resources for food and agriculture.

1. INTRODUCTION AND OBJECTIVES

With growing awareness of the irreversible loss of plant genetic resources for food and agriculture (PGRFA), there has been an immense effort in terms of human and financial resources devoted to the collection and conservation of plant genetic resources and the establishment of an institutional framework at international, national, and local levels. Estimates indicate that there are 6.2 million accessions of 80 different crops stored in 1,320 genebanks and related facilities in 131 countries (FAO, 1998). Deficits of information and uncertainties are, however, hindering an economically efficient approach to optimizing agrobiodiversity conservation. Because of a lack of estimates on (1) the value of PGRFA for global welfare or the cost of their extinction, (2) the rate of PGRFA extinction, and (3) the costs of conservation, investments in PGRFA conservation are most likely sub-optimal at the margin. Additionally, allocative problems such as the imbalance between the shared costs and the benefits of conservation hamper optimal conservation at all levels. For example, some countries with a high amount of unique PGRFA are the poorest countries in the world, where investment in conservation is constrained by very limited resources and other priorities for the use of available funds (von Braun and Virchow, 1997).

Despite the existing uncertainties, the political will, expressed by all governments present at the International Technical Conference (hereafter, ITC) on Plant Genetic Resources for Food and Agriculture in Leipzig in 1996, stressed the importance of genetic resources conservation (FAO, 1996). This lent support to continued conservation of PGRFA, even though long-term conservation activities face strong competition from other, often more short-term development activities for the allocation of financial resources.

Considering these circumstances, uncertainties on the one hand and the political will to conserve PGRFA on the other hand, there is a need for cost-effective and efficient strategies for PGRFA conservation, in addition to further scientific and economic research. Cost-efficient conservation will reduce the risk of losing unique, genetically coded information and reduce the problem of allocating an excessive amount of financial resources to conservation activities. The political and economic discussion has been focused on the value of plant genetic resources and on the issue of "fair and equitable sharing" of the benefits derived from the use of PGRFA. However, an intensive analysis of the costs of conservation activities has been neglected. Therefore, this chapter will analyze the national and international actors' costs of PGRFA conservation activities and will discuss existing and potential collaborations between the actors with the aim of increasing the efficiency of PGRFA conservation. Furthermore, this chapter will highlight the opportunities and limitations of funding of conservation activities, especially applied to regional and international collaborations.

2. THE ACTORS OF PGRFA CONSERVATION

A wide range of different players at local, national, and international levels are involved in maintaining PGRFA. By grouping the actors of the conservation activities according to their activities relating to conservation, one can identify five major groups: farmers, public conservators, regional and international gene banks, private breeding companies, and local conservators.

According to the estimates of Wood and Lenné (1993), 60% of global agriculture depends on the cultivation of farmers' varieties. Even though generally farmers do not maintain these varieties out of conservation motivations, they are conserving them de facto (see Chapter 7). In contrast with their importance in the de facto *in situ* conservation of PGRFA, farmers play only an insignificant roll in *ex situ* conservation. All other actors are mainly involved in *ex situ* conservation, with only a few activities related to *in situ* conservation. Their involvement with *in situ* conservation activities are, however, increasing (FAO, 1998).

As depicted in Fig. 8-1, public conservators at the national level dominate the *ex situ* conservation, storing 83% of all conserved accessions (FAO, 1998). Of these, 34% are stored in public genebanks of developing countries and 49% in public genebanks of industrialized countries (Virchow, 1999a). According to FAO, 15% of all *ex situ* conserved accessions are held in regional and international genebanks (FAO, 1998). The majority of these accessions are stored in the *ex situ* collections of the Consultative Group on International Agricultural Research (CGIAR).[1] Private breeding companies in industrialized countries store approximately 1% of the accessions and the relevant private companies in developing countries roughly 0.2% (Iwanaga, 1993). Finally, it is estimated that less than 0.2% of all *ex situ* conserved accessions are held by local conservators, i.e., farmers supported by NGOs (FAO, 1998).

The leading role of the national public sector in conservation activities is supported by the fact that approximately 85% of all estimated expenditures for PGRFA conservation were spent by the national public sector in 1995 (Virchow, 1999a). Together with the reaffirmation of national sovereignty over genetic resources made at UNCED (UNEP, 1994), these figures indicate that countries are the most important conservators of *ex situ* collections. It is unavoidable that the public sector takes responsibility for the conservation of the national genetic resources particularly in biodiversity-rich countries.

---

[1] The CGIAR is a an informal association that supports a network of 16 international agricultural research centers, primarily sponsored by the World Bank, FAO, and the United Nations Development Program (UNDP).

This responsibility is based on the social benefits which accrue at the national and international level, and which arise from the conservation of genetic resources. Although the discussion concerning the "fair and equitable sharing" of the benefits derived from the use of PGRFA has not yet been resolved, it is important to enable all diversity-rich countries to conserve the existing PGRFA *ex situ* as well as *in situ*. Hence, existing conservation costs have to be analyzed as the first step.

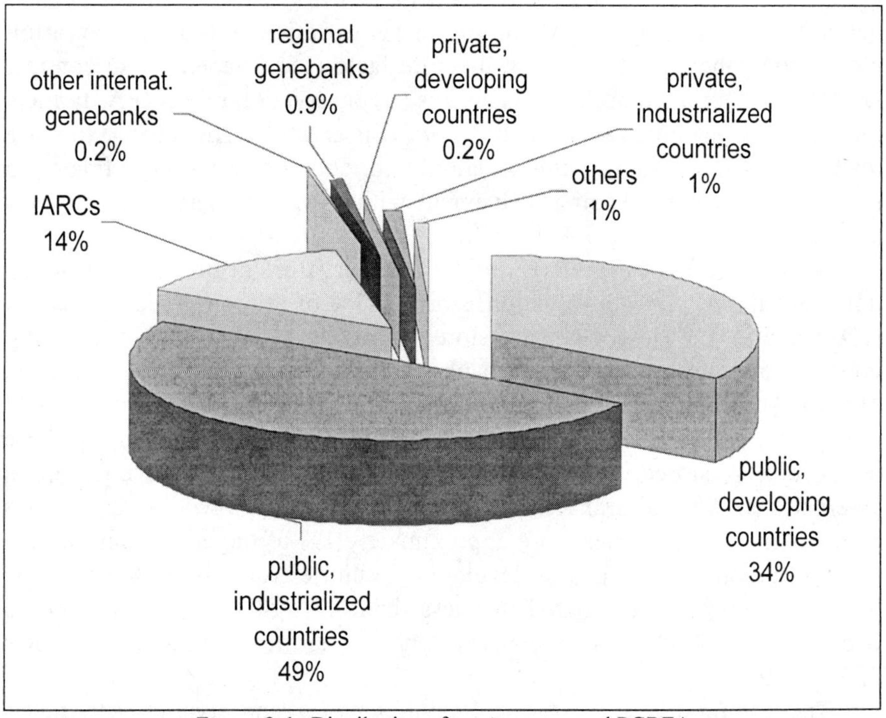

Figure 8-1. Distribution of *ex situ* conserved PGRFA

Source: Calculated based on data from FAO (1998), Virchow (1999a), and Iwanaga (1993).

3. THE COSTS OF CONSERVATION: A GLOBAL OVERVIEW

For an analysis of the efficiency of conservation systems, cost elements have to be analyzed, especially those related to conservation activities at the national level, where the bulk of expenditures are made. Besides an assessment of the costs associated with current global and national conservation activities, whose financial responsibility these activities fall under need to be explored as well.

3.1 Methodology, data, and limitations of current expenditure survey

The overall costs of PGRFA conservation are made up of fiscal costs and opportunity costs (see Fig. 8-2). Fiscal costs include the expenditures arising for PGRFA conservation, which have to be budgeted and invested either on the national or the international level. Fiscal costs include the costs for planning, implementing, and running *ex situ* and *in situ* conservation activities. These costs are determined by specific conservation activities, the depreciation costs of investments, and the costs of institutional and political regulations for access to PGRFA. Additionally, costs for compensation and incentives paid for maintaining PGRFA have to be included. In addition to the fiscal costs, there are also opportunity costs associated with conservation, reflecting the benefit foregone by the country by maintaining the diversity of genetic resources in the field (Chapter 6).

Different approaches may be taken in identifying the specific costs of PGRFA conservation activities. Costs can be identified at different administrative levels as well as in different categories. The latter depend on the conservation methods used: costs of *in situ* and *ex situ* conservation as well as costs for supporting activities and institutional process for the conservation of and access to PGRFA. Considering the players in the conservation activities, costs can arise at the farm level or at national and international levels as well as at the level of conservation activities in the private sector. Consequently, the method of estimating costs depends upon the approach taken.

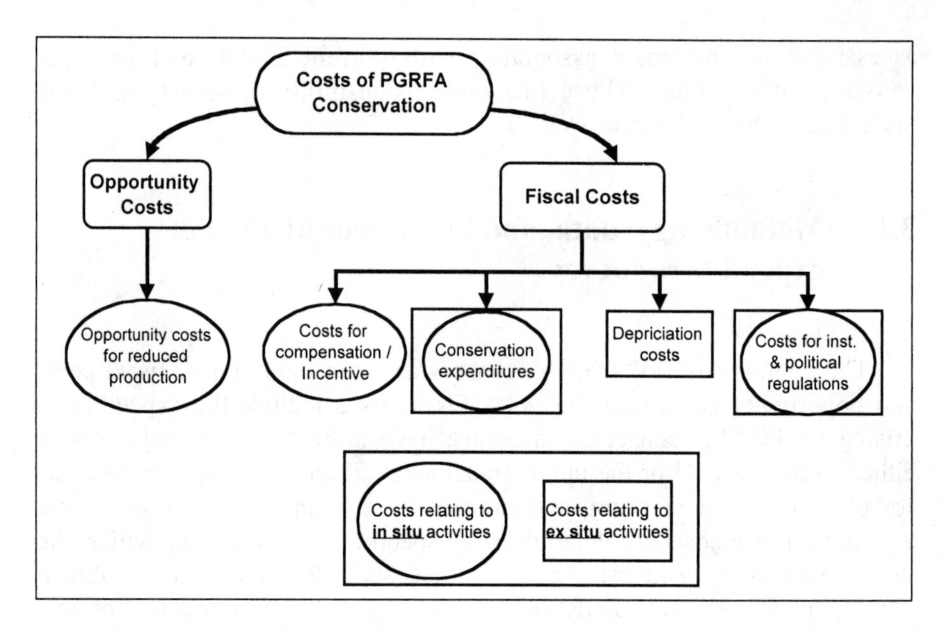

Figure 8-2. Economic concept of the costs of PGRFA conservation

Source: Virchow (1999a).

The main source of information for this study was a survey conducted in 1995/1996. Each country had established a focal point for the preparatory process of the ITC on PGRFA in 1996. These focal points were contacted for the survey. As of June 1996, 28% of all surveyed countries (39 countries) provided data that could be analyzed. Among those responding were countries thought to have substantial programs in PGRFA (inter alia, the United States, France, Germany, Russian Federation, United Kingdom, Japan, China, India, Brazil, and Ethiopia), as well as a number of countries with smaller programs. Because of a lack of information from governments, expenditures for 1995 could not be calculated precisely. National data were estimated, based on available information, in order to obtain an order-of-magnitude estimate of total expenditures at the national level.

It was apparent from the national reports submitted during the preparatory process of the ITC that the available information on the state of PGRFA in the countries and activities for their utilization is vague or even nonexistent in many countries, while few have supplied very precise figures (FAO, 1998). This applies especially to the expenditure data for PGRFA conservation and utilization. Only a few countries have explicit budget lines for these activities. Another problem is that the scope of the conservation and utilization of PGRFA is so broad that activities with other objectives may have a positive impact on the implementation of conservation activities. Consequently, even if a country has a refined cost monitoring system, the actual expenditures may be made up of allocations from different financial resources than those explicitly dedicated to the conservation and utilization of PGRFA.

The first expenditure survey revealed a number of other difficulties with collecting and processing the existing data:

- *Participation of reporting*: Less than 26% of countries involved in the preparatory process actually provided expenditure data.
- *Partial information*: Significant information was not comprehensively provided by countries, even though the expenditure data should have been known to certain agencies in the countries concerned (e.g., national contributions to multi- and bilateral activities related to the conservation and utilization of PGRFA).
- *Homogeneity of reported data*: The proposed reporting format was not followed; consequently, it needed interpretation skills to process the data and to harmonize the data from different sources (e.g., received data were often not disaggregated or sums were given without indicating whether they applied to conservation, utilization, or both).
- *Defining the scope of activities*: There was no unified definition (nor understanding) of the scope of activities related to the conservation and utilization of PGRFA, e.g., some countries included plant-breeding activities, while others only included the conservation of PGRFA in a very narrow sense. Most countries did not clearly define what was covered by expenditure or foreign assistance data. Similarly, some data on financial contributions included only activities closely related to the conservation and utilization of PGRFA. Only a few countries included *in situ* conservation and utilization, while others only provided data on the general national contribution to international organizations;
- *Multiple-impact activities*: Projects or programs often deal with PGRFA conservation and utilization as part of a broader initiative including actions not strictly related to PGRFA. This poses the problem of having to estimate the portion of the program dealing with the conservation and utilization of PGRFA and identifying the expenditures on that portion. This, however, can only be done relatively accurately by those involved in the specific projects and programs.

3.2 National expenditures for PGRFA conservation

National expenditures on PGRFA conservation are difficult to assess, largely because of uncertainties in defining the scope of PGRFA programs. It seems that most countries' national efforts to conserve PGRFA are in the hands of different departments in different ministries. In addition to the complex administrational structure, other parastatal and nongovernmental organizations are involved in the conservation activities as well. Only in some countries are all efforts coordinated by an overall national program.

Hence, the costs involved are not always visible. Furthermore, countries are involved in PGRFA conservation but do not have specially defined budget lines for these activities. For instance, if a genebank belongs to a national breeding institute and its costs are incorporated in the institute's overall budget, it is difficult to assess its specific costs.

Data concerning the national expenditures for conservation of PGRFA can be divided into two different groups:

- Domestic expenditures, which have been spent for conservation activities in the country.
- Foreign assistance contributions, i.e., expenditures, which have been contributed as financial aid for PGRFA conservation in other countries (through bi- or multilateral contributions).

3.2.1 Domestic expenditures

The most important category for national PGRFA conservation is in domestic expenditures. The information received was tabulated into the main two cost categories (*ex situ* and *in situ* conservation activities), wherever possible. Due to the previously discussed difficulties concerning the homogeneity of the data, the comparison of all data received is only possible at a high level of aggregation, i.e., for PGRFA conservation as a total.

Based on the data provided, the order of magnitude of domestic expenditures spent for the conservation of PGRFA by 37 countries amounts to approximately US$475 million for the year 1995 (see Table 8-1). This figure includes the financial assistance of US$17 million, which 15 countries received through bilateral and multilateral contributions (Virchow, 1999a).

Table 8-1. Domestic expenditures for PGRFA conservation in 1995

Country	Domestic expenditures, 1995	Country	Domestic expenditures, 1995
	US$1,000		US$1,000
Germany	113,215	Madagascar*	2,385
France	98,660	Seychelles*	2,322
United Kingdom	70,154	Haiti*	1,896
Spain	33,413	Canada	1,584
Italy	27,208	Russia*	1,526
USA	20,433	Ethiopia*	1,346
South Africa	19,000	Portugal	1,030
Norway	16,208	Suriname*	1,028
Egypt*	11,528	Poland*	656
Greece	10,958	Lesotho*	615
Brazil*	8,000	Romania	408

Table 8-1. (continued).

India*	6,776	Tanzania	187
Japan	6,480	Cyprus*	186
Peru*	4,137	Togo	151
Switzerland	3,825	Belarus	135
Slovak Republic	3,608	Pakistan	120
Czech Republic	3,255	Tonga*	56
China*	2,526	Saint Kitts & Nevis	20
		Austria	10
Total			475,045

Note: *Includes foreign received assistance.
Source: Data according to Virchow (1999a).

3.2.2 Foreign assistance contribution

Of the 37 countries mentioned in Table 8-1, 12 contributed bilateral and multilateral financial assistance of approximately US$50 million (see Table 8-2). It is interesting to note that the amount of foreign assistance contributed by these 12 countries varies widely. Countries like France or Portugal contributed 1% of their total PGRFA conservation expenditures, whereas countries like Switzerland, Canada, and Austria contributed 47%, 69%, and 99%, respectively. Although the results might be biased as a result of the insufficient data, they do show the different levels of international commitment from the various countries.

Table 8-2. Foreign assistance contribution for PGRFA conservation in 1995

Country	Total expenditures on the conservation of PGRFA	Foreign assistance contributed 1995	Foreign assistance contributed as percentage of total PGRFA conservation expenditures
	US$1,000		percent
Germany	131,742	18,527	14
France	99,160	500	1
United Kingdom	87,685	17,531	20
Spain	34,298	885	3
Norway	18,820	2,612	14
Egypt	11,772	244	2
Switzerland	7,225	3,400	47
Canada	5,164	3,580	69
Portugal	1,040	10	1
Austria	1,510	1,500	99
Finland	n.i.[a]	1,180	
Ireland	n.i.[a]	142	
Total		50,111	

[a] No information.
Source: Data according to Virchow (1999a).

3.2.3 Interpretation of the expenditure data

Concluding the analysis of the international expenditures on PGRFA conservation, the survey results of the 39 countries can be summarized as follows: 89% of the PGRFA conservation expenditures by the Organization for Economic Cooperation and Development (OECD) countries surveyed go towards their domestic conservation activities—US$406 million out of a total of US$456 million, mainly for *ex situ* conservation of their PGRFA accessions. 76% of the expenditures for conservation activities in the developing countries surveyed (US$52 million) were funded nationally, while US$17 million, representing nearly one-quarter of the domestic expenditures, were funded through bi- and multi-lateral financial contributions. Although the 16 OECD countries are conserving 53% and the 23 developing countries 47% of their combined conserved accessions, the OECD countries spent 85% of the combined total costs of *ex situ* conservation of US$475 million. Not surprisingly, the contributions for foreign assistance originate predominantly from the 16 OECD countries (Virchow, 1999a).

When the countries are grouped into agrobiodiversity rich and poor countries and furthermore into countries having high and low absolute domestic expenditures on PGRFA conservation, most of the OECD countries analyzed can be categorized as the agrobiodiversity-poor countries with the tendency to higher absolute expenditures.[2] The majority of these genetic resource-poor countries are very interested in building up and maintaining a high level of PGRFA from many other countries and need gene centers to supply their breeding industry with sufficient resources and to ensure long-term sustainable food production. In addition, some developing countries, such as Egypt may be seen as an example of a genetically resource-poor country with a large agricultural sector that has to grow crops under harsh conditions. Consequently, its government must ensure that the need for a sustainable supply of crucial inputs is met. Even if domestic expenditures are expressed as a percentage of the gross domestic product (GDP) per capita, a country like Egypt still has a high ranking in terms of expenditures (see Fig. 8-3).

Not only are some resource-poor countries interested in the conservation of PGRFA, but some agrobiodiversity-rich countries are as well, e.g., India, Ethiopia, South Africa, China and Tanzania. These countries are spending as much on PGRFA conservation in relation to their average income as genetic

---

[2] The concept of agrobiodiversity-rich and -poor countries can be summarized as:
(i) Agrobiodiversity-poor country: The country is not part of a gene center or has less than 10,000 accessions stored *ex situ*.
(ii) Agrobiodiversity-rich country: The country is part of a gene center and has more than 10,000 accessions stored *ex situ* (see Virchow, 1999a, for more detail).

resource-poor countries like Germany, France, and the United Kingdom. Especially in India, Ethiopia, and China, the estimated value for PGRFA conservation turns out to be very high. Indeed, these countries are also playing a leading role in international negotiations on the issue of internalization and compensation with regard to PGRFA conservation in their countries. The countries fall into four groups when measured in terms of the degree of agrobiodiversity and the level of domestic expenditures expressed as GDP per capita (see Fig. 8-3). Of major interest are the two groups with high relative domestic expenditures. They are countries strongly committed to PGRFA conservation, but for different reasons. On the one side (left top) are the demand-driven spenders. These are agrobiodiversity-poor countries, which spend a large amount on PGRFA conservation. The governments of these countries see the need for their breeding industry to safeguard its supply of genetic resources as inputs for breeding. On the other side (right top) are the supply-driven spenders, which are agrobiodiversity-rich countries. These countries invest a great deal in the conservation of PGRFA not only for their own country's breeding efforts but above all to be able to operate as PGRFA suppliers on a market for genetically coded information that is yet to be developed.

On the other hand, there are countries that show a low domestic expenditure level in relation to the national average income regardless of whether they are poor or rich in agrobiodiversity. Countries with high agrobiodiversity like Russia or Pakistan do not invest much in conservation programs in spite of being genetically resource-rich. These countries lack the financial resources to enlarge their conservation activities (e.g., Pakistan) or face a steady decline in these financial resources (e.g., Russia), which undermines their ability to maintain a high quality of conservation. In both cases, the lack of funds and relatively low investment in PGRFA conservation makes the threat of PGRFA loss highest in this group.

Finally, there is a group of agrobiodiversity-poor countries with low financial commitment. This group mainly consists of countries with few or no activities in the breeding and seed industry (e.g., Switzerland, Austria, Poland, and Romania). Other countries (e.g., United States and Canada), however, do not seem to fit into this group due to their intensive activities in the breeding and seed industry. This leads us to the recognition that endowments of PGRFA on its own is a fundamental but not sufficient criterion for characterizing and comparing the efforts made to conserve PGRFA at the national level.

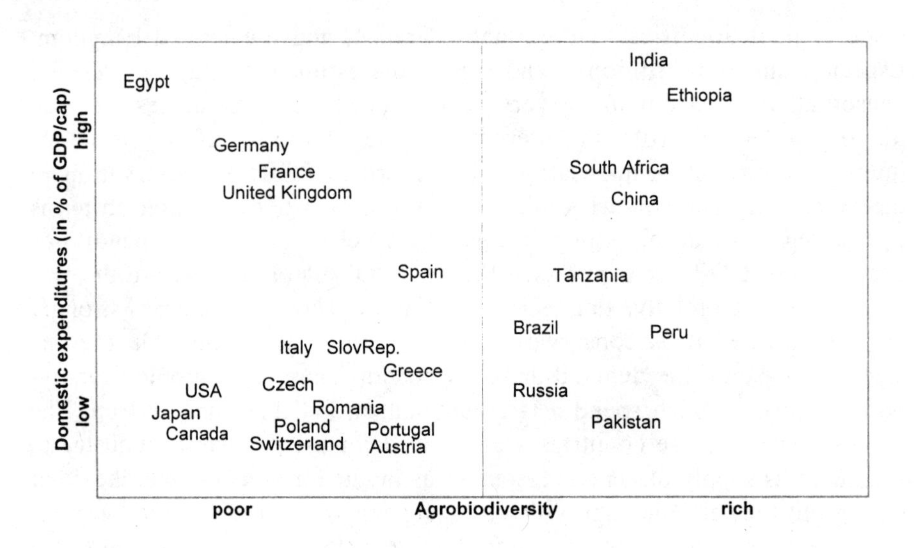

Figure 8-3. Relative domestic expenditures on PGRFA conservation for selected countries[3]

Source: Virchow (1999a).

3.3 International expenditures for PGRFA conservation

In addition to national activities, there are significant efforts taken at the international level to conserve the diversity of PGRFA. A significant amount of international funding and executing agencies are involved in activities relating to the conservation of PGRFA.

3.3.1 The flow of financial resources for the conservation of PGRFA

Fig. 8-4 depicts the flow of financial resources for conservation of PGRFA, which may be divided into two different groups. On the one hand, there are activities by international organizations, e.g., FAO or the CGIAR centers. Single countries or all countries benefit from the output of their work. This may lie in the access to unique accessions, conserved in one of the CGIAR centers' genebanks or it may lie in specific programs and projects implemented by FAO or other implementation agencies in single countries or specific regions. All these activities are financed by the contributions of member countries to the different organizations. Besides the

[3] Low domestic expenditures in % of GDP/cap are less than 200% of GDP/cap for PGRFA conservation. High domestic expenditures in % of GDP/cap is more than 200% of GDP/cap for PGRFA conservation.

international organizations in charge of funds for contributing financial assistance, e.g., Global Environment Facility (GEF) and the regional development banks. Their contributions, either as grants or credits, may be implemented by national organizations or by international implementing organizations. The World Bank and other development banks and funds are major players in agricultural development projects and NARS capacity building. However it is difficult to assess the proportion related to PGRFA. It is estimated that international flows include about US$7 million annually channeled through the GEF for PGRFA-related activities[4] (Virchow, 1996).

Bilateral financial assistance contributed by some countries for specific conservation activities in other countries also plays an important role. To quantify this contribution is difficult. For instance, the new genebank in New Delhi, India, which was inaugurated in November 1996, was financed mainly by the United States. The United States, however, did not mention the US$28 million nine-year program with India in their response to the expenditure survey, even though there were significant expenditures for the project in 1995 (Chandel, 1996).

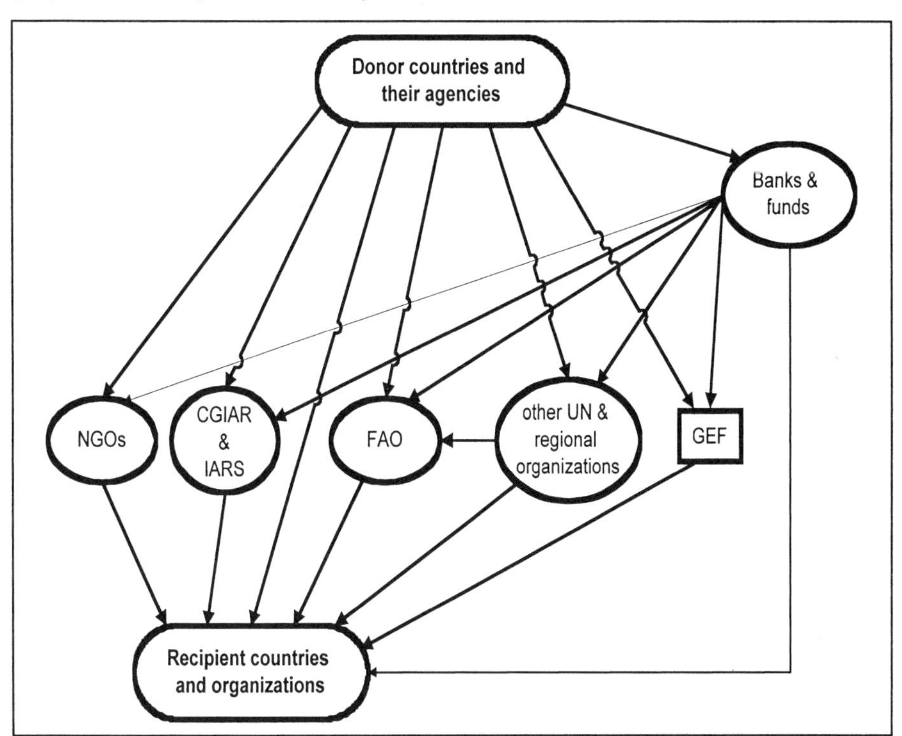

Figure 8-4. Flow of funds for the conservation of PGRFA
Source: Virchow (1999a).

[4] This includes two projects specifically devoted to PGRFA, plus an estimated share (5%) of a number of projects for biodiversity conservation which are likely to contribute to PGRFA conservation.

Bearing the limitations of this survey approach in mind, the rough overall estimate of the expenditures in PGRFA conservation at the international level is summarized in Table 8-3. About US$83 million was channeled through international organizations to activities relating to the conservation of PGRFA in 1995. Most of the money was spent by the above-mentioned international funding and executing agencies for the conservation of PGRFA, mainly as an integrated element of larger development projects or programs. Only a minor part of the entire international expenditure, namely US$7 million, has been spent mainly for technical assistance by FAO, the UNDP, and the U. N. Environment Program (UNEP) financed projects, which represents 8% of all expenditures. The US$83 million (estimate includes the CGIAR's expenditures. The CGIAR's expenditures on PGRFA-related activities represent (with approximately 60%) the largest single expenditure share in all of the international expenditures (including core funds and complementary activities). The expenditures for the CGIAR

Table 8-3. Estimated international expenditures for the conservation of PGRFA (including indirectly related activities)

Origin of expenditures	Expenditures	Expenditures with project/program based estimates added	Shared expenditures according to source (as a share of total international expenditures on PGRFA conservation)
	US$ million		percent
Technical assistance	7	7	8
Funds[a]	7	26	32
Subtotal	14	33	
IARS	50	50	60
Total	64	83	100

[a] IFAD's and the World Bank's contributions were estimated.
Source: Virchow (1996).

for PGRFA conservation, US$26 million, come from fund, such as the World Bank and GEF. Because of the lack of information on the projects and programs of the World Bank, some other development banks, and the International Fund for Agricultural Development (IFAD), it was not possible to include an exact figure derived from the survey in the above totals. Therefore, the expenditures for two of the main funding organizations were estimated. It was assumed that 0.5% of the US$3 billion of WB's agricultural expenditures and 2% of IFAD's $200 million budget were spent for activities related to PGRFA conservation. GEF's spending includes all their projects and programs that had an impact on the conservation and utilization of PGRFA. These were tabulated and calculated according to the above-mentioned method. The same procedure was utilized to calculate the expenditures of FAO, UNDP, UNEP, and the Common Fund for Commodities (CfC).

These data must be carefully interpreted because most agencies do not have a separate budget for funding PGRFA activities. Because these activities make up only a small part of overall development projects and programs, it is difficult for each agency to calculate their expenditures for this task. Consequently, the estimates calculated on the basis of the received data can only be accepted as a rough estimate and are subject to distortions because of the lack of accuracy of the primary data.

3.3.2 Aggregated global expenditures

Concluding the analysis of the international expenditures for the conservation of PGRFA, aggregated global expenditures can be calculated by adding the estimated results of the international expenditures to the national expenditure estimates (see Fig. 8-5).

According to the results, countries contributed US$733 million for national activities related to the conservation of PGRFA in their own countries in 1995. Additionally, US$189 million was spent for multi- and bilateral international activities. OECD member countries have contributed approximately US$162 million of this latter sum. According to the estimation of international expenditures for conservation and utilization of PGRFA (including all technical assistance, funds, and other activities of the international organizations included in the analysis), approximately US$83 million (Table 8-3) of the US$189 million was contributed through multilateral channels. Based upon the information received, approximately 60% (US$50 million in Table 8-2) was contributed to international organizations for the conservation of PGRFA on the international level and approximately 40% (US$33 million) was contributed to international funding and implementing agencies for assistance to countries. While stressing once again the inaccuracy of the data, it is, however, still possible to obtain a tendency and a rough estimation of the current expenditures for PGRFA conservation, which amounts to over US$800 million.

By the same process of estimation, the level of foreign assistance received for conservation of PGRFA is estimated to be approximately US$100 million. Because of the lack of information, there is no reliable estimate for the contribution through bilateral channels. Based upon known information, bilateral contributions are calculated to be at least US$106 million. The estimation of the foreign assistance received by countries would have to be increased by the unknown amount of expenditures through bilateral channels that contribute to the conservation of PGRFA to arrive at the total estimate.

Very little information is available on the conservation expenditures of the private sector and of NGOs. Because of this lack of information, no expenditures were estimated for these two players in PGRFA conservation. It is interesting to note that the bilateral financial assistance seems to be significantly higher than the multilateral assistance. In the future, this bilateral financial assistance may be the key focus point in negotiations and agreements. If the international negotiations on the access to and the benefit sharing of PGRFA make progress, as appears to indeed be the case (see Chapter 9), the likely result may be to provide compensation for access to PGRFA by increased financial assistance. Consequently, the flow of financial contributions will increase, partly via the international organizations, but mainly through the bilateral channels. The main question will be whether additional contributions will proportionally improve conservation efforts on the national level. Besides increased funding, the efficiency of PGRFA conservation will be increased by strengthening all possible collaborations between the different actors involved in conservation activities.

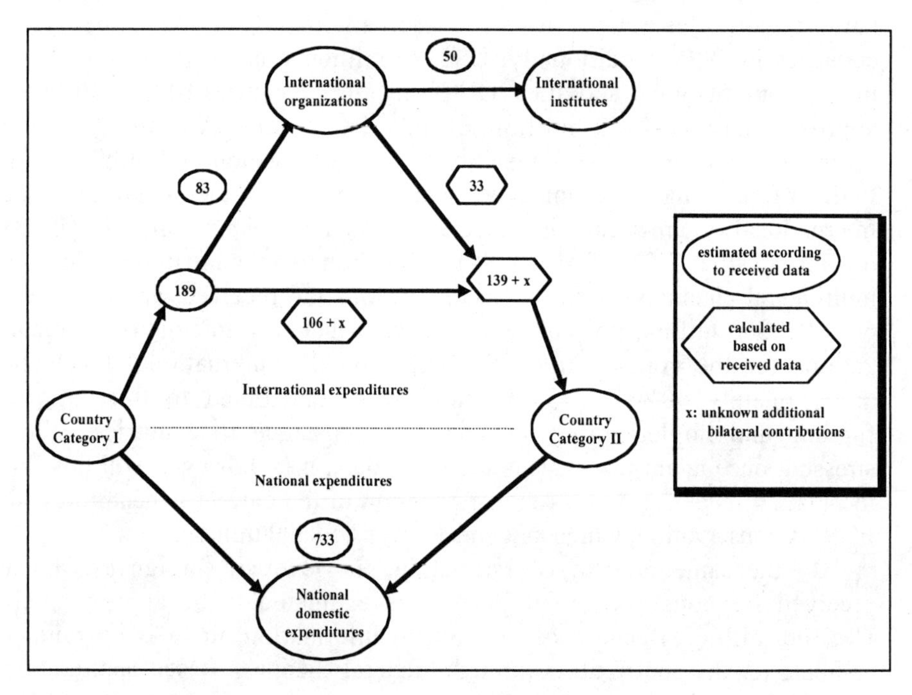

Figure 8-5. Current national and international expenditures for the conservation of PGRFA in 1995 (in US$ million)

Note: Country I: Countries that do not receive (international) financial assistance for their PGRFA conservation.
Country II: Countries that receive (international) financial assistance for their PGRFA conservation.
Source: Virchow (1999a).

4. COLLABORATION AS A MATTER OF PRIORITY

Before the Convention of Biological Diversity (CBD) came into force in 1993, the instruments for the institutional framework of PGRFA conservation management, germplasm exchange and utilization were developed in a rather ad hoc manner, based mainly on national and international codex for research work. Germplasm exchange was regulated according to the transfer of natural resources in research, i.e., free to all bona fide users and based on "pro mutua communatione," the mutual exchange as it is also practiced between botanical gardens (Hammer, 1995). Systematic survey, collection, and conservation of PGRFA have been underway since the beginning of the century. This was undertaken predominantly by the public sector. Today, a complex international and national system for PGRFA conservation is emerging. Engaged in the conservation and utilization of PGRFA since its beginning, FAO has developed some instruments, which are now integrated into FAO's *Global System for the Conservation and Utilization of Plant Genetic Resources for Food and Agriculture*. This global system is the formal framework for the access and exchange of PGRFA since the adoption of the International Undertaking on Plant Genetic Resources (Chapter 9), based on the undertaking's basic concept of a multilateral system.

Since the establishment of the CBD in 1993, however, the international exchange system for PGRFA has experienced some setbacks. In some Eastern European countries, the recent privatization of agricultural research institutes has increased the uncertainty over the continuing free availability of their PGRFA (FAO, 1998). Furthermore, there are signs that the access to PGRFA in some supplier countries is starting to be restricted solely to national utilization[5] (Virchow, 1999a). This may be however, only a transitional phase given the new International Treaty on Plant Genetic Resources for Food and Agriculture (CGRFA, 2002)—hereafter denoted by IT (Chapter 9), because in the long run, the benefit of global exchange of germplasm will exceed all exchange restrictions. Denying access to genetic resources may have a negative impact on those countries themselves when they are not able to participate in the further technological improvements of breeding. It can be assumed that in the future—as suppliers of germplasm—agrobiodiversity-rich but technology-limited countries may be in need of newly developed seed. These are supplied by countries, which are mainly characterized as agrobiodiversity-poor, industrialized countries.

---

[5] For instance, the Chinese genebank and some African countries are restricting the exchange of indigenous germplasm. Furthermore, some other countries restrict the access by establishing bureaucratic obstacles (Hammer, 1995).

Collaboration between countries in conservation activities aims to strengthen national PGRFA programs; to avoid unnecessary duplication of activities in countries within the same region; to promote the exchange of germplasm, information, experiences, and technology related to PGRFA conservation; to promote and co-ordinate collaborative research, evaluation and utilization of conserved germplasm; as well as to identify and promote opportunities for collaboration in training and capacity building. The IT, Agenda 21, the Leipzig Declaration, and the Global Plan of Action for the Conservation and Utilization of PGRFA (GPA) stress the importance of collaboration at different levels for the conservation and sustainable utilization of PGRFA. International collaboration can be differently institutionalized, determined by the objectives of the partners involved. In the following section an overview is given of different types of collaborations.

4.1 International collaborations

At the international level, there is considerable collaboration between international and national organizations for the conservation and utilization of PGRFA. These collaborations are mainly based on the bi- or multi-lateral transfer of technologies and financial resources. These include international collaborative programs of FAO, including the Global System for the Conservation and Sustainable Utilization of PGRFA, CGIAR, and other international organizations, bilateral programs, foundations, and NGOs as well as institutional processes like the International Undertaking on the Plant Genetic Resources or the Convention on Biological Diversity and other international agreements. The compositions of the partners involved in this kind of collaboration are countries and organizations from the PGRFA, on the demand as well as supply side. For example, thirteen maize-breeding countries in the Americas agreed to collaborate on a germplasm project called the **Latin American Maize Project (LAMP)**. Pioneer Hibred provided $1.5 million and technical inputs in support of this project. LAMP has been a highly successful initiative in regional collaboration to improve the conservation and use of maize genetic resources. While the main objective of the program was to evaluate the agronomic characteristics of maize accessions in germplasm banks in Latin America and the United States for future use, several other objectives were also set. These were, inter alia, to determine the exact number of accessions in each bank; to identify the amount and quality of seed in each accession; as well as to list accessions that are in need of regeneration.

In response to the information on regeneration needs, a subsidiary project entitled **Regenerating Endangered Latin American Maize Germplasm** was developed by USAID, USDA, and CIMMYT to salvage maize holdings in Argentina, Bolivia, Brazil, Chile, Colombia, Costa Rica, Cuba, Ecuador, El Salvador, Guatemala, Mexico, Peru, and Venezuela. These 13 countries are participating in the regeneration of nearly 10,000 endangered landrace accessions. Newly regenerated material is conserved in the national collections, with samples duplicated at CIMMYT and/or NSSL (USDA National Seed Storage Laboratory).

FAO also carries out regional activities related to plant genetic resources through various projects. One example is the **Improved Seed Production** project in the CARICOM (Caribbean Community) countries. This project includes training in seed technology, elaboration of a regional seed quality standard, and the establishment of the Caribbean Seed and Germplasm Resources Information Network (CSEGRIN). In addition, FAO coordinates, through its regional office for Latin America and the Caribbean, a Network for Technical Co-operation in Plant Biotechnology (REDBIO) for the exchange of information on tissue culture and other biotechnological techniques.

4.2 Regional collaborations

Regional collaborations are characterized by the common interests and objectives of the partners. One such category of collaboration is between countries from one region with similar plant genetic diversity. These regional networks cover all the different conservation activities that are of relevance to a specific region and, as such, it may include various crops and different programs. These national PGRFA programs often have similar objectives, and regional cooperation usually includes supporting each other's programs and combining similar tasks to increase the efficiency of conservation activities in the region. The main criteria to delimit a regional network are: (1) a common group of indigenous plant genetic resources and agroecological conditions; and (2) the coincidence of country groupings with actual or potential mechanisms for cooperation.

Another type of regional collaboration is defined by the specific crops to be conserved. These crop-specific networks deal with specific conservation and utilization tasks determined by the relevant crop. These networks bring together specialists from different fields on an international and/or regional basis to improve the conservation and utilization efforts for a particular crop's genetic resources. This may include a shared database of all accessions in *ex situ* collections as well as the *in situ* distribution of the crop, and the strengthening of collaboration in collecting and evaluation of germplasm. In addition, regional networks may consist of various crop-related working groups, i.e., crop-specific networks for specific regions.

In each region, there exist some PGRFA networks, some of which are in an advanced stage of development, others being in the stage of establishment:[6]

- The **European Cooperative Program for Crop Genetic Resources Networks** (**ECP/GR**) is the main plant genetic resources network in Europe, currently consisting of 30-member countries, which entirely finance the network. Besides this network, some other collaborative programs among European countries worth mentioning are: the **European Forest Genetic Resources Program** (**EUFORGEN**) and the **Nordic Gene Bank,** which is a centralized regional center for the conservation and utilization of plant genetic resources in the Nordic countries.
- The **West Asia and North Africa Plant Genetic Resources Network** (**WANANET**), as the main plant genetic resources network in the Near East, is strengthening national programs by reinforcing the role of national plant genetic resources committees and by promoting cooperation between organizations within countries as well as programs within the subregion. The operational regional plant genetic resources network in sub-Saharan Africa is the **Southern Africa Development Corporation (SADC) Plant Genetic Resources Center** (**SPGRC**). Its primary objective is to conserve indigenous plant genetic resources within the region, provide training and promote germplasm collection, characterization, documentation, and utilization. A regional plant genetic resources center in Lusaka, Zambia, and a network of national plant genetic resources centers in each SADC-member state characterizes its structure.[7]
- Four regional plant genetic resources networks for Southeast Asia, South Asia, East Asia, and the Pacific are at various stages of development. Of these, the most developed is the **Regional Collaboration in Southeast Asia on Plant Genetic Resources** (**RECSEA-PGR**), which identified a regional network information system and on-farm conservation as its priority working areas.

Three subregional plant genetic resources networks cover South America according to its three agroecological zones. The three networks are **REDARFIT**, the Andean Plant Genetic Resources Network; **TROPIGEN**, the Amazonian Plant Genetic Resources Network; and the network of **PROCISUR**, the Programa Cooperativo para el Desarrollo Tecnológico Agropecuario del Cono Sur. In Central America, the **Red Mesoamericana de Recursos Fitogenéticos** (**REMERFI**) is a well-established network.

---

[6] For more details, see Virchow (1999a), FAO (1998), and Virchow (1996).

[7] SADC-member states are Angola, Botswana, Lesotho, Malawi, Mozambique, Namibia, Swaziland, South Africa, Tanzania, Zambia, and Zimbabwe.

A specific example of regional collaboration is the development of regional or subregional genebanks. This collaboration might provide an alternative to building national genebanks, especially for the conservation of duplicate base collections. National genebanks may give priority to active or working collections, while long-term conservation in base collections might be more effectively carried out at the regional level, as some existing examples show:

- A central subregional base collection that also supports national programs (e.g., SPGRC in Southern Africa).
- The use of existing national genebanks to hold material on behalf of other countries in the subregion with appropriate legal arrangements where necessary (e.g., the role of Ethiopian and Kenyan genebanks in East Africa).
- A network of national genebanks, each one specializing in a particular species or group of species according to mandates agreed upon by the participating countries (proposed for North Africa).

In addition to the conservation of germplasm in single genebanks in individual countries and in regional genebanks, international organizations hold germplasm collections for particular crops, which complement the collections.

4.3 Further potential collaborations

Further cooperation is needed to increase the effectiveness of PGRFA conservation and utilization. Cooperation between the public and the private sector, the public conservation facilities, and the private seed industry, as well as between the professional breeding and farmers in marginalized areas, are possible.

The breeding industry needs support from public sector investment, particularly for processing of information on genetic resources and prebreeding activities. These activities represent long-term and high-risk (i.e., uncertain returns) programs of basic research especially in countries with an emerging breeding industry, like Kenya. Furthermore, these programs compete for resources with other long-term, but basic, and applied research in the private breeding sector (Smith and Salhuana, 1996). Aside from increasing the attractiveness of PGRFA in genebanks by carrying out information processing and prebreeding, this task—carried out by genebanks and financed by the public sector—can be seen as support for all (private, public, and informal) breeding efforts. There are already some collaborative efforts between the private seed industry and national and international public conservation facilities to support regeneration and evaluation, providing seed and information (see section 4.1). More of these collaborations between breeders and genebank managers will eventually lead to a more intensive use of landraces in breeding programs.

Furthermore, the breeding industry could cooperate with farmers in marginal areas. While the private sector's influence is currently rudimentary in marginal areas due to poor infrastructure or because the site-specific needs are of less interest to the private sector, cooperation with the individual breeding of farmers in marginal areas could, however, produce outputs of some interest for the breeding sector in the long run. Consequently, the breeding sector should support these farmers with their conservation activities. This cooperation should be in the interest of the governments, because many countries have to continue and increase the integration of the resource-poor farmers into the market to increase national food security (von Braun and Virchow, 1997).

The final potential collaboration is institutional linkages. *In situ* conservation often involves institutions (e.g., ministries responsible for forestry and environment) other than those that have prime responsibility for *ex situ* conservation (e.g., the Ministry of Agriculture).[8] Additionally, NGOs often play an important role in *in situ* management. These informal conservation efforts, however, are rarely coordinated with public sector activities. Therefore, effective coordination is necessary to strengthen the linkages between all formal and informal organizations and their *ex situ* and *in situ* conservation efforts.

5. FUNDING FOR CONSERVATION ACTIVITIES ON THE INTERNATIONAL AND REGIONAL LEVEL

While many PGRFA conservation and utilization activities are long term and require sustainable long-term funding, international funding is often short term and insufficient. There is a clear need for funding on a planned and sustainable basis. This could be provided both by new and additional funding as well as through a reallocation of existing resources.

5.1 Funding mechanisms for collaborative conservation activities

Although collaborative activities are aimed at increasing the efficiency of the PGRFA conservation efforts partly by outsourcing national activities on the regional or international level, additional funding is required. There are different funding sources, which can be roughly divided into two groups: domestic sources of funding from the public or private sector and external sources of funding from the public and from the private sector (Table 8-4).

---

[8] See Virchow (1999a) for detailed information.

5.1.1 Existing funding mechanisms for collaborations

As can be seen in Table 8-4, there are a vast number of potential funding sources to be utilized for any conservation activity. However, not all possible funds are well adopted to the specific challenges of implementing conservation activities. Two categories of possible funding can be identified: existing funds and new funds. Utilizing existing sources means to reallocate existing funds, to increase the cost efficiency of conservation and utilization activities, and to reduce misallocation.

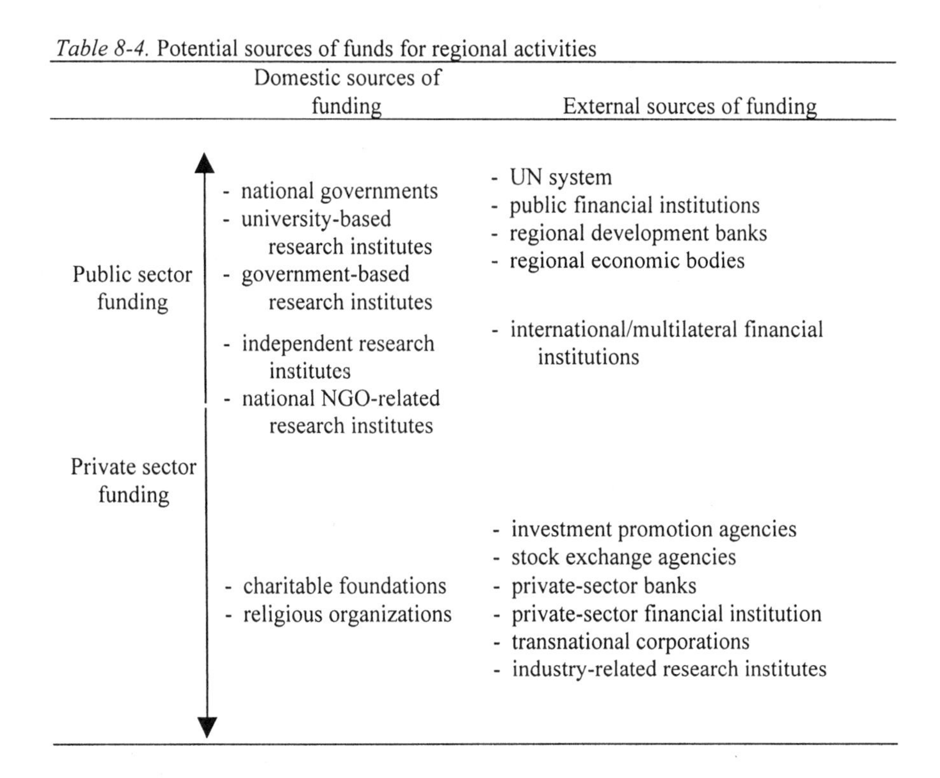

Table 8-4. Potential sources of funds for regional activities

	Domestic sources of funding	External sources of funding
Public sector funding	- national governments - university-based research institutes - government-based research institutes - independent research institutes - national NGO-related research institutes	- UN system - public financial institutions - regional development banks - regional economic bodies - international/multilateral financial institutions
Private sector funding	- charitable foundations - religious organizations	- investment promotion agencies - stock exchange agencies - private-sector banks - private-sector financial institution - transnational corporations - industry-related research institutes

Among existing sources of funding which may be applicable is official bilateral development assistance, the World Bank, the FAO, the UNDP, the UNEP, other specialized UN trust funds, the International Fund for Agricultural Development, regional development banks, NGOs, foundations, and universities and other research institutes.[9]

For countries that do not yet enjoy sufficient access to private flows and in order to assist developing countries more generally in investing in areas

[9] See FAO (1996).

that do not seem "profitable" in the short term, official development finance remains a primary source of external resources. However, with the higher priority in donor countries for fiscal consolidation, there has been a reassessment of official development assistance commitments in several of these countries. Conventional bilateral aid flows now seem set on a downward trend.

5.1.2 Potential new funding mechanisms

Increased attention has been recently given to possible alternative sources of finance. A number of possible new and potentially complementary options for the funding of regional activities can be identified (see Fig. 8-6). As an essential element, an international fund, say, under the auspices of the IT, could be established which would be supported by contributions from various financial sources. The International Fund for the conservation and utilization of PGRFA can be seen as a further component of FAO's Global System for the Conservation and Utilization of PGRFA. This fund could be utilized on the one hand as a financing instrument to provide additional financial resources to national, regional, and international plant genetic resources activities. On the other hand, the International Fund could be used for financial transfers as compensations for countries or specific farmer groups or individuals, which have offered their *in situ* conservation areas to the international community for bioprospecting.

The International Fund is expected to become a key mechanism for sharing benefits and a critical element in ensuring the equitableness of the Global System for the Conservation and Utilization of PGRFA. The International Fund, as the core element of the funds, could be complemented by a "virtual fund." This fund would monitor all financial resources, which are invested for PGRFA-conservation activities, no matter what their source and funding channels. In this way, all private, bi- and multi-lateral investment into implementing any activity related to the PGRFA conservation can be recorded. The International Fund as well as the virtual fund could be based partly on mandatory contributions and partly on voluntary or ex gratia contributions. Besides financial contributions, technology transfer will be the most important alternative resource for the implementation of conservation activities.[11]

An additional fund can also be classified as resources transferred for activities with other objectives, but with a positive impact on the conservation and utilization of PGRFA. These nonrelated expenditures, for instance, development assistance, may be listed in the financing mechanisms due to their positive effect on the conservation and utilization of PGRFA.

---

[11] See the IT's text for detail on funding possibilities (CGFRA [undated]).

All potential funding sources should be invited to contribute or to implement some or more activities related to the conservation of PGRFA. Besides already being involved in the funding of activities, organizations not yet involved as well as all beneficiaries of PGRFA utilization are targeted as funding sources. These sources could contribute to the International Fund as well as to the virtual fund by contributing on a bi- or multilateral level. All contributions could be recorded so as to develop a funding record. This will contain contributions made to the International Fund as well as to the virtual and the additional fund.

All countries, as well as organizations of the public and private sector which act as a funding source to the International Fund, as well as to the virtual fund, will be given access to PGRFA in return for their contributions.

Financial contributions from a country, a private company, or other funding sources should be directed to the International Fund. It may, however, also be contributed by a bi- or multilateral agreement to the activities of a country or to a local, national, regional, or even international organization working in the field of conservation and utilization of PGRFA.

From the discussion above, it can be seen that there are several possibilities for allocating resources for the implementation of conservation activities. On the one hand, a decision is needed as to how to allocate the funds from the International Fund (see, e.g., Chapters 10 and 11). On the other hand, countries, organizations, and all other funding sources can choose their own priorities for fostering specific activities for the conservation and utilization of PGRFA.

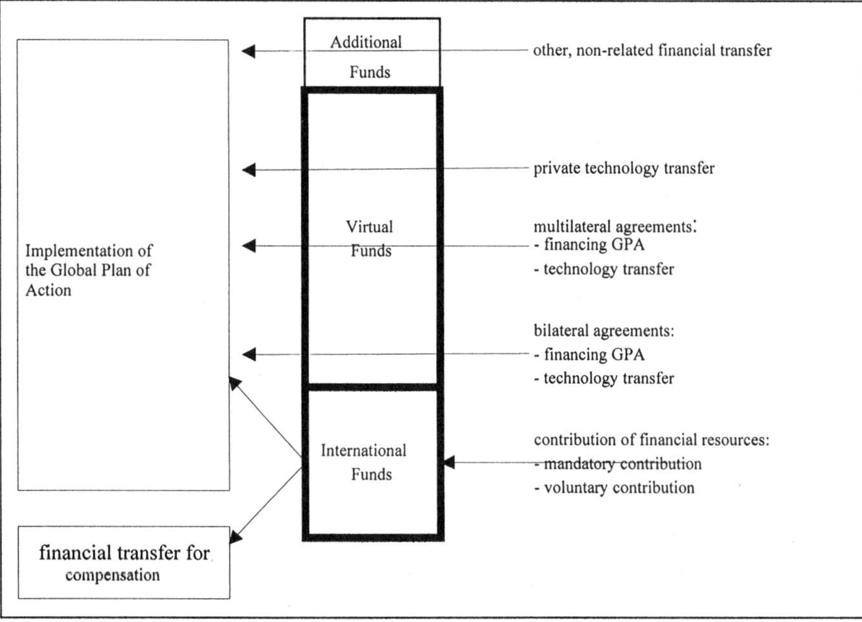

Figure 8-6. The financing mechanisms for conservation activities
Source: Virchow (1999b).

6. PREREQUISITES FOR AN EFFICIENT PGRFA CONSERVATION APPROACH

All of the mentioned actors involved in the conservation of PGRFA will have a role to play in the conservation of plant genetic resources in the future. It will be essential for the future success of increased utilization of improved varieties that the relations between these actors will be productive and some increase in the efficiency of their efforts attained. The main task for the public sector is to set priorities according to the national capacity. Furthermore, it will improve the efficiency of conservation and utilization in a country if collaborations between the actors, especially at the regional and international levels, as well as between the public and the private sectors, are realized, strengthened, and reinforced. These efforts will increase the efficiency of all conservation activities.

A method of strengthening conservation activities worldwide through international and regional collaborations can utilize different synergetic effects based on the principle economies of scale, resulting in an efficient PGRFA conservation and utilization system for the benefit of all involved. The opportunity for collaborations is good, all the more because there is an urgent call for action to improve PGRFA-conservation efforts in the majority of countries.

The backbone of all regional and international collaborations is reliable national programs. It should be emphasized that, in order to be sustainable, collaborations must be based on sound national programs. In other words, international funding does not remove the need for domestic funding. National commitment is essential to provide sustainable funding for national programs and projects through specific funding allocations from governments. Only a regular funding for a national program can guarantee the sustainability of the national program as well as attract sufficient external funding for regional networks or international collaborations.

Different collaborations are based upon varying objectives. Emphasis must be on measures, which improve the efficiency of conservation through the rationalization of efforts, e.g., the rationalization of collections through regional and international collaboration, improved data and information management, as well as the reduced over-duplication of samples. This can be achieved through a joint accessible database with information on the *ex situ* and *in situ* germplasm that are available in the region. Furthermore, this can be achieved through the information gathering and distribution as well as the sharing of research results.

The efficiency of the implementation of PGRFA conservation activities will be increased by including an expenditure survey of all related activities in an overall monitoring-reporting system. In this way, a flexible but objective feedback mechanism for the implementation process can be established. Only by monitoring the expenditures can the effectiveness of the implementation be identified and the allocation of scarce financial resources be optimized. This guarantees the best possible implementation of PGRFA conservation activities. This system may take advantage of synergy effects, reducing the financial and human resources needed for the reporting of the different data. By harmonizing the reporting system with other international monitoring systems, e.g., animal genetic resources or CBD, synergy effects can also be increased.

A final prerequisite for any collaboration on the regional or international level is the maintenance of national sovereignty of those countries involved. Only with their *sovereign rights* over materials such as germplasm are countries willing to place materials in secure storage facilities outside their borders. This includes the implementation of a cost-efficient "Farmers' Rights" system.

REFERENCES

Chandel, K. P. S., 1996, Personal Communication, NBPGR, Director, New Delhi.

Commission on Genetic Resources for Food and Agriculture (CGRFA), 2002, *The International Treaty on Plant Genetic Resources for Food and Agriculture,* FAO (January 23, 2002); http://www.fao.org/ag/cgrfa/itpgr.htm.

FAO, 1996, *Global Plan of Action for the Conservation and Sustainable Utilization of Plant Genetic Resources for Food and Agriculture*, FAO, Rome.

FAO, 1998, *The State of the World's Plant Genetic Resources for Food and Agriculture*, FAO, Rome.

Hammer, K., 1995, Personal communication, Institute of Crop Science and Plant Breeding, Gatersleben, Head of Genebank.

Iwanaga, M., 1993, Enhancing links between germplasm conservation and use in a changing world, *International Crop Science*, **I**:407-413.

Smith, S., and Salhuana, W., 1996, The role of industry in the conservation and utilization of plant genetic resources, Presentation at the Industry Workshop: The Conservation and Utilisation of Plant Genetic Resources for Food and Agriculture, Basel (February 15-16, 1996).

UNEP (United Nations Environment Program), 1994, *Convention on Biological Diversity*, Text and Annexes, Interim Secretariat for the Convention on Biological Diversity, Geneva Executive Center, Switzerland.

Virchow, D., 1996, National and international expenditures for conservation and utilization of plant genetic resources for food and agriculture in 1995, A draft background paper for the document ITCPGR/96/INF/1: *Current Expenditures for the Conservation and Utilization of Plant Genetic Resources for Food and Agriculture*, Prepared for the International Technical Conference on Plant Genetic Resources, Leipzig, FAO, Rome (June, 17–23, 1996).

Virchow, D., 1999a, *Conservation of Genetic Resources: Costs and Implications for a Sustainable Utilization of Plant Genetic Resources for Food and Agriculture*, Springer, Heidelberg.

Virchow, D., 1999b, Financing of The Global Plan Of Action, Background report for FAO, 1999, *Financing of the Global Plan of Action,* CGRFA-9/99/4, Commission on Genetic Resources for Food and Agriculture, Eighth Session, Rome, FAO (April 19-23, 1999).

von Braun, J., and Virchow, D., 1997, Conflict-prone formation of markets for genetic resources: Institutional and economic implications for developing countries, *Quart. J. for Inter. Agri.*, **1**:6-38.

Wood, D., and Lenne, J., 1993, Dynamic management of domesticated biodiversity by farming communities, in: *Proceedings of the UNEP/Norway Expert Conference on Biodiversity*, Trondheim, Norway.

PART III.
Distributional Issues in the Management of Plant Genetic Resources

Chapter 9

THE SHARING OF BENEFITS FROM THE UTILIZATION OF PLANT GENETIC RESOURCES FOR FOOD AND AGRICULTURE*

Joseph C. Cooper
[1] *Deputy Director, Resource Economics Division, Economic Research Service (United States Department of Agriculture), 1800 M Street, NW, Washington, DC 20036-5831*

Abstract: A major issue in international multilateral negotiations is the creation of a fund for the fair and equitable sharing of benefits arising out of the utilization of plant genetic resources for food and agriculture (PGRFAs). This chapter provides a conceptual understanding of the economic value of PGRFAs, identifies proxies for this value that can be used to determine the relative contribution of each country to the benefit-sharing fund, and evaluates the suitability of each proxy to this task.

Key words: agriculture; benefit-sharing; biodiversity; Commission on Genetic Resources for Food and Agriculture; conservation; developing countries; economic value; environmental indicators; International Treaty on Plant Genetic Resources for Food and Agriculture; plant genetic resources.

1. INTRODUCTION

There is growing international consensus on the urgency of slowing the human-induced deterioration of biodiversity, a deterioration that may be coming at high costs to present and future generations. Indeed, within the United Nations System, the adoption of the International Undertaking (IU) on Plant Genetic Resources in 1983, and of the Convention on Biological Diversity (CBD) at the Rio Earth Summit in 1992, was motivated by the goal of maintaining sustainability and diversity of species and ecosystems. In addition to issues regarding wild

* The author thanks Rick Horan, Michigan State University, and Eric Van Dusen, University of California, Berkeley, for their input. The views contained herein are those of the author and do not necessarily represent policies or views of the Economic Research Service or United States Department of Agriculture.

species diversity, the Convention also recognizes the particular importance of biodiversity of relevance for food and agriculture. In 1993 the Food and Agriculture Organization of the United Nations (FAO) adopted a resolution requesting member countries to negotiate—through the FAO inter-governmental Commission on Genetic Resources for Food and Agriculture (CGRFA—the revision of the IU in harmony with the CBD. The Third Conference of the Parties (COP) to the Convention also decided to establish a multi-year programme of activities on agricultural biological diversity with the aims of: (1) promoting the positive effects and mitigating the negative impacts of agricultural practices on biological diversity in agroecosystems and their interface with other ecosystems; (2) promoting the conservation and sustainable use of genetic resources of actual or potential value for food and agriculture; and (3) promoting the fair and equitable sharing of benefits arising out of the utilization of genetic resources (COP, 1997).

Benefit sharing is also called for under the IU's endorsement of the concept of Farmers' Rights, which aims to, inter alia, "allow farmers, their communities, and countries in all regions, to participate fully in the benefits derived, at present and in the future, from the improved use of plant genetic resources." A major observation underlying the negotiations is that agricultural biodiversity "hotspots" tend to be in the developing world, while modern commercial varieties based on plant and genetic resources for food and agriculture (PGRFAs) from these hotspots tend to be developed and marketed by developed countries. As such, many promoters of the Undertaking assert that developed countries are benefiting more from the utilization of PGRFAs from developing countries than do the developing countries themselves, and that these developing countries are not being compensated in return for use of these resources. As touched upon already in Chapter 1, enough concern has developed internationally over the need to conserve agricultural genetic resources that in April 1999, the 161-member nations of the UN-based CGRFA agreed that a multilateral system of access and benefit sharing should be established for key crops, with proposals for payment for conservation of agricultural genetic resources in developing countries. This proposal falls under the auspices of the IU on Plant Genetic Resources for Food and Agriculture, which is the first comprehensive international agreement dealing with PGFRAs. The IU is an evolving international agreement related to the Convention of Biodiversity (COB), yet a separate agreement in its own right (e.g., some countries that are a party to the CGRFA are not a party to the COB). According to the proposal, financing of the Global Plan of Action for the conservation and sustainable development of plant genetic resources will cost the international community an estimated US$155 to $455 million annually (CGRFA, 1999). In November 2001, the 161-member nations of CGRFA agreed that a multilateral system of access and benefit-sharing should be established for key crops and approved the legally binding

International Treaty on Plant Genetic Resources for Food and Agriculture. This international treaty—hereafter denoted by IT—entered into force on June 29, 2004.

The IT's objectives "are the conservation and sustainable use of plant genetic resources for food and agriculture and the fair and equitable sharing of the benefits arising out of their use, in harmony with the CBD, for sustainable agriculture and food security" (CGRFA, undated). Under the IT, countries agree to establish a multilateral system for facilitating access to PGRFAs, and to share the benefits in a fair and equitable way.[1] The multilateral system applies to approximately 60 major crops and forages as listed in the IT's annex. The governing body of the IT, which is composed of the countries that have ratified it, will set out the conditions for access and benefit-sharing in a "Material Transfer Agreement" (MTA). PGRFAs may be obtained from the multilateral system for utilization and conservation in research, breeding, and training. When a commercial product is developed using these resources, the IT provides for payment of an "equitable share" of the resulting monetary benefits, if this product may not be used without restriction by others for further research and breeding. If others may use it, payment is voluntary.

The Treaty provides for sharing the benefits of using PGRFA through information-exchange, access to and the transfer of technology, capacity-building, through the sharing of monetary benefits as mentioned above, and through financial contributions (Articles 13.2 and 18.4). The IT foresees that funds received under the multilateral system "should flow primarily, directly and indirectly, to farmers in all countries, especially in developing countries, and countries with economies in transition, who conserve and sustainably utilize plant genetic resources for food and agriculture" (Article 13.4). While the general funding principles are laid out in the treaty (articles 18 and 19), specifics of who pays, and how much, are not laid out.[2]

This chapter discusses the concept of the economic value of the contribution of PGRFAs to commercial and other uses of plant genetic resources, identifies proxies for this measure that can be used to determine the relative contribution of each country to the benefits-sharing fund, and evaluates the suitability of each proxy to this task. Given political realities and the lack of existing data on benefits, the level of total annual contributions to this fund will be determined through a multilateral negotiation process that is independent of any estimates of the value of PGRFAs. Hence, a goal of this chapter is to discuss how, given the existing data, each country's relative contribution to this fund

---

[1] PGRFAs consist of the diversity of genetic material contained in all domestic cultivars as well as wild plant relatives and other wild plant species and plant matter (germplasm) that are used in the breeding of new varieties either through traditional breeding or through modern biotechnology techniques.

[2] See CGFRA (undated) for further details.

can be as highly (and positively) correlated as possible with each country's benefits from utilizing a defined set of PGRFAs as well as satisfy equity considerations. Since country contributions will be of a monetary form, the focus of this chapter will be on the economic value of PGRFAs.

2. ECONOMIC INDICATOR FOR BENEFIT SHARING

If, for the sake of argument, the idealized goal is to set total country contributions to the benefit-sharing arrangement proportional to the benefits derived from the commercial and other uses of plant genetic resources obtained from outside sources, then the question is how to derive this value.

Fig. 9-1 is a stylized graphical representation of the economics benefits associated with an agricultural output-enhancing set of PGRFAs, denoted as ***I*** (possible definitions of this set are discussed below).[3] In the figure, equilibrium crop quantity produced and supplied without availability of *I* (denoted as " $\backslash I$") is denoted by Q_1 and P_1, respectively. Assuming that set *I* served as an input to a crop breeding program that increased the crop supply by a fixed amount at any given price, and that in this case, the equilibrium crop quantity produced and supplied with availability of *I* is denoted by Q_0 and P_0, respectively. The shaded area in the figure is the loss in economic benefits (defined in terms of consumer and producer surplus) of researchers and crop breeders not having access to this set *I*, say, due to loss of these crop varieties over time.

The basic difficulty in arriving at the economic value of PGRFAs, such as that depicted in Fig. 9-1, is largely due to the structure of the market for PGRFAs. One can assume that PGRFAs are valuable: Breeders need them as input to producing new varieties and, furthermore, to the extent that downstream producers as well as consumers benefit from these new varieties, they benefit from PGRFAs as well. However, the number of suppliers of PGRFAs is high enough that the market price for PGRFAs is essentially driven to zero. Because many farmers may grow the same PGRFA, they are all potential suppliers. Plus, the number of

---

[3] This theoretical section covers some of the same ground as in Chapter 4, but the former is more general and oriented towards institutional issues and the latter is oriented towards motivating an actual valuation exercise.

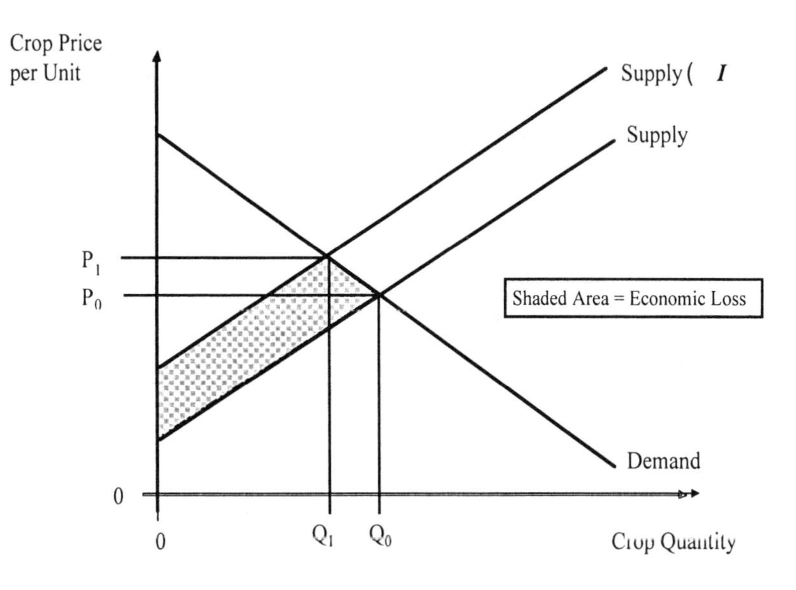

Figure 9-1. Economic losses stemming from loss of PGRFAs set *I*.

suppliers may cut across country boundaries. Furthermore, for most PGRFAs, the level of substitutability by other PGRFAs appears to be quite high.[4]

Given the divergence between the social and the market value of PGRFAs, a related difficulty in establishing the value of genetic resources is that it is extremely difficult to separate the value of the raw PGRFA input from the value of the research used to produce a new variety. With enough information on the plant breeders' PGRFA choice set and their process of selecting between PGRFAs, it may be possible under certain conditions to separate these values, perhaps in some manner analogous to bioprospecting models for pharmaceuticals (e.g., Simpson, Sedjo, and Reid, 1996), at least in some narrowly focused studies. However, the necessary information is largely proprietary and/or costly to collect. The contingent valuation and hedonic methods may come to mind as short-cuts to estimating values of PGRFAs, but have some serious shortcomings.

[4] That the market price is zero is not to say that the value of PGRFAs is zero, but that suppliers cannot capture this value. The only way suppliers could obtain this value in the marketplace would be if these suppliers formed an effective cartel in access to their PGRFAs. However, even with a single supplier, the willingness to pay for a "marginal" PGRFA is likely to be close to zero. Some genetic resources may be quite valuable in the market, but most will not be. A cartel might function more profitably by not charging the monopoly price on all genetic materials, but rather, by attempting to appropriate the buyers' entire surplus by charging a price discriminator-based fee for any access to genetic materials. In practice, however, there are too many potential suppliers, and a cartel cannot be formed through private actions.

The general concept of economic value in period j of a set of PGRFAs, can be formally stated in a simple manner. Ignoring for ease of exposition the lags in the development of new products that use $\boldsymbol{I}$ as input, the value of $\boldsymbol{I}$ summed across all stakeholders in period j is

$$V_j(I) = \left[\left(W_j\right) - \left(W_j \setminus I\right)\right]$$

where W_j is the value of the agricultural sector in the PGRFA-utilizing countries. This value can be defined as consumer plus producer surplus at the retail level, and $W_j \setminus I$ (i.e., W_j without $\boldsymbol{I}$) is the value of the agricultural sector assuming that it could not obtain PGRFA set $\boldsymbol{I}$. If suppliers of $\boldsymbol{I}$ had some form of market power of the supply of set $\boldsymbol{I}$ in period j, then they would capture a portion of $V_j(\boldsymbol{I})$.[5]

For an illustration of how $V_j(\boldsymbol{I})$ can be formally represented, take the seed industry as an example and assume for simplicity that $V_j(\boldsymbol{I})$ represents only benefits to that stakeholder. In a highly stylised form that considers as beneficial only the number, and not the quality, of varieties in set $\boldsymbol{I}$, the seed industry maximises the following profit function:[6]

$$\pi_I = p(S)S - c(S, N)$$

where S is seed sales, N is the number of varieties in set $\boldsymbol{I}$, p (S) is the demand function for S, and $c(S, N)$ is the cost function, and $S = S(N)$. While plant breeders usually do not pay for N, because research costs are a function of N, N enters into the cost function directly as well as indirectly. Totally differentiating π with respect to N yields

$$\frac{d\pi}{dN} = \left(P + \frac{\partial p}{\partial S}S - \frac{\partial c}{\partial S}\right)\left(\frac{\partial S}{\partial N}\right) - \frac{\partial c}{\partial N}$$

If the industry is producing S at profit-maximising levels, then marginal cost equals marginal revenue and the difference in the first set of brackets is zero, yielding

---

[5] For making intercountry assessments of the importance of imported PGRFAs, $V_j(I)$ can be re-expressed as $R_j(I) = 1 - \left[\left(W_j\right) - \left(W_j \setminus I\right)\right] / \left[\left(W_j\right)\right]$ Since $\left(W_j \setminus I\right) \leq \left(W_j\right)$, the ratio $R_j(I)$ ranges from 0 for no dependency on set I to 1 for complete dependency on set I.

[6] While noting that set I can be more generally conceived of as germplasm, it is defined as varieties for ease of exposition.

$$\frac{d\pi}{dN} = -\frac{\partial c}{\partial N}$$

which says that the increase in profits for a marginal increase in N equals the negative of the decrease in costs associated with a marginal increase in N. If we assume that research costs fall with increases in N, then $d\pi/dN > 0$ and profits increase with increases in N. If the seed industry were to be taxed for contributions to the benefit-sharing fund in the amount $t(N)$, so that $\pi = p(S)S - C(N, S) - t(N)$, where $\partial t/\partial N > 0$, then $d\pi/dN$ would be positive only if $-\partial c/\partial N > \partial t/\partial N$. In other words, the marginal tax rate on N must be less than the decrease in marginal cost of N for breeding firms to benefit from increases in N. Similar analysis can be used evaluate the impacts of N and of $t(N)$ on consumer and producer surplus (Cooper, 1998).[7]

Of course, $V_j(\boldsymbol{I})$ is a function of the scope of PGRFAs included in $\boldsymbol{I}$ as well as the time period over which benefits are to be evaluated. For instance, is $\boldsymbol{I}$ the set of all varieties of wheat, or even all major crop species in existence at time j, or $\boldsymbol{I}$ is the set of varieties obtained by plant breeders from suppler countries in j. If it is the former, $V_j(\boldsymbol{I})$ will be enormous. Or is $\boldsymbol{I}$ measured at the gene level? With regards to the time dimension, in quantifying total benefits for the purpose of determining contributions to the benefit-sharing pool, is the present value of past benefits to be included in addition to current benefits? If so, how far back? While scope and time dimensions have not been set in the multilateral negotiations, some definitions have been discussed. For example, the African Region states that the contribution to the fund should be some percentage of "the value of the commodity produced using intellectual property rights [IPR] material..." (CGRFA, 1998), which is similar to a definition set out in a proposal by Malaysia. Since Intellectual Property Rights (IPR) for PGRFAs are a fairly recent concept, this definition addresses the time dimension as well as the scope of PGRFAs to be considered in determining benefits.

Even if $\boldsymbol{I}$ is precisely defined, a complicating factor in estimating $(Wj \setminus \boldsymbol{I})$ is that the substitutability of other inputs for set $\boldsymbol{I}$ must be considered, given that $(Wj \setminus \boldsymbol{I})$ in the case where some substitutability of inputs is possible will be higher than $(Wj \setminus \boldsymbol{I})$ in which no other inputs can substitute for set $\boldsymbol{I}$. Substitutability may be in the form of technological innovations. If advances in biotechnology make the availability of PGRFAs for plant breeding less necessary, then $V_j(\boldsymbol{I})$ may decrease.

[7] In the equation for $d\pi/dN$, it would be more accurate, but less eloquent, to explicitly consider the stochastic aspects of N with respect to producing new commercial varieties. If so, the expectation of the function is not the same as the function of the expectation. The point of the expression given in the text is to convey the essence of the valuation argument $V(I)$ and to serve as a serviceable first approximation to the exact, but much more complicated, expression.

Because of substitution possibilities between PGRFAs, one would expect diminishing marginal benefits to be associated with adding another randomly chosen PGRFA to $\boldsymbol{I}$. The more varieties are contained in $\boldsymbol{I}$, the smaller the difference in $V_j(I)$ with and without any randomly chosen variety. For example, if $\boldsymbol{I}$ is a set of PGRFAs currently in publicly owned gene banks, given that most of these accessions are never examined, then the market value portion of $V_j(\boldsymbol{I})$ changes little when a randomly chosen PGRFA is added to the collection (though, as discussed in the next section, the value added to society may be greater). Also, if one wants to consider the value associated with $\boldsymbol{I}$ supplied from any one country i, or $V_j(\boldsymbol{I}_i)$, then the extent of geographic substitutability must be considered in estimating this value. The value $V_j(\boldsymbol{I}_i)$ will be higher the more PGRFAs in $\boldsymbol{I}_i$ are also grown in other countries other than i, and higher the less easily varieties from other countries can substitute for varieties in I_i.

3. COMPONENTS OF VALUE OF PLANT GENETIC RESOURCES FOR FOOD AND AGRICULTURE

The previous section discussed the general concept of benefit-sharing. In this section, we discuss the composition of $V_j(\boldsymbol{I})$. From an economic standpoint, several value categories can be ascribed to PGRFAs. The most concrete one is its *use value*, i.e., the value associated with the direct and indirect benefits resulting from the use of PGRFAs by plant breeders, farmers, food processors, and consumers. For plant breeders, PGRFAs are inputs to producing more productive or disease-resistant varieties. To a large extent, this use value is a function of the breeding technology and of the income achievable from productive use of the improved seed. Improvements in breeding technology, through biotechnology for example, may increase breeders' demand for germplasm and thus raise its value and market price. A share of the economic benefits of improved varieties goes ultimately to consumers in the form of lower food prices and another to farmers in the form of greater revenues due to higher yields or higher quality products.

The second-value component, the *option value* (Arrow and Fisher, 1974; Henry, 1974), is the value to society (their willingness to pay) of avoiding irreversible decisions on the conservation of PGRFAs. In particular, the loss of native landraces (or "traditional" varieties) is irreversible. As first noted by Hanemann and Fisher (1986) in the context of option value, other varieties may be close, but not perfect, substitutes, so once a particular landrace is extinct, its value for future plant breeding will remain unknown. In this context, option value suggests that a premium exists that is associated with actions that preserve flexibility. In other words, it is an option to be able to consume the product in the future. However, this premium is held by society, and not necessarily by private industry, which has little incentive to maintain a *in situ* conservation program outside of the firm's own private lines. The reasons are that the likelihood that any particular PGRFA currently *in situ* will yield a useful input to the breeding of a new variety is very low and, given its characteristics of nonrivalness and nonexcludability as a breeding input, one firm's conservation of a PGRFA will not necessarily exclude any other firm from conserving the same PGRFA.[8] If the PGRFA-utilizing firm could be assured that the same PGRFA it has bought rights to will not be sold by the supplier to another firm as well, the price it is willing to pay could have a positive option value component to it. Depending on the structure of the market for PGRFAs, even in the case where suppliers of PGRFAs could control access, the social value of conserving PGRFAs may be higher than the private value. In other words, even in the case of a hypothetical single-market supplier for PGRFAs, the option value may still not be fully revealed in market prices for PGRFAs.

If society has a value for conservation actions that assure the potential for future use of the PGRFAs, then this value is part of the total benefits from PGRFAs. However, this option value makes sense as a part of the contributions to the fund only if distributions from the fund require conservation activities on the part of the recipients. Otherwise, it would not be equitable to make option value part of a country's contribution to the fund. The possibility of an option value component demonstrates that, at least in principle, how the money is to be distributed affects the level of contributions. At any rate, the discussion of the exclusion or inclusion of option value in contributions is an academic one at this point, given that no data exist on this value.

A third component of the value of PGRFAs is its *existence value*, i.e., the value one holds for a variety or set of varieties just for its own sake or for some moral or cultural reason. With respect to loss of PGRFAs, the main threat facing agriculture is the loss of intra-species diversity, and not inter-species diversity—species that may be lost are not major players in agriculture. Hence, pure existence values for conserving PGRFAs are likely to be only a small portion of their total economic value. Almost certainly, people hold a larger existence value for knowing a certain species exists than they hold for individual varieties within that species.

---

[8] Unlike the use of PGRFAs as food, PGRFAs (as embedded in native landraces) as an input to breeding research are generally nonrival and nonexcludable. In this case, what is being purchased is the knowledge contained, and not the physical product itself. Because so little of any PGRFA is needed by a breeder as a research input and because it is a renewable product, for all practical purposes, most PGRFAs exist in sufficient quantities that all breeders could use it as a research input without bidding up its price. Native landraces are generally nonexcludable because property rights over them are not assigned and because there are usually many potential suppliers of the same landrace. Note that these two properties do not apply to a breeder's private lines, given that the breeder can effectively restrict the distribution of these lines.

4. PROXY INDICATORS FOR BENEFIT-SHARING

In a world with full information, contributions towards a benefit-sharing arrangements would be based on $V_{jl}(\boldsymbol{I})$, $l = 1,\ldots,L$ countries, where $\sum_{l=1}^{L} V_{jl}(\boldsymbol{I}) = V_j(\boldsymbol{I})$, given that the benefit-sharing fund requires contributions from each of the L countries endorsing the IU. Of course, for the reasons discussed earlier, $(W_{jl} \setminus \boldsymbol{I})$ is unknown for almost any definition of $\boldsymbol{I}$, and hence, $V_{jl}(\boldsymbol{I})$ is unknown. Hence, it appears that the best one can do is to base contributions on just the W_{jl} portion of $V_{jl}(\boldsymbol{I})$.[9] While existing indicators cannot be explicitly based on benefits, given that a benefit-sharing fund will be created, the best one can do is identify existing indicators that are politically acceptable and that appear to have some connection to the benefits associated with PGRFAs.

Table 9-1 presents a list of potential country level indicators for benefit sharing that are at least somewhat within the realm of feasibility. No claim is made that this list is all-inclusive, but it does attempt to cover a broad range of classes. Indicators 1-3 are easily obtainable for almost all countries. While the collection of indicators 4-13 is technically feasible, they cannot be found for more than a few countries and/or crops, and hence, may be of limited policy usefulness, regardless of their correlation with the value of PGRFAs. Given that the negotiations on setting the total size of the Fund will essentially be independent from the negotiations that determine each country's contribution, a country's

Table 9-1. Potential country level indicators for benefit sharing

Indicator	Observability[a]	Equity[b]	Correlation[c]
Value of agricultural output	+	-	-
Gross Domestic Product	+	+	-
Agricultural Gross Domestic Product	+	-	-
Seed industry profits and/or revenues	-	+	+
The value of agricultural commodities produced using intellectual property rights (IPR) material	-	+	+
Value of commodities produced using improved (e.g., "green revolution" varieties)	-	+	+
Royalties earned on agricultural patents	-	+	+

[9] While there may exist a few contractual relationships between farmers and breeders that purport to be benefit-sharing arrangements, in practice, these appear to be simply contracts that cover the cost of collecting PGRFAs.

Table 9-1. (continued).

Agricultural research and development expenditures by country	-	+	-
Plant protection titles issued	-	+	-
Number of landraces used in agriculture	-	-	-
Domestic-origin patents used in the agricultural and food sectors	-	+	-
Matrix of varietal exchange and matrix of parental exchange	_	_	_
Diversity measures	-	-	-

Notes: The scale "-" / "+" are used to construct relative ranking within the group of the 13 potentially feasible indicators.

[a] Ease in obtaining the data for all countries (and/or crops), where "+" = obtainable for all countries and "-" = not obtainable for all countries.

[b] In terms of income (re)distribution from developed to less developed countries within the group, "+" = most equitable and "-" = least equitable (assumes that the higher a country's value for an indicator, the more it contributes to the fund).

[c] Subjective ranking of correlation between the true value of PGRFAs and the listed indicator. As this is a relative ranking, no implication is made that any of these indicators are anywhere close to the true value, merely that "+"coded indicators may be closer to the true value than "-" coded indicators.

share of contributions to the Fund can be set to its share of the world total value of the indicator.[10] Analytically, country l's contribution to the Fund can simply be $C_l = (Total\ Fund\ Value) *$, $\left(Indicator_l \Big/ \sum_{l}^{N} Indicator_l \right)$ where $Indicator_l$ is the value of the indicator for that country, and $l = 1, \dots, N$ countries. Due to the continuous cycle of imports and exports of varieties between regions, in some sense every country is a net user of PRFAs (Wright). Under this formula, every country makes a contribution to the Fund.[11]

The first indicator in the table is the value of agricultural output (VAO). This indicator has several desirable characteristics. First, this indicator is readily available for each country and, among the indicators

---

[10] This assumption of independence is somewhat of a simplification, given each country's position in the negotiations over the total value of the Fund is probably influenced by suppositions regarding what it may have to pay.

[11] It is noted in passing that negative values are possible for the indicators. For instance, it is possible to consider an indicator based on net flows of genetic resources. In fact, Gollin and Evenson (1998) have a table of international flows of rice landrace ancestors for selected countries. This table shows some large developing countries to be net borrowers (e.g., India) and shows the United States to be a net lender. It is not certain how to use this indicator in determining a country's contributions to the fund, but certainly, it would result in the United States being a net recipient and India being a net contributor to the fund. Whether or not this allocation is equitable, it is unlikely to satisfy the political motivations underlying the creation of the fund.

listed in Table 9-1, is most tied to the value of agricultural activities as it is the value of agriculture production at the farm gate. Second, it is equitable in the sense that countries with greater agricultural value pay more, which is analogous to a progressive income tax on agriculture. For instance, if the U.S. payment to the Fund were based on its share of total average annual worldwide value of agricultural production, its contribution would represent 12.7% of the total value of the fund. Since most developing countries represent only small shares of the total value of agricultural production, their contribution would be small.[12]

The justification for the use of Gross Domestic Product (GDP), Agricultural Gross Domestic Product (AGDP), and Seed Industry Profits and/or Revenues indicators are similar to those for VAP. GDP is the value-added at each stage of production of all goods and services during one year. A potential benefit of VAP over GDP from a negotiation standpoint is that an indicator based on the former is at least directly tied to agricultural production, if not to the value of PGRFAs. AGDP, which is the value-added only of agricultural goods and services, is more closely tied to agriculture than is GDP but includes the value of services that are not directly related to agricultural production. Its use as an indicator would explicitly acknowledge that consumers and food processors as well as farmers and seed companies benefit from PGRFAs. Therefore, when GDP or AGDP is used as the basis for a funding mechanism, countries who benefit on the consumer side but are without an agricultural production base will contribute to the Fund. Similar to VAP, the large size of AGDP relative to any realistic prediction for the total size of the Fund suggests that a benefit-sharing payment at the retail level is unlikely to produce significant market effects. In practice, the indicators VAP, GDP, and AGDP are all highly correlated with one another and a country's share of the contributions to the Fund would be similar using any of these as a basis.

As a proportion of total revenues, commercial seed producers are probably the stakeholders most dependent on access to PGRFAs, and therefore, seed industry profits and/or revenues by country could be, among the class of measures that may be obtainable, the one most highly correlated with the value of PGRFAs. However, this measure is not currently available for all countries, but with data available for 39 countries, it has better coverage than some other indicators. Given the International Association of Plant Breeder's (ASSINEL) definition of "commercial seed producer," the value at the world level of the commercial seed industry is $30 billion per year (ASSINEL, 1998).[13] If

[12] The relative size of VAP with respect to GDP tends to be higher for developing countries than for the industrialized northern countries. This may decrease the acceptance of VAP as an indicator since some of the larger developing countries, such as China, India, Brazil, and Indonesia will pay more than most OECD countries, which may be viewed as inequitable.

[13] Worldwide, the value of seed export turnovers is 15% of the internal commercial seed

industry profits are around 10%, then a $50 million benefit-sharing tax levied on the industry represents 1.7% of profits, and a $300 million tax represents 10% of profits. While the extent to which this increase in cost would be passed on to downstream stakeholders depends on supply and demand conditions in each stakeholder's market, the relatively large ratio of the tax to total profits (or revenues) suggests that a tax aimed only at commercial seed producers would produce more notable price impacts, at least in the market for seed, than a tax at the farm gate or retail level.

The indicator "Value of Agricultural Commodities Produced Using Intellectual Property Rights (IPR) Material" is more narrow in scope than the previously mentioned indicators. This indicator is available for only a few countries, making it difficult to develop a benefit-sharing mechanism. The indicator "Value of Agricultural Commodities Produced Using Improved Varieties" has a slightly broader scope than the IPR indicator, but the term "improved" must be clearly defined. Under many possible definitions, one could expect that all varieties grown in developed countries are improved. If this is the case, this indicator would be the same as the VAP for these countries, but for developing countries the values would be difficult to estimate given the lack of data.

The indicator "Royalties Earned on Agricultural Patents," along with the indicator "Seed Industry Profits and/or Revenues," focuses on stakeholders who may be the most dependent on PGRFAs relative to other stakeholders. Again the caveat must be noted that the value of the PGRFA contribution cannot be easily separated from the value of the research contribution. At any rate, the royalty data tends to be proprietary, and hence, little of it is available.

The indicator based upon "Agricultural Research and Development Expenditures" reflects a country's capacity for upstream use of PGRFAs. It is likely that the higher a country's expenditures are, the greater its relative financial benefits from the use of PGRFAs. This expenditure reflects the R&D industries' expected share of the total value-added in the marketplace resulting from their development of new products, i.e, the presumption is that the industries will not spend more than they expect to earn. However, the measure does not include the share of the benefits received by users (both producers and consumers) of the new products. A practical disadvantage is that it is not available for many countries and that private and public expenditures may have to be disentangled.

Unlike the first eight indicators listed in Table 9-1, indicators 9-13 are nonmonetary measures. These nonmonetary measures share the same drawback as the available monetary ones, that is, the connection between

---

market value (ISF, undated). For the 17 countries for which it was possible to construct these ratios, Japan had the lowest percentage (1.6%) and The Netherlands the highest (206.7%). Other examples include the United States at 15.7% and Chile at 62.5%.

them and the value of PGRFAs is unknown. The International Union for the Protection of New Varieties of Plants (UPOV) charts certificate applications and titles issued for plant variety protection, and has made the data available for 28 member countries (UPOV, 1995).[14] The indicator "Number of Landraces used in Agriculture" is available for rice for select countries from the International Rice Research Institute, along with a tabulation by country of own and borrowed rice landraces. Data on "Domestic-Origin Patents Used in the Agricultural and Food Sector" are available only for some countries. The indicator "Matrix of Varietal Exchange and Matrix of Parental Exchange" tracks international exchanges of varieties and is available for rice for selected countries.

With nonmonetary measures such as these, one essentially has the problem of comparing apples and oranges. For example, one country could hold 20 plant protection titles, but the sum of their values could be less than one plant protection title held by another country. Most likely, a large proportion of the total value of improved varieties is ascribed to a small percentage of improved varieties. This asymmetry is apparently the case for U.S. university-held patents, in which only a few patents (and universities) account for most of U.S. university royalties on patents (AUTM, 1993). Given the small number of commercially successful varieties relative to the total number of available varieties in genebanks (FAO, 1998), there is reason to believe that this asymmetry is also true for the economic value of PGRFAs. That said, monetary measures as a basis for contributions have the advantage that contributions tend to be equitable (contributors with higher incomes often pay more) and may be rationalized from a development aid standpoint.

The final category in the table (Diversity Measures) covers indicators of PGRFA diversity. Many different measures are available, and no single measure can satisfy all aspects of diversity (see, e.g., Ferris and Humphrey, 1999). These diversity measures are most useful for tracking changes in diversity over time and for assessing the impact of conservation measures. As with the other indicators, the relationship between this indicator class and the value of the PGRFAs is unclear. Unless the diversity itself is of value (which may be the case for wild species and ecosytems), the degree of crop diversity in a country is somewhat irrelevant as a proxy for the value of PGRFAs. Instead, it is their current or future contribution to agricultural production and to protecting agricultural production (e.g., from disease) that are of value, and not the resources themselves (Cooper, 1998). For instance, while two countries may have the same genetic diversity index, there is little reason to assume that their respective contributions to the value of agriculture are equal.

After weighing the advantages and disadvantages of each of the listed indicators, the VAP and AGDP appear to be "superior" in the case of

---

[14] Note that WTO rules require signatories to WTO to also be members of UPOV.

PGRFA conservation. If each countries' contribution to a benefit-sharing fund is the ratio of its indicator value to the total world value of the indicator, then application of either of these indicators will be progressive, primarily with respect to equity considerations, but also with respect to efficiency considerations. That is to say that although all countries contribute, those countries with higher VAP and AGDP benefit from PGRFA more than those with lower values and also pay more.

5. CONCLUSION

If a multilateral environmental agreement (MEA) requires member countries to meet specific environmental targets or to fund the provision of a global public good, then environmental indicators become necessary. An environmental indicator measures environmental quality, whether as a measure of the physical quantity itself or of the monetary impact of that quantity. For agriculture, in particular, indicators generally include measures of land-use changes between agriculture and other land uses, both on-farm and off-farm impacts of soil erosion, total agricultural water use, nutrient balances, pesticide use, and water quality (Organization for Economic Cooperation and Development [OECD], 2000). Obviously, some decision on the choice of indicators is necessary if one is to make inter-regional comparisons of the environmental benefits from the MEA.[15]

The OECD has developed a "core set . . . of commonly agreed indicators for OECD countries and for international use, . . ." (OECD, 2001). According to the OECD, the purpose for developing this set of core indicators is to:

- Allow countries to track environmental progress.
- Ensure integration of environmental concerns into sectoral policies (e.g., agriculture).
- Ensure integration of environmental concerns into economic policies.
- Measure environmental performance.
- Determine whether countries are on track towards sustainable development.

Among the criteria for selecting a core set of indicators are that the indicators are policy relevant, analytically sound, measurable, and easily interpreted (OECD, 2000).

This set of OECD indicators could be used to assess whether or not countries are meeting environmental commitments specified under an MEA. Environmental indicators can also be used to determine a

[15] Of course, environmental indicators can be important in the domestic policy setting as well. For example, in the United States, the Environmental Benefits Index (EBI) uses environmental indicators to help prioritize land for inclusion into the Conservation Reserve Program.

country's funding obligations to an MEA, as well as disbursements from an MEA to member countries, say, for conservation efforts. With respect to the agricultural sector, the OECD is seeking to develop indicators to better integrate environmental and economic policies with the goal of sustainable agriculture in mind. While developing such indicators may be a relatively easy task for an MEA seeking to address some fairly concrete externality, such as loss of forest cover, as pointed out earlier, it is not as easy in the case of conserving agricultural genetic resources.

From an economic standpoint, it seems reasonable to tie a country's contribution to the benefit-sharing fund to the benefits it receives from its use of PGRFAs. Unfortunately, as discussed in this chapter, these benefits cannot be quantified, except perhaps in limited case studies. Hence, an alternative can be to appeal to indicators that take equity and development considerations into account in determining contributions. Furthermore, to ease the process of multilateral negotiations, these indicators must be available for all (or almost all) the countries participating in the negotiations. Among indicators satisfying these conditions, the VAO and AGDP appear to be the most applicable. Both are progressive in the sense countries with higher values pay more. A potential advantage of VAO over AGDP in multilateral negotiations is that the former is explicitly focused on agricultural interests. Use of the latter in the negotiations is appropriate if the benefits of PGRFAs to consumers are to be accounted for in setting country contributions to the Fund. Furthermore, the latter is somewhat more equitable (in terms of income distribution) than VAO. For all practical purposes, given that these two indicators are highly correlated, the choice between the two will not have much impact on the relative size of each country's contributions. Indicators such as the value of commercial seed production focus better on stakeholders most dependent on access to PGRFAs, but in addition to data limitations, their downside is that they ignore downstream benefits to farmers, food processors, and consumers.

It is worth mentioning in closing that no indicator exists that will be more than an imprecise guide for how to distribute the cost of any benefit sharing fund in an economically efficient fashion. Obviously, the ideal indicator relies on fully observable environmental and economic benefits, and thus increases the possibilities for efficient and equitable distribution of any and all benefit sharing funds. However, at least having some reasonable indicators available as a guide may reduce the potential for the funds being distributed through an opaque process.

Finally, as is discussed in the "Components of value" section, how the funds are to be distributed has some bearing on what is included in the valuation of the PGRFAs. However, some proponents of benefit-sharing assert that since PGRFA suppliers are due compensation in return for utilization of their genetic resources by others, they should not be restricted to what use they put the compensation to. In fact, it is the political considerations regarding this view that are the primary reason why multilateral negotiations have not simultaneously discussed contributions to, and distributions from, the fund. However, data limitations aside, the often-blurred distinction between users and suppliers of PGRFAs, as well as other complications, make explicit compensation impossible. Hence, unrestricted distribution from a benefit-sharing fund would appear to be a simple income transfer. To help ensure that a benefit-sharing fund is maintainable over time, the major contributors must perceive that the funds are being redistributed in a constructive fashion. Earmarking the fund for conservation and sustainable use of PGRFAs or related food security activities will help the fund distinguish itself from other forms of development aid.

REFERENCES

Arrow, K., and Fisher, A., 1974, Environmental preservation, uncertainty, and irreversibility, *Quarterly J. of Econ.* **88**:312-319.

ASSINSEL (International Association of Plant Breeders), 1998, ASSINSEL: Position on Access to Plant Genetic Resources for Food and Agriculture and the Equitable Sharing of Benefits Arising from their Use, Position Statement, ASSINSEL, Nyon, Switzerland (June 5, 1998).

Association of University Technology Managers, Inc. (AUTM), 1993, AUTM Licensing Survey: Fiscal Years 1991–1993.

Commission on Genetic Resources for Food and Agriculture (CGRFA), undated, International Treaty on Plant Genetic Resources for Food and Agriculture (Official text), Food and Agricultural Organization of the United Nations Rome, Italy; ftp://ext-ftp.fao.org/ag/cgrfa/it/ITPGRe.pdf and www.fao.org/ag/cgrfa/itpgr.htm.

Commission on Genetic Resources for Food and Agriculture (CGRFA), 2000, Item 2 of the Draft Provisional Agenda: Third Inter-Sessional Meeting of the Contact Group, CGRFA/CG-3/00/2, Food and Agricultural Organization of the United Nations, Rome (April, 2000).

Commission on Genetic Resources for Food and Agriculture (CGRFA), 1999, Report of the Eight Extraordinary Session of the Commission on Genetic Resources for Food and Agriculture, CGRFA-8/99/4, Food and Agricultural Organization of the United Nations, Rome (April 19-23, 1999).

Commission on Genetic Resources for Food and Agriculture, 1998, Report of the Fifth Extraordinary Session of the Commission on Genetic Resources for Food and Agriculture, CGRFA-Ex5/98/Report, Food and Agricultural Organization of the United Nations, Rome (June 8-12, 1998).

Conference of the Parties to the Convention on Biological Biodiversity, 1997, Report of the Third Meeting of the Conference of the Parties to the Convention on Biological Diversity, Buenos Aires, Argentina, November 1996, UNEP/CBD/COP/3/38 (February, 1997).

Cooper, J. C., 1998, The economics of public investment in agro-biodiversity conservation, in: *Agricultural Values of Genetic Resources*, R. Evenson, D. Gollin, and V. Santaniello, eds., CABI International, Wallingford, U. K.

FAO, 1996, State of the World's Plant Genetic Resources for Food and Agriculture, background documentation prepared for the International Technical Conference on Plant Genetic Resources, Leipzig, Germany, FAO, Rome (June 17-23, 1996).

Fisher, A., and Hanemann, M., 1986, Option value and the extinction of species, *Advances in Applied Micro-Economics* **4**:169-190.

Gollin, D., and Evenson, R., 1998, Breeding values of rice genetic resources, in: *Agricultural Values of Genetic Resources*, R. Evenson, D. Gollin, and V. Santaniello, eds., CABI International, Wallingford, UK.

Henry, C., 1974, Investment decisions under uncertainty: The irreversibility effect, *American Economic Review* **64**:1006-1012.

International Seed Federation, undated, Seed statistics, http://www.worldseed.org/~assinsel/stat.htm.

Organization for Economic Cooperation and Development (OECD), 2001, *Key Environmental Indicators*, OECD, Paris.

Organization for Economic Cooperation and Development (OECD), 2000, Background Information on Future OECD Work on Agri-Environmental Indicators, COM/AGR/CA/ENV/RD(2000)131, OECD, Paris (March 27, 2000).

Reid, W., McNeely, J., Tunstall, D., Bryant, D., and Winograd, M., 1993, *Biodiversity Indicators for Policy-makers*, World Resources Institute, Washington, D. C.

Simpson, R., Sedjo, R., and Reid, J., 1996, Valuing biodiversity for use in pharmaceutical research, *J. Polit. Econ.* **104**(February):163-185.

UPOV, 1995, Plant Variety Protection Statistics for the Period 1990-1994, document number C/27/7, UPOV, Geneva (October 1995).

Wright, B., 1998, Intellectual property and farmers' rights, in: *Agricultural Values of Genetic Resources*, R. Evenson, D. Gollin, and V. Santaniello, eds., CABI International, Wallingford, U. K.

Chapter 10

ECONOMIC CRITERIA FOR THE MULTILATERAL DISTRIBUTION OF AGRICULTURAL BIODIVERSITY CONSERVATION FUNDS*

Joseph C. Cooper
Deputy Director, Resource Economics Division, Economic Research Service (United States Department of Agriculture), 1800 M Street, NW, Washington DC 20036-5831

Abstract: Enough concern has developed internationally over the need to conserve agricultural genetic resources that a multilateral system of access and benefit sharing with regards to key crops has been proposed by the United Nations, including suggestions for funding conservation of agricultural genetic resources in developing countries. This chapter addresses the question of what is the most economically efficient method of distributing the funds among countries and world regions, given the available data. An overview of the economics of investment in agricultural genetic resources conservation along with a theoretical dynamic model is presented. Next, a proxy indicator for regional allocation of conservation funds is developed. Finally, mechanisms for *in situ* conservation are proposed.

Key words: agricultural production; agriculture; biodiversity; centers of diversity; current value Hamiltonian; *in situ* conservation; International Treaty on Plant Genetic Resources for Food and Agriculture; plant genetic resources for food and agriculture.

1. INTRODUCTION

Chapter 9 focused on indicators for funding conservation activities. This chapter discusses the economics of the distribution of funds from the International Treaty on Plant Genetic Resources for Food and Agriculture—hereafter denoted by IT—discussed earlier in the book. While economic benefits accruing from the distribution of these funds for conservation

* The views contained herein are those of the author and do not necessarily represent policies or views of the Economic Research Service or United States Department of Agriculture.

activities are almost impossible to ascertain, political pressures motivate this distribution to be made. The question then becomes what is the most economically efficient method of distributing the funds among countries or among world regions, given the available data. The rest of this chapter is organized as follows. Section 2 is an overview of the economics of investment in Plant Genetic Resources for Food and Agriculture (PGRFA) conservation. In Section 3, a proxy indicator for regional allocation of conservation funds is developed. Section 4 addresses mechanisms for *in situ* conservation. The last section is a discussion and conclusion.

2. PRINCIPLES OF VALUATION OF THE CONSERVATION INVESTMENT

A reduction in biodiversity matters for two reasons (Chapter 5): (1) the vulnerability of a crop and its current productivity; and (2), the potential for farmers to cease cultivating a variety with untapped, potential use value. Any variety, whether bred with conventional methods or with techniques of genetic transformation, will be widely grown by farmers if they view it as superior to those they currently cultivate. With respect to the first concern, if that variety, or a set of varieties, is uniform with respect to certain genes conferring biotic resistance, then cultivation of these varieties over a widespread area increases the probability of a mutation in the disease pathogen that overcomes the source of genetic resistance. Once that occurs, widespread cultivation of varieties with that same source of genetic resistance also contributes to more rapid spread of infection. In other words, uniformity with respect to resistance genes can make a crop more vulnerable to economically meaningful crop losses—but this would be true for conventional and genetically modified seed, as well as for traditional varieties. Though traditional varieties or landraces are typically composed of more heterogeneous populations or mixtures, historically there are important cases of epidemics in landraces such as in the Indian subcontinent for the rusts of wheat.[1] In fact, these epidemics were part of the motivation for early scientific plant breeding programs. Furthermore, diversity in genetic backgrounds and other resistance mechanisms (not confined a single gene) are often very important in explaining different disease reactions among varieties.

---

[1] Otherwise known as farmers' or traditional varieties, landraces are the product of farmers' breeding and selection carried out over many generations. They tend to be more heterogeneous populations that are selected for traits conferring local agronomic adaptation and value rather than commercial value (Melinda Smale, undated).

Depending on how modern breeding techniques are used, they can maintain or decrease genetic diversity. These techniques may contribute to maintaining diversity in systems dependent on traditional varieties if they enable the insertion of traits to overcome specific disadvantages into landraces that developing country farmers value for their consumption traits or agronomic performance. If so, the relative economic value of these landraces to local farmers may be enhanced (Smale et al., 1999). On the other hand, if a company with breeding skills purchases a seed company and inserts the desired traits only in a subset of varieties that become dominant, then biotechnology may reduce the diversity of genetic materials used in plants (Yarkin, 1998).[2]

The economic value of PGRFA stems from its value as an input to the agricultural production process. Genetic erosion is the loss of genetic diversity, including the loss of individual genes, and the loss of particular combinations of genes such as those manifested as locally adapted landraces. The term can be used in a narrow sense, i.e., the loss of genes or alleles, and in the wider sense, including the loss of varieties (for example, see FAO, 1998, Annex 1.1; Reid et al., 1993). The main cause of genetic erosion in crops is the replacement of local varieties by improved or exotic varieties and species (FAO, 1998). Erosion can occur when a smaller number of varieties replace a larger number of older varieties and/or the newer variety has a different gene base from the old one. This chapter defines genetic erosion as the loss of genetically distinct varieties.

Chapter 9 provides the general economic background behind the economic valuation of the genetic resources for agriculture. The appendix to this chapter presents a more detailed, and hence, more complex, analytical model of the economics of the conservation of PGRFA. In the model in the appendix, an increase in accessions to a gene bank is a function of conservation investment, and the change in agricultural supply is a function of the change in accessions. According to the model, a reasonable criterion for choosing the optimal level of conservation investment is to choose the level of investment that maximizes producer plus consumer surplus in a dynamic context.

The comparative statics of the model show that the marginal value of an additional dollar of conservation is a function of the marginal change in varieties conserved per additional dollar of investment as well as being a direct function of the change in welfare due to a change in accessions. The economically efficient allocation of conservation funds among regions would be the one in which the marginal value of conservation investments is

---

[2] In production systems with both modern and traditional varieties, or ones that are still dominated by traditional varieties, it is not necessarily the case that traditional varieties will be entirely replaced. There are many empirical examples where both coexist because it suits the interests of the farmers who grow them (Brush, 1995).

equated across regions. The marginal value of an additional dollar of conservation investment is not only a direct function of the marginal change in welfare due to a change in accessions, but it is also a function of the marginal change in cultivars conserved per additional dollar of investment. The latter is certainly higher in areas of high diversity than in areas of low diversity. Hence, the analysis in the appendix can be seen as making a case for concentrating limited conservation resources in areas of high diversity, especially when little is known about the quality of the diversity with regards to agricultural uses. Note that it is likely easier in principle to generate data on the marginal change in the number of cultivars conserved per additional dollar of investment than on the marginal change in welfare due to the change in the number of these accessions.

3. PROXY INDICATORS OF THE VALUE OF DIVERSITY

This section describes a proxy indicator for the importance of a region as a primary center of diversity. With the exception of a perhaps a case study or two, figures on crop production by variety are unavailable. The probability of variety loss is unknown, as is the marginal economic value of adding (subtracting) an additional variety (to) the set of varieties that make up a species. In practice, the best one can do is to estimate a third-best proxy for the optimal allocation of the conservation funds. On a global scale, all we know is roughly where areas of high diversity are located for various species, and the best we can do is to rank regions by a proxy for the importance of their agri-biodiversity to the global agricultural economy. Doing so can provide a rough index for targeting conservation funds.

For the purpose of this chapter, and given the state of the available data, there is no reason to strongly promote any particular measure of diversity. For policymaking purposes, it is useful to divide the world into major regions of diversity, hoping to use some expert advice to assign as objectively as possible an area's relative importance to the world's major crops. Whether one follows the approaches of Vavilov, Harlan, or Zeven is not a point for this economics paper to make. The basic goal of this discussion in this section on centers of diversity is that there appears to be enough research done on the subject of centers of diversity to make a plausible case that some botanical expert(s) could make diversity assignments to the world's major regions.

Our indicator is based on the principle that plant genetic variability is not uniformly distributed throughout the world. In the 1920s, the Russian geneticist Vavilov (1926, 1949/50) noted that the level of inter- and intra-specific genetic variability varies geographically across the world. He identified the geographic areas with the highest genetic variability in cultivated food crops.

Vavilov thought that areas of maximum genetic diversity represented centers of origin and the origin of a crop could be identified by the simple process of analyzing variation patterns and plotting regions where diversity was concentrated. Although his proposed centers of origin are widely accepted even today (Harlan, 1992), his notion is somewhat simplistic and it turns out that even though many crops do exhibit centers of diversity, these centers of diversity have little to do with centers of origin (Harlan, 1992; Smith, 1995). Domesticates can, and did, originate in one region and then develop much of their diversity in another. However, while this distinction is of importance for anthropological and other reasons, it is of little practical import to the purposes of this chapter. For conservation of landraces—as opposed to wild relatives—we assume that centers of diversity are more important than centers of origin. However, centers of origin might be of importance in wild species conservation, at least in the case of species to which the concept is relevant.[3] Conservation of wild varieties has a separate set of policy mechanisms from that for domesticated varieties that, by definition, require the intervention of man. Mechanisms for wild species conservation include the set-asides, nature preserves, etc., discussed in the usual context of wild species conservation.

The term "region of diversity" is currently used to refer to the variability generated by crops during their dispersal from point of origin. Thus, a plant population can be described at any point in its evolution by frequency of genes and genotypes, which illustrates its historical evolution. The regions of crop diversity are areas with high variability in number of alleles and genotypes. The genetic composition of these populations represents varying adjustments to ecological and social imperatives (Palacios). Given that these data on regions of diversity are the only relevant genetic data we have on a world scale, the best we can do is develop an indicator for the world (or OECD countries) value of agricultural production ascribed to primary regions, or centers, of diversity.

The first columns of Table 10-1 present geographic distribution of centers of diversity (FAO, 1998) derived from Zeven and de Wet on the basis of the centers identified by Vavilov. The fourth column presents the price per metric ton (in international dollars) for each commodity, and the fifth and sixth columns present the total world (and OECD) production data for each crop. The rational for the OECD figures is that if the OECD

---

[3] Just because some regions use domesticates first cultivated elsewhere does not mean that it is low on wild ancestors' diversity. It just may mean that in that region it was more convenient to adopt varieties domesticated elsewhere than to go through the great efforts needed to breed local wild ancestors to the point where they are useful domesticates (Diamond, 1999).

countries are paying for the lion's share of this conservation effort, they may be most interested in focusing on the crops of interest to them. Given the price and quantity data, the total value of each crop is calculated. This value is then assigned to the geographic region corresponding to the center of diversity for each crop, with the total value ascribed to each center of diversity being the sum of the values of the crops for which the region is a primary center. This process produces the last column to the right in Tables 10-2 and 10-3. However, because a crop may have more than one geographic center of diversity, this measure has only an ordinal interpretation. The alternative is to normalize the measure to sum to one by dividing each instance of a primary center for a crop by the number of primary centers for that crop (i.e., for a given crop, the weight assigned to a primary center falls as the number of primary centers increases). Based on the assumption that each center provides an equal contribution as a center of diversity (which is all that can be done given the lack of data), imposing this normalization provides the measures in the last column in Tables 10-2 and 10-3 with a cardinal interpretation, and the normalized indicator, when expressed as a fraction of total agricultural value, sums to one.[4]

Table 10-4 ranks the regions in descending order according to their value as primary centers of diversity based on the values in Tables 10-2 and 10-3. Comparing the normalized indicator for the OECD with that for the world, we see that for the OECD, Central Asia and then West Asia rank highest as primary centers, while for the world, Southeast Asia and then South Asia are the highest. The result is due to Central and West Asia being primary centers for wheat, which is of greater importance to OECD countries than rice, while Southeast and South Asia are primary centers for rice. Not surpris-ingly too, the Mediterranean and European centers rank higher in the OECD value than they do in the overall world value. The low rankings associated with Central Africa, North America, and the Caribbean are the same for both the OECD and the world as whole. These results suggests that, with no other information being available on the general state of PRGFA, the OECD might be inclined to allocate a high share of funds to conservation efforts in higher ranked regions such as Central Asia, but if their view was more magnani-mous, they may give Southeast Asia higher consideration. In other words, if the OECD is paying for the conservation activities, they may allocate funds differently than world ranking would dictate. However, if their goals include global food security, they may well want to follow world values. Central America ranks highest in the normalized ranking primarily because it does not have to share its title as center of diversity for maize with any other region.

---

[4] The normalization process could be more precise for the case of particular crops such as wheat and rice since certain regions are more important and data for these crops are likely to be more available than for the aggregate level I examine here.

Table 10-1. Crops, their primary centers of diversity, and international price and production

Crop	Primary centers of diversity	$/Mt[a]	World production	OECD production[b]
Rice	E./S.E./S. Asia/W. Africa	190	577,349,526	31,298,772
Wheat	W. & C. Asia	144	591,632,321	262,804,453
Sugar:				
- Cane	S.E. & S. Asia/Pacific	34	1,253,253,810	121,072,256
- Beet	Mediterranean/Europe	17	260,886,055	192,686,066
Maize	C. America	124	614,003,156	321,448,834
Soybeans	E. Asia	234	159,822,505	79,520,120
Potatoes	S. America	110	293,377,361	111,438,719
Cassava	S. America (Brazil-Paraguay) & C. America	68	162,072,352	11,096
Sorghum	Africa	124	61,044,434	21,358,847
Millet	Africa (excl. C. Africa)/ S.E./S./E. Asia	158	28,606,826	260,298
Barley	W. & C. Asia/ Mediterranean	114	138,454,668	96,457,199
Sweet potatoes	S./C. America	85	138,914,332	2,143,844
Oil palm	W. Africa	165	96,894,224	35,000
Rape/mustard	Mediterranean/Europe/E. Africa	328	36,423,345	21,907,635
Beans		445		
Phaseolus	S. & C. America	445	21,196,276	5,151,141
Vicia	C. Asia	445	5,581,953	1,294,213
Groundnut	S. America	491	33,750,854	2,098,819
Banana	S.E. & S. Asia/Indian Ocean	154	55,988,655	2,233,423
Plantain	S.E. & S. Asia/Indian Ocean	96	29,969,764	0
Cotton	S. & E. Africa/C. Asia/S. & C. America	175	51,566,253	13,726,239
Coconuts/copra	Pacific/S.E. Asia	106	47,398,344	1,302,500
Yams	S.E. & S. Asia/Africa	85		
Oranges	E. Asia	202	35,753,627	181,000
Grapes	Mediterranean/W. & C. Asia	n.a.[c]	63,140,901	22,582,779
Apples	Europe/C. Asia	n.a.		
Sesame	S. & C. Asia/E. Africa	626	2,515,592	85,500
Olives	Mediterranean	n.a.		
Oats	Mediterranean/Europe	n.a.		
Rye	W. Asia	n.a.		
Tomato	S. America	n.a.		
Cocoa	S. America	663	2,914,830	43,968
Sunflower	N. America	292	24,833,755	7,703,127
Date	Mediterranean/W. Africa	n.a.		
Grapefruit	S.E. Asia	n.a.		
Pea	W. Asia/E. Africa	n.a.		
Onion	C. Asia	n.a.		
Paprika	Caribbean	n.a.		
Pineapple	S. America	n.a.		

Table 10-1. (continued).

[a] 1989-1991 average of International Commodity prices (most recent available): These prices, derived from the Geary-Khamis equations for the agricultural sector, have been introduced to avoid the use of exchange rates. The "prices" are expressed in so-called "international dollars." This method assigns a single "price" to each commodity (e.g., 1 ton of wheat has the same price in whatever country it is produced).

[b] Production figures are in metric tons for 1998.

[c] Data not available.

Table 10-2. Total number of primary centers per world region and world value of agricultural production ascribed to primary centers of diversity (expressed as a fraction of total world value of the agricultural commodities)[a]

Region	Number of primary centers[a]	Fraction of world value ascribed to primary center[b] (non-normalized)[c]	Fraction of world value ascribed to primary center[b] (normalized)[c]
Central Africa	2	0.019	0.004
East Africa	6	0.069	0.016
Southern Africa	4	0.044	0.008
West Africa	5	0.257	0.084
Indian Ocean	5	0.049	0.012
South Asia	7	0.316	0.086
Southeast Asia	7	0.322	0.090
East Asia	4	0.301	0.143
Pacific	2	0.087	0.031
Mediterranean	3	0.059	0.021
West Asia	2	0.185	0.087
Central Asia	5	0.209	0.096
Europe	2	0.030	0.011
South America	7	0.168	0.126
Central America	5	0.215	0.172
Caribbean	0	0.000	0.000
North America	1	0.013	0.013
			Total: 1.000

[a] Total commodity value is $546,967,706,426 (1989-1991 international dollars, 1998 production).

[b] Only crops for which international price data are available are considered. Hence, some relatively minor crops, such as paprika, are excluded.

[c] Normalized to sum to one by dividing each instance of a primary center for a crop by the number of primary centers for that crop (i.e., for a given crop, the weight assigned to a primary center falls as the number of primary centers increases). Imposing this normalization provides the measure with a cardinal interpretation.

Table 10-3. OECD value of agricultural production ascribed to primary centers of diversity (expressed as a fraction of total OECD value of agriculture at the farm gate)[a]

Region	Fraction of OECD value (non-normalized)[a]	Fraction of OECD value (normalized)[b, c]
Central Africa	0.017	0.003
East Africa	0.079	0.022
Southern Africa	0.033	0.006
West Africa	0.055	0.013
Indian Ocean	0.019	0.004
South Asia	0.067	0.019
Southeast Asia	0.068	0.019
East Asia	0.186	0.157
Pacific	0.027	0.009
Mediterranean	0.137	0.049
West Asia	0.312	0.144
Central Asia	0.331	0.151
Europe	0.067	0.026
South America	0.116	0.096
Central America	0.286	0.265
Caribbean	0.000	0.000
North America	0.014	0.014
		Total: 1.000

[a] Total OECD commodity value is $156,660,455,522 (1989-1991 international dollars, 1998 production).

[b] Only crops for which international price data are available are considered. Hence, some relatively minor crops, such as paprika, are excluded.

[c] Normalized to sum to one by dividing each instance of a primary center for a crop by the number of primary centers for that crop (i.e., for a given crop, the weight assigned to a primary center falls as the number of primary centers increases). While the non-normalized measure does not impose this assumption, doing so provides a cardinal measure).

4. CONSERVATION MECHANISMS

In situ conservation may be achieved through several possible mechanisms that are discussed in the chapters in the first section of this book, as well as in the final two chapters. Hence, possible mechanisms are discussed only briefly here. One possibility is to directly pay farmers to undertake conservation activities (perhaps through mechanisms like the Conservation Reserve Program in the United States), but doing so will probably not work well, particularly in developing countries. Indirect methods appear more feasible. For example, programs that encourage farmers to adopt or continue

Table10-4. Regions ranked in descending order according to their value as primary centers of diversity

	Region: OECD value		Region: World value	
Rank	(Non-normalized)	(Normalized)	(Non-normalized)	(Normalized)
1	Central Asia	Central America	Southeast Asia	Central America
2	West Asia	East Asia	South Asia	East Asia
3	Central America	Central Asia	East Asia	South America
4	East Asia	West Asia	West Africa	Central Asia
5	Mediterranean	South America	Central America	Southeast Asia
6	South America	Mediterranean	Central Asia	West Asia
7	East Africa	Europe	West Asia	South Asia
8	Southeast Asia	East Africa	South America	West Africa
9	South Asia	Southeast Asia	Pacific	Pacific
10	Europe	South Asia	East Africa	Mediterranean
11	West Africa	North America	Mediterranean	East Africa
12	Southern Africa	West Africa	Indian Ocean	North America
13	Pacific	Pacific	Southern Africa	Indian Ocean
14	Indian Ocean	Southern Africa	Europe	Europe
15	Central Africa	Indian Ocean	Central Africa	Southern Africa
16	North America	Central Africa	North America	Central Africa
17	Caribbean	Caribbean	Caribbean	Caribbean

sustainable farming practices may increase the competitiveness of landraces versus modern varieties. Decentralized breeding programs, niche market development, trademark or other schemes for tying quality or certain traits to source are some of the policy options. The diversity of a production system dominated by modern varieties can be enhanced through policies designed to encourage the release of more genetically diverse materials at a higher rate, where higher rates of release of modern varieties contribute to "diversity in time" (Duvick, 1984). Other policies might enhance the spatial diversity in modern systems by favoring greater numbers of more evenly distributed, genetically different varieties (Smale et al., 1999; Heisey et al., 1997).

Optimal investment strategy would require that precedence in the distribution of the funds be given to countries that do the best job at cataloging their varieties with information of use to breeders, given that the value of germplasm increases with the information associated with it. Of course, for this approach to be undertaken, less-developed countries (LDCs) would require technological assistance with database development, etc.[5] In terms of mechanisms for distribution, the fund can support specific *in situ* conservation projects that are chosen by competition from an open call for conservation proposals.

5. CONCLUSION

As discussed in section 2, the economically efficient allocation of conservation funds across regions is, not surprisingly, the one that equates the marginal value of an additional dollar of investment across the regions. When using centers of diversity to target conservation dollars, one question is whether the marginal value of an additional dollar of conservation investment is higher in an area with high diversity than in an area with low diversity. Given the principle of diminishing marginal returns, it is possible that the marginal economic value of an additional variety conserved is higher in an area with low diversity. However, the marginal value of an additional dollar of conservation investment is also a function of the marginal change in varieties conserved per additional dollar of investment. The latter is likely higher in areas of high diversity than in areas of low diversity.[6] Hence, the case is made for concentrating limited conservation resources in areas of high diversity.

From the idealized economic model presented in section 2, section 3 steps down to assessments that can be made using available data. The basic genetic diversity data we have on a global scale are the association of geographic regions with primary centers of diversity for certain crops. This information allows us to generate a basic guide to the value of agricultural production associated with these centers of diversity. While this information does not incorporate the risks associated with genetic erosion, and the concept of primary centers may lead to imprecise measures, it does give us a measure of the economic importance of crops associated with these regions as centers of diversity. As such, it may be useful as a rough guide for the IT in making allocations of the conservation funds to world regions.

The current data on centers of diversity do not identify areas at the country level or lower. Hence, other indicators are needed to make the allocation from world region levels to country levels. Given the lack of

---

[5] It may be possible, at least up to a certain level that evaluation of germplasm can be done at lower costs in LDCs, with potential benefits for DC breeders. This capacity for database development can be targeted through the conservation mechanism.

[6] This statement is somewhat of a generalization, as costs of conservation can vary across regions according to differences in economic conditions that affect conservation costs (as related in Chapter 6 and in the appendix to this chapter).

scientific information, the IT will mostly likely use the subsidiarity principle in allocating the funds, i.e., they will allocate the funds to the world regions, and each region will decide how to allocate the funds to individual countries.

Instead of directly transferring funds down to the country level, given that there are no indicators and mechanisms available for transferring the funds in anything near a first-best fashion, perhaps the most efficient approach would be to ask countries to submit concrete proposals for use of the funds in conservation activities. These proposals could then be prioritized for funding based upon their merits. Of course, such a mechanism could be used directly by the IT, and it could evaluate conservation project proposals directly. However, regions may argue over their shares of the conservation funds, and some regional indicator, such as the one developed in this chapter, will be necessary to the IT in making some judgments in the parceling out of the funds among regions.

The indicators of centers of diversity discussed in the third section of this chapter are broad enough that they can only serve as a rough guide for conservation targeting. Smaller areas, known as "microcenters," might be especially useful to concentrate limited conservation dollars on. Harlan (1992) defines microcenters, for either crops or wild plants, as relatively small regions, 100 to 500 kilometers across, which may be packed with high variation of one to several crops in comparison to adjacent areas. Some of these have been destroyed, and the rest are threatened with replacement with modern cultivars (*ibid*). A global indicator based on microcenters is not feasible until they are studied on a more systematic basis. However, the analytical framework developed here could easily be desegregated by micro-centers if such data became available. Chapter 13 provides more concrete suggestions for mechanisms for distributing funds at the country level, as well as for policy mechanisms for conservation programs.

The final allocation of the conservation funds will be affected to some extent by the relative bargaining power of actors, both at the regional level and at the country level. Furthermore, as implied in The Global Plan of Action's call for the sharing of benefits to be "fair and equitable," there may be significant demand to distribute at least a portion of the funds on equity grounds to countries most in need of development aid. The next chapter addresses in detail the qualitative and policy-relevant aspects of bargaining power and equity in the allocation of these funds.

APPENDIX
Analytical model of the economics of the conservation of PGRFA

This appendix presents an analytical model of the economics of the conservation of PGRFA. The goal of this model is to link together the economic concepts underlying the conservation of agricultural genetic resources, and to demonstrate the data needs for conducting a first-best economic assessment of the allocation of conservation funds. Let A_t be the number of the distinct cultivars collected (accessions) in period t. It is drawn from the number of uncollected cultivars in time 0 that are still extant in time t (i.e., those that have not been lost before collection), denoted as A_t^u, and put into *ex situ* collections in period t.[7] In addition, A_t includes cultivars not existing at period 0 that may be developed by cultivar selection activities of indigenous farmers in the program target area. The accessions are functions of the following variables:

$$A_t = A[I_t, S_t, A_t^u(I_t, S_t, A_{t-1}^u), A_0, A_1, \ldots A_{t-1}], \tag{1}$$

where I_t = investment by the United Nations fund in agri-biodiversity conservation, enhancement, and collection activities in program target areas, and the vector of socioeconomic and other factors in the program area that are causing the number of existing uncollected cultivars to decrease over time is denoted as S_t. The expected signs are $\partial A_t / \partial I_t > 0, \partial A_t^u / \partial I_t > 0$, $\partial A_t / \partial S_t < 0$, and $\partial A_t^u / \partial S_t < 0$, and it is assumed that $\partial A_t / \partial t = 0$. To limit biases associated with aggregation, assume that A_t^u and A_t represent varieties within a species category, although other grouping based on physical or regional characteristics can be formed. One would expect accessions to be stochastic variables.[8]

---

[7] While the goals of CGRFA may be more than just limited to cultivars that are uncollected—*in situ* conservation can have other values, such as dynamic evolution and response to changes in pest and other environmental conditions (Brush, Taylor, and Bellon, 1992)—we limit the scope of the model for the sake of tractability.

[8] Given annual accessions A_t, the total size of the *ex situ* collection at period t is denoted as TA_t and is simply

$$TA_t = \sum_{k=0}^{t} A_i + \bar{A}_0 - \sum_{k=0}^{t} L_k .$$

where $\bar{A}_0$ is the carryover stock at the beginning of period 0 and L_t is the loss of prior levels of TA_t due to spoilage and other factors each period, and where "*ex situ* collection"

Crop breeding, and in return crop supply, is a function of the number of cultivars available for research. The general form for a supply response function for agricultural crop output with respect to gene bank size can be written as:

$$X_t^s = f(\boldsymbol{P}_t, \boldsymbol{Z}_t, \boldsymbol{h}_t, \tau_t[A_{t-s}(I_{t-s}), A_{t-s-1}(I_{t-s-1})..., A_0(I_0), \boldsymbol{h}_t], \boldsymbol{U}_t) \text{ for } t = s,...,T, \quad (2)$$

where X_t^s is output given the vector of expected prices $\boldsymbol{P}_t$, vector of inputs $\boldsymbol{Z}_t$, vector of environmental and other uncontrolled effects $\boldsymbol{U}_t$, vector of state of technology (excluding breeding research technology) $\boldsymbol{h}_t$, state of crop breeding research technology τ_t, and s, the lag before accessions have impact. Note that $\boldsymbol{h}_t$ appears outside the brackets as well as inside as much research on PGRFA is done using only the private collections of the breeding firm. For τ_t, $\partial\tau_t/\partial A_{t-s-i} \geq 0$, and $\partial X_t/\partial t_t \geq 0$, $i = 0, ..., t - s$. Assuming that crop breeding research technology exhibits diminishing marginal returns with respect to A, $\partial^2\tau_t/\partial^2 A_{t-s-i} \leq 0$. At the farm level, crop breeding research related technical change can be modeled using the common specification (Alston, Norton, and Pardey, 1995) for the profit function for perfectly competitive farm unit j as $\pi_j = g[\boldsymbol{P}_t(\tau_t), \boldsymbol{Z}_{tj}(\tau_t), \boldsymbol{h}_{tj}, \boldsymbol{U}_{tj}]$. For this specification, τ_t can represent either output or input augmenting technical change. Crop breeding research can be either.

In the literature on measuring gains to agricultural research, a common practice is to assume that agricultural research induces shifts in the supply function, which then translates into changes in producer plus consumer surplus (de Gorter, Nielson, and Rausser, 1992; Alston, Norton, and Pardey, 1995). While the policymaker may have objectives other than maximizing producer plus consumer surplus, and while no measure is completely objective, this criterion has the benefit of being reasonably general and objective. Fisher and Hanemann (1990) were the first to apply this criterion in the context of agri-biodiversity. In a two-period model for measuring the option value of saving an identified native landrace, they model the benefits of saving a native corn landrace by assuming this a priori identified landrace impacts the corn supply curve through the intercept. A reasonable criterion for choosing the optimal level of conservation investment is to choose the level of investment that maximizes producer plus consumer surplus in a dynamic context, in which the increase in accessions is a function of conservation investment, and the change in agricultural supply is a function

refers to the aggregate of all generally accessible *ex situ* collections. In practice, L_t can be treated as some fraction of particular age cohorts, but for simplicity's sake, its form is left unspecified in this analysis. As the basic analysis does not change if we consider TA instead of A, for the sake of brevity, we stick to using A in the text.

of the change in accessions, while allowing for both biophysical and economic uncertainty.

Suppose in this formulation that due to the new biodiversity conservation investment I'', the supply curve either shifts to the right or rotates downward, such that γ_0 decreases to γ_1, i.e., $\gamma_1 = f[\tau_t(A_t | I_t = I'')]$ and $\gamma_0 = f[\tau_t(A_t | I_t = I')]$, where γ_1 and γ_0 are either the intercepts resulting from the summation (whether vertical or horizontal) of the supply functions of each farm unit maximizing π_j, or are the new and old slopes of the supply function, respectively, and where I' is the base level and $I'' > I'$, and γ_1 and $\gamma_0 > 0$. The change in welfare in period t when γ_0 decreases to γ_1 is denoted as ΔW_t. The net present value is

$$NW = \sum_{t=0}^{T-s} \frac{\Delta W_{s|t}(\gamma_{s|t.1}(A_t | I_t = I''; I_0, \ldots, I_{t-1}); \gamma_{s+t.0}(A_t | I_t = I'; I_0, \ldots, I_{t-1}))}{(1+r)^{s+t}} - \sum_{t=0}^{T-s} \frac{I_t}{(1+r)^t}, \tag{3}$$

where r is the discount rate for the project. The summation over ΔW_{s+t} in Eq. (3) is the value associated with keeping alive the option of being able to use existing uncollected PGRFA that would be lost without the investment plus the value of new PGRFA that are developed due to the investment.

Eq. (3) can be maximized in a discrete time current value Hamiltonian framework that accounts for the uncertainty and irreversibility effects. Given 1 ..., J *in situ* conservation regions, maximizing the equation with respect to I_{jt} subject to a budget constraint, $\sum_{t=1}^{T} \left(\sum_{j=1}^{J} I_{jt} \right) / (1+r)^t \leq 1$,

produces the first-order condition for the marginal value of incremental increase in conservation expenditures

$$\frac{\partial NW}{\partial I_{jt}^*} = \sum_{k=0}^{T-s-t} \frac{1}{(1+r)^{t+s+k}} \frac{\partial \Delta W_{t+s+k}}{\partial A_{jt}} \left(\frac{\partial A_{jt}}{\partial I_{jt}^*} + \frac{\partial A_{jt}}{\partial A_{jt}^u} \frac{\partial A_{jt}^u}{\partial I_{jt}^*} \right) - \frac{\lambda_I}{(1+r)^t} = \frac{1}{(1+r)^t}, \tag{4}$$

for $t = 0,...,T - s$, where λ_I is the shadow price giving the value of increasing I by a small increment.[9] Eq. (4) simply says that at the optimum, the present value of the marginal increase in welfare of expenditure I_{jt} minus the shadow price of I must equal the present value of an additional dollar of investment. As the equation demonstrates, the value of an incremental increase in I_{jt} is sum of the returns from periods $t + s$ to T.[10] According to Eq. (4), the marginal value of an additional dollar of conservation is a function of the marginal change in varieties conserved per additional dollar of investment as well as being a direct function of the change in welfare due to a change in accessions.

If we explicitly acknowledge the stochastic aspects of the model, using expectations, Eq. (4) becomes

$$\frac{\partial NW}{\partial I^*_{jt}} = \sum_{k=0}^{T-s-t} \frac{1}{(1+r)^{t+s+k}} \left[E_0\left\{ \frac{\partial \Delta W_{t+s+k}}{\partial A_{jt}} \frac{\partial A_{jt}}{\partial I^*_{jt}} \right\} + E_0\left\{ \frac{\partial \Delta W_{t+s+k}}{\partial A_{jt}} \frac{\partial A_{jt}}{\partial A^u_{jt}} \frac{\partial A^u_{jt}}{\partial I^*_{jt}} \right\} \right] - \frac{\lambda_I}{(1+r)^t} = \frac{1}{(1+r)^t} \tag{4'}$$

which, noting the relation cov(A,B) = E{A*B} - E{A}*E{B}, and E{A*B} = cov(A,B) + E{A}*E{B}, can be expanded to

$$\begin{aligned}\frac{\partial NW}{\partial I^*_{jt}} = \sum_{k=0}^{T-s-t} \frac{1}{(1+r)^{t+s+k}} \Bigg[& E_0\left\{ \frac{\partial \Delta W_{t+s+k}}{\partial A_{jt}} \right\} E_0\left\{ \frac{\partial A_{jt}}{\partial I^*_{jt}} \right\} \\ & + E_0\left\{ \frac{\partial \Delta W_{t+s+k}}{\partial A_{jt}} \right\} E_0\left\{ \frac{\partial A_{jt}}{\partial A^u_{jt}} \right\} E_0\left\{ \frac{\partial A^u_{jt}}{\partial I^*_{jt}} \right\} + \mathrm{cov}\left\{ \frac{\partial \Delta W_{t+s+k}}{\partial A_{jt}}, \frac{\partial A_{jt}}{\partial I^*_{jt}} \right\} \\ & + \mathrm{cov}\left\{ \frac{\partial \Delta W_{t+s+k}}{\partial A_{jt}}, \frac{\partial A_{jt}}{\partial A^u_{jt}} * \frac{\partial A^u_{jt}}{\partial I^*_{jt}} \right\} + E_0\left\{ \frac{\partial \Delta W_{t+s+k}}{\partial A_{jt}} \right\}\end{aligned} \tag{4''}$$

---

[9] The increment is $\frac{\partial \Delta W}{\partial A} \approx \frac{\Delta(\Delta W)}{\Delta A} = \frac{\Delta W(A_0 + \Delta A_0) - \Delta W(A_0)}{\Delta A}$ where $\Delta A = A_1 - A_0$.

[10] For simplicity, the impact of A on ΔW is modeled as a continuous process. However, in reality, given that only a relatively small fraction of accessions will ever be used to develop new useful varieties, it is likely that the impact of accessions on ΔW is a process that makes infrequent but discrete jumps.

$$+\mathrm{cov}\left\{\frac{\partial A_{jt}}{\partial A_{jt}^{u}},\frac{\partial A_{jt}^{u}}{\partial I_{jt}^{*}}\right\}\Bigg]\Bigg]-\frac{\lambda_I}{(1+r)^t} \quad = \quad \frac{1}{(1+r)^t},$$

where E_0 denotes the expectation operator over all the stochastic or inherently unknown variables in the model and ΔW in Eq. (4) is replaced with $E_0\{\Delta W\}$. If these covariance terms are positive, then the change in net welfare is higher with the stochastic model (4") than the deterministic model in (4').

The economically efficient allocation of conservation funds among regions would be the one in which the marginal value of conservation investments is equated across regions. Given Eq. (3), at the optimum, resources should be allocated between *in situ* conservation areas such that

$$\frac{\partial NW}{\partial I_{1t}^{*}}=\frac{\partial NW}{\partial I_{2t}^{*}}=\ldots=\frac{\partial NW}{\partial I_{Jt}^{*}}. \tag{5}$$

Eqs. (4) through (4") demonstrate that the marginal value of an additional dollar of conservation investment is not only a direct function of the marginal change in welfare due to a change in accessions, but it is also a function of the marginal change in cultivars conserved per additional dollar of investment. The latter is certainly higher in areas of high diversity than in areas of low diversity, *ceteris paribus*. Hence, the analytical analysis can be seen as making a case for concentrating limited conservation resources in areas of high diversity, especially when little is known about the quality of the diversity with regards to agricultural uses. However, if $\partial A_{jt}/\partial I_{jt}$ can also vary by conservation region according to differences in the farmers' opportunity cost of adoption of modern varieties (i.e., the argument in Chapter 5), generalizations become more difficult to make. Note that it is likely easier in principle to generate data on the marginal change in the number of cultivars conserved per additional dollar of investment than on the marginal change in welfare due to the change in the number of these accessions.

REFERENCES

Alston, J., Norton, G., and Pardey, P., 1995, *Science Under Scarcity: Principles and Practice for Agricultural Research Evaluation and Priority Setting*, Cornell University Press, Ithaca and New York.

Brush, S. B., 1995, In situ conservation of landraces in centers of crop diversity, *Crop Science* **35** (2):346-354.

Brush, S. B., Taylor, J. E., and Bellon, M. R., 1992, Technology adoption and biological diversity in Andean potato agriculture, *J. of Dev. Econ.* **39**(2):365-387.

de Gorter, H., Nielson, D. J., and Rausser, G. C., 1992, Productive and predatory public policies: Research expenditures and producer subsidies in agriculture. *Amer. J. of Agr. Econ.* **74**(February):27-37.

Diamond, J., 1999, *Guns, Germs, and Steel: The Fates of Human Societies*, W.W. Norton & Company, New York.

Duvick, D. N., 1984, Genetic diversity in major farm crops on the farm and in reserve, *Econ. Botany* **38**(2):161-178.

FAO, 1998, *State of the World's Plant Genetic Resources for Food and Agriculture*, FAO, Rome.

Fisher, A., and Hanemann, M., 1990, Information and the dynamics of environmental protection, *Scandinavian J. of Econ.* **92**:399-414.

Harlan, J. R., 1992, *Crops and Man*, 2nd ed., American Soc. Agronomy, Madison, Wisconsin.

Heisey, P., Smale, M., Byerlee, D., and Souza, E., 1997, Wheat rusts and the costs of genetic diversity in the Punjab or Pakistan, *Amer. J. of Agr. Econ.* **79**:726-737.

Palacios, X. F., undated, Contribution to the estimation of countries' interdependence in the area of plant genetic resources, Background Study Paper No. 7 Rev. 1, Commission on Genetic Resources for Food and Agriculture, FAO, Rome.

Reid, W., McNeely, J., Tunstall, D., Bryant, D., and Winograd, M., 1993, *Biodiversity Indicators for Policy-makers*, World Resources Institute, Washington, D. C.

Smale, M., undated, personal communication.

Smale, M., Meng, E., Brennan, J. P., and Hu, R., 1999, Using economics to explain spatial diversity in a wheat crop: Examples from Australia and China, CIMMYT Economics Working Paper 99-13, CIMMYT, Mexico City.

Smith, B. D., 1995, *The Emergence of Agriculture*, 2nd ed., Scientific American Library, New York.

Valvilov, N. I., 1926, *Studies on the Origins of Cultivated Plants*, Inst. Appl. Bot. Plant Breed, Leningrad.

Valvilov, N. I., 1949/50, The phtyogeographic basis of plant breeding, in: *The Origin, Variation, Immunity and Breeding of Cultivated Plants*, transl. by K. Starr Chester, Chronica Botanica, Waltham, Massachusetts.

Yarkin, C. J., 1998, Challenges and opportunities: Pesticides, regulation and innovation, Ph.D. dissertation, University of California, Berkeley.

Zeven, A. C., and J. M. J. de Wet, 1982, *Dictionary of Cultivated Plants and their Regions of Diversity*, Centre for Agricultural Publishing and Documentation, Wageningen, The Netherlands.

Chapter 11

MODELING THE IMPACTS OF BARGAINING POWER IN THE MULTILATERAL DISTRIBUTION OF AGRICULTURAL BIODIVERSITY CONSERVATION FUNDS*

Frederic Chantreuil[1] and Joseph C. Cooper[2]

[1]Institute National de la Recherche Agronomique, Economie et Sociologie Rurales, 4 allée Adolphe Bobierre, CS 61103, 35011, Rennes cedex, France; [2]Deputy Director, Resource Economics Division, Economic Research Service (United States Department of Agriculture), 1800 M Street, NW, Washington, DC 20036-5831

Abstract: The previous chapter addressed the question of what is the most economically efficient method of distributing the agricultural biodiversity conservation funds from the International Treaty on Plant Genetic Resources for Food and Agriculture among countries and world regions. This chapter uses game theory to extend the analysis to take into account the possibilities for players, i.e., countries receiving the funds, to form coalitions with respect to obtaining the funds. The analysis applies the Shapley value concept of an n-person cooperative game to determining distribution of the funds at several levels of the negotiating process, e.g., at the country, world region, and fund administrator levels. Using this approach, the impacts of players' bargaining power on the resulting allocations can be empirically assessed. Furthermore, the approach allows us to explicitly account for potentially competing interests of the players, thereby introducing some equity to the allocation.

Key words: agricultural production; agriculture; bargaining power; biodiversity; centers of diversity; coalition structures; equity; game theory; *in situ* conservation; International Treaty on Plant Genetic Resources for Food and Agriculture; plant genetic resources for food and agriculture; Shapley value.

---

* The views contained herein are those of the authors and do not necessarily represent policies or views of the Economic Research Service or United States Department of Agriculture. The views expressed herein are those of the authors, and not necessarily of the Economic Research Service or the United States Department of Agriculture.

1. INTRODUCTION

With the continuing trend towards globalization of trade and investment, international environmental agreements are becoming increasing important. Examples of agreements with significant environmental components include the Convention on Biological Diversity, the United Nations Framework Convention on Climate Change (UNFCCC), the Montreal Protocol on Substances that Deplete the Ozone Layer, the Basel Convention on the Control of Transboundary Movements of Hazardous Wastes and Their Disposal, and the Convention on International Trade in Endangered Species of Wild Fauna and Flora. Given the incentives to free ride associated with the environment being a public good, the possibilities for the forming of coalitions among subsets of players, and the potential for spillover effects between environmental goods as well as economic activities, a large number of economic issues are opened in these agreements. Game theory naturally plays a major role in analytically modeling a negotiations process. Given the large number of potential players in these agreements, as well as the framers' desire that cooperative solutions be taken to achieving these agreements, cooperative game theory is an especially appropriate technique for modeling these agreements. Nash bargaining and other noncooperative solutions, for instance, are not appropriate game theory tools for models of more than two players when free-riding or other coalition manipulations are possible. Instead, n-person cooperative game theory can be used as an analytic framework for assessing the impact of bargaining power on the division of any economic "pie" or burden. Although n-person cooperative game theory has received relatively extensive coverage in some economic fields, few examples exist within the environmental and resource economics literature. This chapter demonstrates how n-person cooperative game theory can be applied in environmental and natural resource economics, with a numerical application to determining the "fair and equitable sharing" of the benefits arising from the use of plant genetic resources for food and agriculture (PGRFA), using as a starting point, the regional allocations of conservation funds in Table 10-3 of the previous chapter.

In this chapter, we allow for collective action in determining the allocations of the funds from the International Treaty on Plant Genetic Resources for Food and Agriculture—hereafter abbreviated as IT. The difference over the two-person case is that coalitions are possible. That means that two players can manipulate the outcome of the game by acting together against the third. Specifically, to take into account the possibilities for players to form coalitions, this chapter applies the Shapley value concept of an n-person game to determining allocation of the PGRFA conservation funds at several levels of the negotiating process, e.g., at the country, world region, and IT administrator levels. Using this approach, the impacts of players' bargaining power on the resulting allocations can be empirically assessed. Furthermore, the approach allows us to explicitly account for potentially competing interests of the players, thereby introducing some equity to the allocation.[3] Our model is easily generalizable to other applications in which shares of a "pie" or burden are allocated among heterogeneous groups.

Our focus is on efficiency and equity considerations in allocating the funds. An allocation based on the economic efficiency criteria may not be particularly relevant at the international level, where equity considerations are a prime concern (Hagem and Westskog, 1998; Rose et al., 1998; Cline, 1992). For example, in an international environmental treaty, heavy use of side payments may not be a politically acceptable instrument for distributing aid. Section 2 presents the theoretical models for the allocation rules. In section 3, we assess the impact of these allocation rules on the distribution of the funds, whether in cash or in project money, using numerical illustration of several scenarios regarding the level of bargaining power among the players. As for the choice of indicators for making the allocations at the world region level, we use a proxy measure for PGRFA diversity value developed in Chapter 11. At the country level, indicators of the relative importance of country-level PGRFA diversity are not available—and at any rate, may not be applicable to equity considerations—and indicators such as amount of land in agriculture or agricultural GDP must be used.

2. THE MODEL

Table 10-3 in Chapter 10 provides the indicator ("OECD Value of Agricultural Production ascribed to Primary Centers of Diversity") that could be used to distribute PGRFA conservation funds between regions. However, this indicator, or any other, can only be a rough guide. For example, players may attempt to utilize bargaining power and form blocks in order to receive a larger allocation than suggested by this indicator. We present an analytical framework for modeling this process.

To determine the allocation of initial allowances that explicitly accounts for heterogeneity across the players, we propose a payoff function, denoted as φ, that is derived from an n-person cooperative games construct. Application of this construct to the world regional level implies that the

[3] In the manner of Rose et al. (1998), equity in our context refers to the distributional justice of the initial allocation of the conservation funds across countries, i.e., international equity.

determination of such an allocation can be achieved if all regions agree to delegate their decision power—over the definition of the agenda and the type of PGRFA benefits distribution system that can be implemented—to an international institution. In other words we suppose that all world regions accept the fact that the IT has to allocate the initial endowments to each world region. Hence, the problem of the allocation of conservation funds can be set in the form of a simple game in characteristic function form (N, U), i.e., with side payments, where N is the set of players (i.e., world region) and U is the characteristic function of the game.[4] In the context of the negotiations over the distribution of PGRFA conservation funds, this side payment assumption expresses the possibility for world regions to form blocks in receiving the aggregate funds of the world regions forming the block. The characteristic function sums up all possible utility sets of every coalition $S \subseteq N$. For example, if population is the relevant criterion, we need to be able to calculate it for all possible coalitions. If $PO(S)$ represents the population of coalition S, we have the following simple game with the characteristic function $U(S)$:

$$\text{For all} \quad S \subseteq N, \quad U(S) = \begin{cases} 1 & \text{if } PO(S) \geq \beta.PO(N) \\ 0 & \text{otherwise} \end{cases} \tag{1}$$

with $0 < \beta < 1$. For example, a value of $\beta = 0.5$ means that a coalition S can obtain the total conservation fund if its primary center of diversity value is equal or greater that 50% of the value of the IT membership, $PO(N)$. In other words, β reflects a hypothesis regarding the level of bargaining power coalitions can achieve. With the function above, while noting that bargaining power is a function of the initial allocations, bargaining power is decreasing in β.

Given this framework, the Shapley allocation approach is particularly useful to solving the model (Roth, 1998; Mas-Colell, Whinston, and Green, 1995). This normative concept attempts to describe a fair way to allocate gains from cooperation, given the strategic realities captured by the characteristic form. The Shapley value in our case represents the final distribution of initial allocation of the conservation funds between all world regions.

While a lengthy discussion of Shapley values is outside the scope of this chapter, the concept is briefly described here. The Shapley value summarizes

---

[4] The side payment assumption implies the existence of commodities that are linearly transferable. In other words, the utility functions for the individuals can be chosen so that the rate of transfer of utility among any two of them is 1:1. Hence, the total utility obtainable by a coalition S can be divided among the members of this coalition in any number of possible ways.

the complex possibilities facing each player i (i.e., world region) in game (N,U) by a single number $\varphi_i(N,U)$ representing the value to i of playing the game. Thus, the value of game (N,U) is an $n \times 1$ vector in which each element represents the "expected value" to a player of playing the game, where $\varphi_i(N,U)$ represents the expected payoff to player i under a randomization scheme on all coalitions S she can join. For a simple two-player case, the egalitarian solution is represented by the expression $\varphi_1(N,U) - \varphi_2(\{1\},U) = \varphi_2(N,U) - \varphi_1(\{2\},U)$, where $\varphi_1(N,U) + \varphi_2(N,U) = U(N)$. This relationship states that in the egalitarian Shapley value solution, player 1 gets the same utility out of the presence of player 2 as the latter gets out of the former. Extending this concept to a multiple-player game, the general formula is (see, e.g., Mas-Colell, Whinston, and Green, 1995):

$$\varphi_i(N,U) = \sum_{\substack{i \in N \\ i \notin S}} \frac{s!(n-s-1)!}{n!} \left[U(S \cup \{i\}) - U(S) \right] \tag{2}$$

where s is the number of players in a coalition S. The basic principle (marginality principle) behind the share $\varphi_i(N,U)$ is that when a player joins a coalition, she receives the marginal amount $\left[U(S \cup \{i\}) - U(S)\right]$. The probability that a random ordering of coalition $S \subset N$ forms as the union of i and its predecessors equals the probability that i is in the sth place, which is $1/n$ multiplied by the probability that $S - \{i\}$ forms when we randomly select $s-1$ members from $N - \{i\}$, which is $s!(n-s-1)!$. For any given random ordering of players, we calculate the marginal contribution of every player i to its set of predecessors in this ordering.

Consider the simple game denoted in (2) between five world regions (a,b,c,d,e) whose population levels are, respectively, 5, 15, 20, 25, and 35. If $\beta = 0.5$, then the Shapley value of this game is $\left(\frac{1}{30}, \frac{7}{60}, \frac{1}{5}, \frac{1}{5}, \frac{9}{20} \right)$.[5] Note

[5] Please see the appendix to this chapter for an example that highlights how the Shapley value of this simple game is calculated.
This table gives the list of winning coalitions for this game and 1 means that the corresponding world region is decisive for the corresponding coalition while 0 means that the corresponding world region is not decisive. Considering the world region a, its marginal contribution is positive for one and only one coalition: (a,c,d). Then, applying the formula given in (3) for world region a, we have

$$\varphi_a = \frac{(3-1)!(5-3)!}{5!} = \frac{1}{30}.$$

that the Shapley value is the same for world region *c* and *d*, even though the population of region *d* is larger. This result describes a fundamental concept behind the Shapley value: since player *d* has no greater opportunity than c to form a minimal winning coalition, he must have the same share as player *c* in a bargaining game. In contrast, the "vote vector" result—payoffs strictly proportional to each player's share of the total population, i.e., no world region has majority power on his own—is $\left(\frac{5}{100},\frac{15}{100},\frac{20}{100},\frac{25}{100},\frac{35}{100}\right)$.

While the Shapley value can address the allocation of initial endowments between world regions, it cannot address the chain of allocations from the world regional levels down to country and subcountry levels given the allocations, each level of which has its own players with varying levels of bargaining power. Methodology is developed here to capture the bargaining activity through several administrative or other political power levels. To account for negotiations over conservation of PGRFA being held at several levels, we assume that fictitious delegates are elected to represent each world region, each country, and one subcountry level, which can be states, provinces, lobbyist groups, or even firms. The set of players at the lowest level is denoted as $N=\{1,...i,...,n\}$. The world regions and the countries they belong to are denoted by the level structure $\mathrm{B}=\{B^1,B^2\}$, where $B^1=\{R_1,...,R_m\}$ is the set of all world regions and $B^2=\{C_1,...,C_g\}$ the set of all countries that describe N's a priori coalition structure. Given this coalition structure framework, if we assume that in each world region j, $R_j \in B^1$, a delegate is selected to represent the coalition, the bargaining situation is a problem of how to divide the conservation funds and can be formally represented by the quotient $(M,V)=(N,U)/B^1$, where (M,V) is a game with a set of players (world region representatives) $M=\{1,...,j,...,m\}$ and its characteristic function is given by:

$$V(S)=U\left(\bigcup_{j\in S} R_j\right) \text{ for all } S\in M \tag{3}$$

Considering the world region *b*, its marginal contribution is positive for three coalitions: (*b,e*), (*a,b,c*), and (*b,c,d*). Then applying formula given in (3), we have $\varphi_b=\frac{(2-1)!(5-2)!}{5!}+\frac{(3-1)!(5-3)!}{5!}+\frac{(3-1)!(5-3)!}{5!}=\frac{1}{20}+\frac{1}{30}+\frac{1}{30}=\frac{7}{60}$.

Applying the formula for all world regions, the Shapley value of this game is thus $\left(\frac{1}{30},\frac{7}{60},\frac{1}{5},\frac{1}{5},\frac{9}{20}\right)$.

Thus, a reasonable expectation for the jth world region is the amount $\varphi_i(M,V)$, which is the element in the Shapley value corresponding to player j in game (M,V), and would be the value normally expected by this world region.

Given the allocations of the permits among world regions, subregional allocations take place. Obviously, the payoff a player (i.e., a firm) receives depends crucially on the definition of the bargaining relationships between countries and the lower level players, which we can denote as "firms" for convenience. To address this topic, we propose in the following subsections three bargaining and payoff principle scenarios that are based on the same bargaining relationship among world regions, but on different bargaining relationships among countries and among firms. The scenarios depend on the capacities of threat (bargaining power) of some countries over other countries within the same world region or on the capacity of threat of a firm toward other firms within the same country. Section 2.1 presents a payoff function Ψ_1 for a base scenario in which we consider only the amount each lowest-level player can obtain on its own. Section 2.2 presents a payoff function Ψ_2 that captures the "lobbying game" between countries as played out in the IU. Section 2.3 defines a payoff function Ψ_3 motivated by the subsidiarity principle. In these three scenarios, to determine the final distribution of initial endowments in a manner that takes into account interest firms of various types, various countries, and various world regions, we assume a three-step process. In the first step, the world regions bargain with each other to determine the division of the surplus. In the second step, countries belonging to the same world region bargain with each other over the allocation received by their world region. Finally, the firms of a given country divide among themselves the share the country receives. However, because the Shapley value formula in Eq. (3) does not take into account the fact that the lowest-level players are organized a priori into level structure **B**, the model must be extended for determining the sharing of conservation funds.

2.1 Null threat and egalitarian principle

Under the "null threat," a player has no power to negotiate with other players in the same coalition. Hence, in this minimum information game, we need only account for the amount a player can obtain by himself (that is, his value in the characteristic function), and the amount the coalition to which he belongs to can obtain. According to the egalitarian principle, no player has a greater opportunity than another player to form a minimal winning coalition. Hence, every player has the same power, and the symmetry axiom of the Shapley value implies that all players will obtain the same share. For example, if the United States is not a winning coalition (i.e., receives no payoff) at the international level, the value of the characteristic function for the U.S. regions is always the same, and the United States receives an egalitarian division of the PGRFA funds.[6] In sum, the allocation rule proposed in this section supposes that each country [and each firm] passively accepts the funds proposed by the IU.

Formally, we model this bargaining procedure within world regions by only considering the amount $U(K)$ that every coalition of countries K (such as $K \subseteq R_j$, $K \in B^2$) can obtain on its own. For this game, we can define the sub-game (N_{R_j}, W_{R_j}) on the world region R_j, where N_{R_j} represents the set of countries in world region R_j. The characteristic function of this sub-game is given by :

$$W_{C_j}(\varnothing) = 0 \tag{4}$$

$$W_{C_j}(K) = U(K) \text{ for all } K \subseteq R_j,\ K \in B^2 \tag{5}$$

$$W_{C_j}(R_j) = \varphi_j(N,V) \tag{6}$$

Eq. (4) says that the empty set is worth nothing and (6) says that the amount a coalition K can itself obtain in the sub-game, defined on the world region that K belongs to, is the same amount K can obtain in the initial game. Given these equations, a reasonable expectation for the country $C_h \subset R_j$ is the amount $\varphi_h\left(N_{R_j}, W_{R_j}\right)$, which is the Shapley value of player h in the sub-game (N_{R_j}, W_{R_j}).

We can use the same reasoning to divide the amount received by the country $C_h \subset R_j$ among the lowest-level players—we can called these "firms" for convenience—considered in the model. Thus, each firm $i \in C_h \subset R_j$ can obtain the amount $\varphi_i\left(N_{C_h}, W_{C_h}\right)$, which is the Shapley value of firm i in the sub-game defined on the country $C_h \subset R_j$.

---

[6] In our simple game in Eq. (2), a losing coalition is one that has less than $\beta\%$ of the world population. Hence, every subcoalition of this coalition is also a losing coalition. However, this situation does not imply that a loosing coalition will obtain nothing. Instead, the outcome depends on the definition of the game. For instance, two losing coalitions may become winners if they act together.

For this bargaining game, we can define a payoff function Ψ^1, in which the share a firm $i \in C_h \subset R_j$ will receive is (see Chantreuil, 2000):

$$\Psi_i^1(N,U) = \sum_{\substack{G \subset C_h \\ i \notin G}} \frac{g!(c_h - g - 1)!}{c_h!}\left[U(G \cup \{i\}) - U(G)\right]$$
$$+\frac{1}{c_h}\left[\sum_{\substack{K \subset R_j \\ R_h \notin K}} \frac{k!(r_j - k - 1)!}{r_j!}\left[U(K \cup C_h) - U(K)\right] - U(C_h)\right] \quad (7)$$
$$+\frac{1}{c_h}\left[\frac{1}{r_j}\left[\sum \frac{s!(m-s-1)!}{m!}\left[U(S \cup R_j) - U(S)\right] - U(R_j)\right]\right]$$

where m denotes the number of world regions; s, k, and g are the number of players in every coalition of world regions S, every coalition of countries K, every coalition of firms G, respectively; r_j represents the number of countries in the world region R_j; and c_h is the number of firms in the country C_h.

Consider the example of the five world regions using the population criterion. If we take into account the fact that players a and b act together, the payoff vector using the allocation rule Ψ^1 will be $\left(\frac{1}{12}, \frac{1}{12}, \frac{1}{6}, \frac{1}{6}, \frac{1}{2}\right)$.

2.2 Strong threat and the quotient game lobbying principle

The payoff function in this section uses more information than in the first, given that here we need to know the amount a country can obtain by forming a coalition with other world regions, an amount that is different for each country. Hence, the degree of bargaining power varies among the countries, which produces the demand for lobbying activities. This bargaining procedure within world regions consists in accounting for the amount $U(K)$ that every coalition K (such as $K \subseteq R_j$, $K \in B^2$) can obtain itself and the amounts $U\left(K \cup R_q \cup ... \cup R_m\right)$ that K could obtain if it would replace world region R_j and form a coalition with one or more of the remaining world regions of B^1. A country can threaten to leave a coalition on the basis of being able to gain more elsewhere. Even thought this threat is never carried out, it can be used to compute the relative power of each country in a given world region, and to capture much of the "lobbying game" of countries at the international level.[7]

Formally, for any $K \subset R_j$, and K' its complement relative to R_j, we define a restricted game $\left(M\middle|R_j / K, V_{R_j/K}\right)$ as representing the quotient game (M,V) when K replaces the world region R_j in B^1. This restricted game is the formal representation of a bargaining situation which involves K and the other world regions. Its characteristic function is given by

$$V_{R_j/K}(S) = U\left(\bigcup_{q \in S} R_q - K'\right) \text{ for all } S \subseteq M\middle|R_j / K \tag{8}$$

where $M|R_j / K$ represents the set of players when world region R_j is replaced by K. Given this function, a reasonable expectation for K is the Shapley value $\varphi_K\left(M\middle|R_j / K\right|)$ of player K in this restricted game. This amount also represents the relative payoff, in game (N, U), of K if it would replace the world region R_j and bargain with the $m-1$ other world regions.

Using the measure of the relative payoff of each country of the world region R_j, we can define the sub-game $\left(N_{R_j}, W_{R_j}\right)$ of world region R_j by its characteristic function given in Eqs. (4), (6) and by:

$$W_{R_j}(K) = \varphi_K\left(V_{R_j/K}\right) \text{ for all } K \subseteq R_j\,,\ K \in B^2 \tag{9}$$

Thus, each country $C_h \subseteq R_j$ can obtain the amount $\varphi_h\left(N_{R_j}, W_{R_j}\right)$, which is the Shapley value of player h in the sub-game $\left(N_{R_j}, W_{R_j}\right)$. To determine the division of the amount received by the country C_h among the firms of this country, we use the same argument. Each firm $i \in C_h \subset R_j$ can obtain the amount $\varphi_i\left(N_{C_h}, W_{C_h}\right)$, which is the Shapley value of player i in the sub-game defined on country C_h. Given this sub-game, we can define a

---

[7] The possibility for a country to threaten the coalition allows us to compute its relative power and, hence, capture the lobbying force of this country.

payoff function Ψ^2 in which the share a firm $i \in C_h \subset R_j$ will receive is given by the following formula (Chantreuil, 2000):

$$\Psi_i^2(N,U) = \sum_{\substack{S \subset M \\ R_j \notin S}} \sum_{\substack{K \subset R_j \\ C_h \notin K}} \sum_{\substack{G \subset C_h \\ i \notin G}} \frac{s!(m-s-1)!k!(r_j-k-1)!g!(c_h-g-1)!}{m!r_j!c_h!} \left[U(S \cup K \cup G \cup \{i\}) - U(S \cup K \cup G)\right]. \quad (10)$$

With the five-world region game example, if we take into account the fact that players a and b act together, the payoff vector using the allocation rule Ψ^2 will be $\left(\frac{1}{24}, \frac{3}{24}, \frac{1}{6}, \frac{1}{6}, \frac{1}{2}\right)$.

2.3 The subsidiarity principle

Finally, we can define a third payoff function that assumes that the problem of the IT is only to determine the allocation of conservation funds to each world region. This assumption can be motivated by the principle of subsidiarity, which essentially means that the IT does not make decisions for the lower levels. Instead, the world region itself has to solve the allocations problems for the lower level(s). This is the most likely scenario to become practice. For example, Central Asia and South America blocs can be given allocations, and they in turn must decide how to allocate these blocks among their member countries. In this case, as discussed in section 2, the payoff vector we are looking for corresponds to the Shapley value of the quotient game (M,V) played by the delegates of each world region. The share every world region $j \in M$ will receive is given by:

$$\varphi_j(M,V) = \sum_{\substack{S \subset M \\ j \notin S}} \frac{s!(m-s-1)!}{m!} \left[V(S \cup \{j\}) - V(S)\right] \quad (11)$$

Once this payoff vector is defined, the IT considers the problem solved at the international level. Then, with respect to the principle of subsidiarity, each world region can choose its own conservation fund allocation process regardless of what the others do. We can set this problem of the division of the gain among the countries of a given world region in the form of a simple game. Hence, the share a country will obtain corresponds to the Shapley value of the game defined on the world region of interest.

Using the same argument for the allocation of conservation funds between competing conservation proposal within each country, we can define a payoff function Ψ^3 in which the share a firm i, such as a conservation agency, $i \in C_h \subset R_j$ will receive is given by (see Mathurin, 1997):

$$\Psi_i^3(N,U) = \sum_{\substack{G \subset C_h \\ i \notin G}} \frac{g!(c_h - g - 1)!}{c_h!}\left[V_{C_h}\left(G \cup \{i\}\right) - V_{C_h}(G)\right]$$

$$+ \frac{1}{c_h}\left[\sum_{\substack{K \subset R_j \\ C_h \notin K}} \frac{k!(r_j - k - 1)!}{r_j!}\left[V_{R_j}\left(K \cup C_h\right) - V_{R_j}(K)\right]\right]$$

$$+ \frac{1}{c_h}\left[\frac{1}{r_j}\left[\sum \frac{s!(m - s - 1)!}{m!}\left[V\left(S \cup R_j\right) - V(S)\right] - V_{R_j}\left(R_j\right)\right] - V_{C_h}\left(C_h\right)\right]$$

where $\left(N_{C_h}, V_{C_h}\right)$ is the simple game defined on the country C_h and $\left(N_{R_j}, V_{R_j}\right)$ is the simple game defined on the world region R_j. With the five world regions game example, if we take into account the fact that players a and b act together, the payoff vector using the allocation rule Ψ^3 will be $\left(0, \frac{1}{6}, \frac{1}{6}, \frac{1}{6}, \frac{1}{2}\right)$.

3. NUMERICAL ILLUSTRATION

The numerical illustration makes use of data from Chapter 10 (Table 10-3, second column for the region allocation, while country level data are taken from the fourth column in Table 11-1 below. For the sub-region level allocation, we chose the example of Southeast Asia. Tables 11-2 and 11-3 present the final allocations in the presence of bargaining power and coalition building. In other words, these represent the payoff function (Ψ_1, Ψ_2, and Ψ_3), or power indices, results using the values in the second column in each table as the basis for the initial allocations. Table 11-2 presents the power indices at the world region level that feed into the construction of Ψ_3 in Table 11-3. When power parameter $\beta = 0.5$, the countries with the largest base allocations (Mediterranean, West Asia, Central Asia, and Central America) gain at the expense of the other regions. Equity plays a stronger role when $\beta = 0.95$, with the largest countries losing notable shares and some of the smaller ones gaining.

Table 11-3 uses agricultural GDP from Table 11-1 to determine the country level allocations. Under Ψ_1 the final allocation in Southeast Asia is an egalitarian one, whatever the value of power parameter β is. Small countries (Brunei, Cambodia, Laos), with respect to their weight in the total GDP of Southeast Asia, prefer an allocation based on Ψ_1. Moreover, this preference remains valid whatever the value of power parameter β. For other countries, however, the picture is different. When the value of power parameter β is 0.50 or 0.75, large countries (Indonesia and Philippines) prefer an allocation based on Ψ_3, but prefer an allocation rule based Ψ_2 when the value of power parameter rises to 0.95. Thailand and Vietnam prefer an allocation based on Ψ_2 when the value of power parameter β is 0.50 or 0.70, and an allocation based on Ψ_3 when the value of power parameter β is 0.95. Finally, Malaysia and Burma prefer an allocation based on Ψ_1 when the value of power parameter β is 0.50 or 0.70, and an allocation based on Ψ_3 when the value of power parameter β is 0.95. In practice, one would expect both the egalitarian extreme under Ψ_1 and the subsidiarity principle (Ψ_3) when β is 0.95 to be untenable. However, the Ψ_3 allocations under the other two power levels seem feasible.

Table 11-1. Country level agricultural indicators for Southeast Asia

Country	Agricultural land	Rural population	Agricultural GDP	Agricultural labor force
	(km. sq.)		(x 1 billion $US)	
Brunei	105	--	0.3	50,000
Cambodia	22,948	9,882,542	4.0	1,402,500
Indonesia	310,495	131,826,090	113.2	35,670,000
Laos	6,924	4,217,813	3.4	115,200
Malaysia	49,463	9,405,469	28.0	1,763,580
Myanmar (Burma)	105,238	35,099,350	33.1	12,257,600
Philippines	93,000	34,118,699	54.1	12,457,400
Thailand	205,600	47,881,146	44.3	17,604,000
Vietnam	69,208	61,848,968	37.7	21,255,000

Source: *CIA Country Factbook* (except for Rural Population: The World Bank).

Table 11-2. Final allocations (%) of conservation funds using OECD value of agricultural production ascribed to primary centers of diversity

World region	Base allocations[a]	Final allocations		
		$\beta = 0.50$	$\beta = 0.75$	$\beta = 0.95$
Central Africa	0.94	0.81	0.85	0.92
East Africa	4.36	4.08	4.22	5.37
Southern Africa	1.82	1.60	1.64	2.51
West Africa	3.03	2.70	2.77	2.77
Indian Ocean	1.05	0.89	0.90	0.94
South Asia	3.69	3.38	3.49	3.94
Southeast Asia	3.75	3.38	3.49	3.94

Table 11-2. (continued).

East Asia	10.25	10.00	10.28	12.04
Pacific	1.49	1.39	1.34	2.51
Mediterranean	7.55	7.25	7.90	12.04
West Asia	17.20	18.31	17.57	12.04
Central Asia	18.25	19.62	19.06	12.04
Europe	3.69	3.36	3.49	3.94
South America	6.39	6.03	6.47	12.04
Central America	15.77	16.47	15.77	12.04
Caribbean	0.00	0.00	0.00	0.00
North America	0.77	0.75	0.77	0.92
Total	100.0	100.0	100.0	100.0

[a] The base allocations are the values from the second column of Table 3, Chapter 10, scaled down (by a factor of 1.814) to sum to one, and multiplied by 100.

Table 11-3. Final allocations in Southeast Asia under each of the three payoff functions, and three values of β using column 2, Table 10-3 in Chapter 10 as the basis for the regional values and column 4 in Table 11-1 above for the country level values

Countries		Final allocations (%)								
		$\beta = 0.5$			$\beta = 0.75$			$\beta = 0.95$		
	Base[a]	Ψ_1	Ψ_2	Ψ_3	Ψ_1	Ψ_2	Ψ_3	Ψ_1	Ψ_2	Ψ_3
Brunei	0.09	11.11	0.12	0.00	11.11	0.12	0.00	11.11	0.12	0.00
Cambodia	1.26	11.11	1.20	1.18	11.11	1.17	0.95	11.11	1.05	0.00
Indonesia	35.59	11.11	35.88	43.10	11.11	35.11	40.24	11.11	31.12	16.67
Laos	1.07	11.11	0.14	1.18	11.11	0.91	0.95	11.11	0.81	0.00
Malaysia	8.80	11.11	9.00	7.86	11.11	8.72	5.95	11.11	7.76	16.67
Philip-pines	17.01	11.11	17.14	17.86	11.11	16.70	20.24	11.11	22.76	16.67
Thailand	13.93	11.11	14.11	13.10	11.11	13.79	13.57	11.11	12.26	16.67
Vietnam	11.85	11.11	11.95	7.86	11.11	13.18	10.24	11.11	14.89	16.67
Total	100.00	100.00	100.00	100.00	100.00	100.00	100.00	100.00	100.00	100.00

[a] The base allocations are the values from the fourth column of Table 11-1 above, scaled-down to sum to one, and multiplied by 100.

4. CONCLUSION

The concepts we use to model the distribution of PGRFA conservation funds between world regions and countries are all characterized by what Rose and Stevens call "consensus equity". Unfortunately, no consensus exists on the best definition of equity (Rose and Stevens, 1998). While efficiency-equity tradeoffs receive substantial attention in the mainstream economics literature, equity concerns receive little attention in the resource economics literature. Equity tends not to be a strong mechanism for making allocations, and the traditional economist prefers notions of efficiency and allocations that maximize consumer plus producer surplus.[8] Realistically, however, political realities can require equity concerns to be of important, or even prime, concern. Methodological rules for interactions between efficiency and equity theory have yet to be worked out. Within the class of equity allocations at least, the Shapley value format we use as the basis for our cooperative games solutions is efficient.[9]

In this chapter, we considered three rules, denoted as Ψ_1, Ψ_2, and Ψ_3, for allocating the biodiversity conservation funds among regions. The difference between the three allocation rules proposed here depends on the possibilities for bargaining. The outcome of the first allocation rule Ψ_1 (the null threat and egalitarian principle in section 2.1) is characterized at the domestic (country) level by the egalitarian criterion, while the outcome of payoff function Ψ_2 (the strong threat and the quotient game lobbying principle in section 2.2) depends exclusively on the bargaining power of each country.

The third allocation rule Ψ_3 (the subsidiarity principle in section 2.3) is particularly appropriate as part of a flexible mechanism for biodiversity conservation as it allows the use of different types of control mechanisms at different levels of negotiation process. In other words, conservation fund targeting at the world region level can be based on an efficient and equitable allocation rule and, as each world region is then free to distribute the conservation funds within the region, each world region and each national program can be based on a different set of rules tailored to regional and national characteristics that address the requirement of efficiency and equity.

The possibility for parties to form blocks in competing over the distribution of the conservation funds motivates the cooperative game theory backdrop to this chapter. Perhaps surprisingly, in this framework, the notion of equity and power are very similar. The key concept behind the allocation rules presented here is that the outcome a player can obtain depends on her bargaining power.

---

[8] Some economists may argue that equity and fairness concerns are really a veneer for hiding self-interests. To someone who has that view, we note that the model presented here may be of interest as a tool for analytically expressing self-interests.

[9] For a discussion of the efficiency of the Shapley value in making allocations, see Eatwell, Milgate, and Newman (1989), p. 24 and p. 213.

APPENDIX
Example of how a Shapley Value is constructed

As denoted by Eq. (2) in the main text, consider the simple game between five world regions (a,b,c,d,e) whose population levels are, respectively, 5, 15, 20, 25, and 35. If $\beta = 0.5$, then the Shapley value of this game is $\left(\frac{1}{30},\frac{7}{60},\frac{1}{5},\frac{1}{5},\frac{9}{20}\right)$. The following table can be used to highlight how the Shapley value of this simple game is calculated. In this game, five world regions (a,b,c,d,e) whose population levels are, respectively, 5, 15, 20, 25, and 35, are considered. If $\beta = 0.5$, a winning coalition, is a coalition whose population level is 50 or more.

Winning coalitions	a	b	c	d	e
b,e	0	1	0	0	1
c, e	0	0	1	0	1
d, e	0	0	0	1	1
a, b, e	0	1	0	0	1
a, c, d	1	0	1	1	0
a, c, e	0	0	1	0	1
a, d, e	0	0	0	1	1
b, c, d	0	1	1	1	0
b, c, e	0	0	0	0	1
b, d, e	0	0	0	0	1
c, d, e	0	0	0	0	1
a, b, c, d	0	0	1	1	0
a, b, c, e	0	0	0	0	1
a, b, d, e	0	0	0	0	1
a, c, d, e	0	0	0	0	0
b, c, d, e	0	0	0	0	0
a, b, c, d, e	0	0	0	0	0

REFERENCES

Chantreuil, F., 2000, Axiomatics of level structure values, *Homo Oeconomicus* **17**(1/2):177-191.

Cline, W., 1992, *Tshoulhe Economics of Global Warming*, Institute for International Economics, Washington, D. C.

Eatwell, J., Milgate, M., and Newman, P., eds., 1989, *The New Palgrave, Game Theory*, W. W. Norton, New York.

Hagem, C., and Westskog, H., 1998, The design of a dynamic tradable quota system under market imperfections, *J. of Environ. Econ. and Manage.* **36**:89-107.

Mas-Colell, A., Whinston, M., and Green, J., 1995, *Microeconomic Theory*, Oxford University Press, Oxford.

Mathurin, J., 1997, Application de la théorie des jeux à l'analyse économique de la coopération internationale: Illustrations par l'étude du cas de la politique agricole commune, thèse [Dissertation] de l'Université des Sciences Sociales de Toulouse, France.

Rose, A., and Stevens, B., 1998, Will a global warning agreement be fair to developing countries? *Inter. J. Environ and Pollution* **9**(2 and 3):157-178.

Rose, A., Stevens, B., Edmunds, J., and Wise, M., 1998, International equity and differentiation in global warming policy: an application to tradeable emission permits, *Environ. and Res. Econ.* **12**(July):25-51.

Roth, A., 1998, *The Shapley Value: Essays in Honor of L.S. Shapley*, Cambridge University Press, Cambridge.

PART IV.
Biotechnology: Concepts, Values, and Management

Chapter 12

AGRICULTURAL BIOTECHNOLOGY: *CONCEPTS, EVOLUTION, AND APPLICATIONS*

Maria José de O. Zimmermann[1] and Enrico Porceddu[2]
[1]*FAO-OCD, Via delle Terme di Caracalla, 00100 Roma, Italia;* [2] *Genetica Agraria, Universita della Tuscia, Viterbo, Italia.*

Abstract: Without being exhaustive, this chapter covers the main principles that are the basis of all methods of genetic improvement and relates these principles to the modern methods of plant biotechnology. It explains why the modern procedures are an extension of the old principles and how they fit together. Besides it discusses evolution and its relation with plant improvement and with biotechnology. It also discusses most of the regular biosafety concerns that relate to genetically modified organisms and shows that in many cases these concerns can be dealt with by the knowledge of the biology and evolution of the species that was modified. It also shows that some of the biosafety concerns are true for any new genotype, be it genetically modified or not. It ends by assessing the current status of agricultural biotechnology and by addressing the implications drawn for its application to developing countries.

Key words: biosafety; biotechnology; evolution; genetically modified organisms; plant improvement.

1. INTRODUCTION

Given that the term "biotechnology" is generally applied quite loosely, it may be useful to commence this discussion of biotechnology with a definition of the term. Semantically, the word is formed from the union of bios (life) with technology (techniques), and it encompasses all technologies and processes involving living beings. The Convention on Biological Diversity (UNEP/CBD/94/1, 1998) defines biotechnology as "any technological application that uses biological systems, living organisms, or derivatives thereof, to make or modify products or processes for specific use." Interpreted in this broad sense, the definition covers many of the tools and techniques that are commonplace in agriculture and food production.

While the "official" definition of biotechnology is quite broad, biotechnology is generally considered by the public to be applicable on a much narrower sense, one which restricts itself to applications of DNA technology, genomics, modern reproductive techniques, most of which were developed over the last 30 years, whereas their application to agriculture dates to the last 10 or 15 years or so.[1] Because of the novelty of this set of techniques, and the controversies over its potential impacts and applications, the subject can stir passionate debates. Controversies over the uses of biotechnology aside, this set of tools has potentially important applications, especially when used in conjunction with other technologies, for the production of food, agricultural products and services, applications that may be of significance in meeting the food supply, and nutritional needs of an expanding and increasingly urbanized world population, not to mention other applications in areas that go beyond agriculture.

The goal of this chapter is to provide a basic overview of the key agronomic and biologic concepts and applications of biotechnology that can supplement the economic issues and perspectives addressed in the rest of the book. This chapter will begin with a discussion of the evolution of agriculture. Next it will present and examine some of the key concepts related to the applications of modern biotechnologies to agriculture, agricultural production, and for biosafety. Finally, the current state of agricultural biotechnology will be assessed, implications drawn for its application to developing countries, and future prospects of its development and application will be addressed.

2. DOMESTICATION, AGRICULTURE, AND PLANT AND ANIMAL BREEDING

About 130,000 years ago, while people were still hunters and gatherers, they began developing the knowledge of plants and animals that set the stage for the beginnings of agriculture (Harlan, 1975). They began to recognize which plants and animals could be eaten and which could not due to toxicities or unpalatabilities. People began to learn which plants needed to be cooked to be edible, and those that could be eaten raw. Later, they also discovered how to use some plants for treating illnesses. Even so, at this stage, humans were heavily dependent on the available food in the

---

[1] DNA (deoxyribonucleic acid) is the long chain of molecules in most cells that carries the genetic message and controls all cellular functions in most forms of life. It is the information-carrying genetic material that comprises the genes. RNA (Ribonucleic acid) is molecule derived from DNA that may carry information (messenger RNA (mRNA)), provide subcellular structure, transport amino acids, or facilitates the biochemical modification of itself or other RNA molecules.

surroundings of their dwellings, and their average life expectancy was relatively short.

Considering the natural division of tasks at the time, the male was the hunter and forager and the female likely gathered fruits and grains, close to the setting, to feed both herself and the children. At some stage, Man began selecting plants for their ease of use (e.g., cereals with larger kernels whose ears did not shed the grains easily and plants with less toxic compounds) and began to cultivate them. People began the act of planting simply as the process of throwing seeds of fruits around their dwellings, noticing that those seeds would develop into plants that were easier to find at the time of harvest. The step from understanding that seeds were the reproductive part of a plant to promoting plant growth near their dwellings was a short one. The selection of the most suitable types for further growing was the logical next step that marked the beginnings of agriculture. This fact happened around 10,000 years ago at several different locations around the globe (Harlan, 1975).

As humans started nurturing plants and animals, they protected and modified the growth environment and, as a consequence, these organisms were gradually modified to become ever more adapted to the new conditions. Domestication is the process that resulted in inherited modifications in plants and animals that made them more suited to human needs (Harlan, 1975; Futuyma, 1979).

Wild plants have many traits that make them different from their domesticated counterparts. Wild wheat collected by the original aboriginal societies in South West Asian countries were very different from cultivated wheat, in which a single ear has many grains which mature simultaneously and are not shed. The loss of wild traits was not fortuitous nor was it due to deliberate selection pressure. When a plant with the convenient combination of characteristics appeared and was recognized, it allowed man to transform a species into a cultigen (a category of plants found only in cultivation). Probably the next discovery was that seed germination could be improved if seeds were thrown in a soil that was not too hard or if seeds were not left uncovered under the hot sun. It was also noticed that seeds needed water to grow and early growers began nurturing plants in order to increase the probability of adequately feeding the family. Plants were additionally selected for better traits or to give a better response to the care provided by the humans.

Even though there are about 250,000 to 300,000 plant species on Earth, only a small number of them sustain humans (The Crucible Group, 1994). The bulk of calories (80%) in the human diet are furnished by only 30 crop species. Wild types of these species are so different from domesticated types that some may think that they are not related, and many have compounds that are toxic to humans. The process of domestication not only reduced or eliminated such compounds, but it also reduced to a much lower level the amount of genetic variation present in those plants. Crop plants have been highly modified to the point that they are unique and dependent upon humans for their survival.

A similar pattern can be found in the domestication of animals. Initially, animals were simply hunted. At some point in time, they began to live around human settlements, where they discovered that they could get leftovers to eat or could steal food. The proximity of animals meant more abundant and easier hunting and the process of encouraging this proximity became deliberate. For both plants and animals, such practices modified the genetic structure of the populations of wild plants and animals that were targeted, restricting their variability to the types more adequate to the human needs and making some of those species dependant on the humans for survival. Domestication was an important factor of plant and animal evolution (Futuyma, 1979).

2.1 Plant and animal breeding

According to Vavilov (1951), breeding is the human-controlled "improvement" or modification of plants and animals to suit human needs (Allard, 1960). While the origins of breeding of plants and animals goes back over 100,000 years (more or less coinciding with the origin of the domestication process), scientific breeding gained momentum in the 20th Century with the rediscovery of the genetic laws that had been first described by Mendel at the end of the 19th Century. All scientific breeding procedures pass through two steps: (1) the creation and or release of variability; and (2) the identification and selection of the useful genotypes (the genetic constitution, or gene makeup, of an organism). In this section we will first give a brief overview of both steps.

2.1.1 Creation and release of variability

The creation and release of variability can be made by induced mutations or by the production of new recombinations among existing genotypes.[2] The induction of mutations creates a new form (allele) of an element (gene) that was not formerly present in a population. Natural mutations created most of the existing variations and forms. Spontaneous mutations take place at low rates, and in most cases mutants are disadvantageous for the genotype survival and or for reproduction. Hence, mutant individuals tend to be naturally eliminated from the population or are

[2] In classical genetics, recombination (crossing-over) is done through traditional breeding techniques, e.g., Parents: AB/ab and ab/ab produce recombinant offspring: Ab/ab. In molecular genetics, it refer to the process that yields a molecule containing DNA from different sources. The word is typically used as an adjective, e.g., recombinant DNA.

present in very low frequencies. The very rare ones that are advantageous or have become advantageous in some environments are the few that tend to predominate in the populations and, because of their numbers, have a larger probability of surviving (Falconer, 1960; Futuyma, 1979).

Mutagenic agents are chemical or physical means that increase the mutation rate. A number of them have been identified and are exploited by scientists for inducing mutations with greater or lower efficiency in plants, animals, and microorganisms (Allard, 1960). Unfortunately, mutations occur at random in the genome, and there are no ways of directing the induction. Molecular genetics allows a better understanding of the action of some mutagenic agents; however, mutation breeding is a time-consuming, expensive, and largely random procedure that most breeders avoid in their programs.

Polyploidy is the natural state of some species or individuals who have more than two complete sets of chromosomes in their somatic cells. Induction of polyploidy is a random process (Allard, 1960), and not all genotypes in a species will survive the treatments that have that effect (i.e., colchicine, heat treatment), although the reason why this happens is not yet clear. Polyploidy, although being of great importance for the evolution of many species of plants and to some extent of animals, is not used very much for breeding purposes.

Formation of new recombinants is usually obtained by sexually crossing two compatible genotypes belonging either to the same biological species or to closely related species. In general, recombination can only be obtained among individuals with similar genetic background due to close evolutionary histories and thus similar traits. Most, if not all, breeding progress has so far been based on the production of new recombinations by crossing followed by selection. Breeding was, and will continue to be highly successful based on these procedures, but it is well known that some characteristics cannot be improved, due to the lack of adequate variability among sexually compatible genotypes. Interspecific (i.e., interspecies) crossing is usually very difficult and only in a few cases within each genus can it be performed successfully and/or produce fertile offspring. Intergeneric crosses are usually impossible, as the isolating mechanisms are very strong between two different genus.

Another way of producing distant hybrids through bypassing sexual barriers is by protoplast (plant cells whose cell wall have been removed) fusion, and plant regeneration, e.g., "in vitro" procedures. Cells are submitted to a treatment that degrades cell walls, making them amenable to fusion, and plants from the fused protoplasts are regenerated. The process takes place in the "in vitro" cell culture medium. After being submitted to selection for some generations, they may end up by being regenerated as hybrid genotypes. Although theoretically all plants could be submitted to this procedure, protocols have been developed only for a few species

(Nakano and Mii, 1990; Ohgawara et al., 1989). Furthermore, irregularities of cell division during the subsequent growth (and mitotic divisions) make the process much less useful and reliable than it may sound. Protocols have recently been developed that improve the process involved, and this process will likely have greater use in the future. Almost certainly it will be a useful tool in combination with other techniques to help in genetic engineering procedures, for instance, to obtain transient gene expression within the hybrid protoplasts or cells.[3]

The existence of useful variation for a given trait has always been the main issue in plant breeding, and all possible steps have been taken by breeders to increase its amount. Induction of mutations and interspecific crosses are two methods, but these require large expenditure in terms of time and money. Hence, use of these approaches is often avoided. Furthermore, the potential rate of success in terms of the number of commercial genotypes is extremely low, the results in terms of the gene affected are largely random, and several generations of careful selection have to follow in order to obtain a useful product. Frequently, the desired mutation is obtained together with many other undesirable traits, which have to be selected against and eliminated in a number of generations of selection.

It is also known that wild relatives of a crop species have much higher variability than domesticated types. When a genotype possessing a desirable trait is identified in a related species, it is never known a priori if that genotype can be crossed. Once the cross is made, the hybrid usually has viability problems and is fully or partially sterile. Many backcrosses are needed before the offspring is fully viable, fertile, and possesses the desirable trait introduced in the genetic background of the recurrent parent. There are a few really successful cases that have resulted in excellent commercial varieties that have contributed to solving some pressing agricultural problems in specific areas and, as such, they are a good justification for breeding programs to continue investing in these inter-species crossing techniques. On the other hand, *intra*-specific crosses have always been the main way by which breeders produce new variability, but unfortunately, the variability for agronomic traits, and especially for disease and pest resistance, is many times limited.

Most breeders, when they cannot identify the desired trait in the available genetic material, usually abandon their breeding objective because it is considered an impossible task. In some cases the problem may be bypassed by cultural practices, such as the use of chemical inputs. In other

---

[3] Genetic engineering produces changes in the genetic constitution of cells (apart from selective breeding) resulting from the introduction or elimination of specific genes through modern molecular biology techniques. This technology is based on the use of a vector for transferring useful genetic information from a donor organism into a cell or organism that does not possess it.

cases, the problem may be so serious that the only solution for farmers is to change the crop. That has been the case in some areas of Brazil, for common bean (*Phaseolus vulgaris*), in spite of the fact that it represents a staple crop in Brazil and is the main protein source for the poorer population strata. However, due to the golden mosaic virus affecting beans, it cannot be cultivated anymore in some areas of the country (Farias et al., 1996).

In most such cases where the desired trait is not identified among the available genetic material, biotechnology, or more precisely, genetic engineering techniques, can help breeders to incorporate specific desired traits into the otherwise good genotypes. Namely, if the suitable gene or genes can be found in any other organism (plant, animal, or microorganism), that gene can be isolated using the proper laboratory procedures, the gene can be cloned and inserted into a bacterium or virus that will act as a vector to transfer it to the organism of interest (plant or animal). The isolated gene can also be introduced into the plant, animal or microorganism cell (depending on the case), using "biolistic" procedures (techniques which shoot DNA- coated micro-particles into cells).

Besides allowing the transfer of genes that would otherwise be impossible due to sexual barriers, the other advantage of genetic engineering over the other mentioned procedures is that it is specifically targeted. One single gene (or a few genes) is introduced into a genotype that is usually composed of thousands of other genes; therefore, it is a very small change and the genotype will continue to have the same genetic background, except for the additional characteristic. The total re-shuffling of the genome that happens when any type of crossing is made (intraspecific, with wild types or interspecific) will not occur, and the general behavior of that genotype is expected to be very similar to that of the same genotype before the introduction of the alien gene.

2.1.2 Identification and selection of useful genotypes

All known breeding strategies are based on selection (Allard, 1960). Selection procedures are only efficient when genetic variability is present in the material to be selected. Obviously, the variability caused by environmental factors cannot be fixed by selection. One difficulty breeders have always been confronted with is that selection acts on the phenotypic expression. However, a phenotype (the visible appearance or set of traits of an organism resulting from the combined action of genotype and environment) is the outcome of the genotype as well as of the environment action. Consequently, selection is often a complicated and lengthy process that limits the rate of success of breeding programs (Falconer, 1960; Ramalho, Dos Santos, and Zimmermann, 1993).

A number of characteristics are difficult to measure, and the recognition of the amount and type of genetic variations may require complicated manipulations that limit the number of individuals who can be evaluated and, by consequence, the rate of progress that can be realized. That is the case for many physiological traits, e.g., rate of plant photosynthesis, root growth, disease resistance, cold and/or drought tolerance, physiological efficiency, rate and efficiency of nutrient uptake, and many others. In general, plant and animal improvements have relied on indirect measures and on evaluations of some final products. Also, traditional breeding procedures are often expensive, as they require a somewhat large number of plants, with replications and grown at different locations, before a few superior individuals are recognized.

Researchers in quantitative and population genetics have developed theories and procedures for evaluating progenies and understanding breeding value of genotypes, but many of these are difficult to apply as they refer to one or a few genes, and extension to several genes becomes cumbersome and unrealistic. The net result is that breeding is largely "a numbers' game" in which the most successful programs are those that have more resources and are capable of evaluating larger populations from many crosses at many different locations. Many potentially very useful populations are discarded after a few generations simply because the breeder was not capable of identifying interesting genotypes at an early stage. Therefore, in every breeding program there is a large waste of potentially good genetic material.

The application of molecular marker technology to breeding provides new opportunities to improve selection procedures. Researchers can use molecular markers to detect variations either at the level of DNA sequences or of polypeptides (storage proteins, enzymes), which are direct products of genes. Those markers can also be part of, or closely linked to, a gene of interest. When the isolation of these molecules is an easy, nondestructive, and not very expensive process, large populations can be screened in a short period and individual genotypes can be unambiguously identified. The application of such technology can accelerate the breeding programs, improve their precision, and reduce the number and size of populations that must be planted at each generation. Selection, based on molecular markers, is named marker-assisted selection (McCouch et al., 1988).

The *efficiency of selection* is another important issue to breeding programs and is necessary in identifying genotypes to be crossed and later selected in the offspring. No cross can be superior if the parents are not well selected for true genetic superiority and complementarity. This selection procedure can be difficult to implement. It requires tests to be performed in a wide range of environments and using the proper experimental designs.

It is also very important to utilize all possible procedures that allow the correct identification of a sought characteristic, such as inoculating a pathogen for obtaining disease resistance, inducing drought or cold, etc. Some of the procedures may be very sophisticated, but they are necessary given that nature does not assure the occurrence of the selective atmospheric phenomena.

Breeders can use molecular markers to produce linkage maps to which agronomic traits should be added, and linkage relationships among the markers and the traits established. Based on the knowledge of the linkage relationships, selection can be applied over molecular markers in order to change some linked characteristics that are difficult to measure or to visualize in individuals. This is the case, for example, of root traits in plants. As an example of the use of molecular markers, take the case of rice. Markers were identified in rice that relate to root diameter and to drought resistance (Champoux et al., 1995). Using prior techniques, crosses were made using selected parents, and the segregating population had to be selected for root diameter in a nondestructive manner, and before harvest time, when roots were already degenerating. Using molecular markers, it was possible to take one or a few leaves per plant, extract its DNA, and identify the presence or absence of the marker. The selection was done without any influence of the environment in which plants were growing.

There are several reasons why these marker procedures are not widely incorporated in breeding programs, some of which are expected to change in the near future. One of them is that molecular marker linkage maps are yet not available for many species and, even when the maps exist, the location of useful genes has still to be determined. Other limitations are due to the fact that most breeders do not have access to adequate laboratory facilities, molecular markers are still rather expensive, and the pleiotropic effects of single genes are entirely unknown. Pleiotropy is the property of many genes by which a particular gene has a recognizable effect on several different traits. An area of knowledge that is developing and will help significantly in the application of molecular markers to breeding is bio-informatics.[4]

These difficulties do not change the fact that molecular markers offer the only hope for breeders to be able to begin purposely selecting for characteristics that so far have been almost impossible, such as those linked to some physiological traits. The potential results of molecular marker applications are new genotypes similar to those that could be obtained by traditional procedures. However, they will be obtained at a much faster rate, and more precisely even for characteristics that formerly were considered impossible to be individually recognized and selected.

As with traditional breeding techniques, molecular biotechnology can be used to improve the performance of plants and animals in different

---

[4] Bio-informatics is the use and organization of information of biological interest. In particular, it is concerned with organizing bio-molecular databases, in getting useful information out of such databases, in utilizing powerful computers for analyzing such information, and in integrating information from disparate biological sources.

agricultural conditions and even for harsh environments. With the latest techniques, plants can also be bred to recover degraded environments (bio-remediation) utilizing the genes of plants and microorganisms that are able to live on soils or water that contains heavy metals or other toxic compounds. The resulting products could be used to detoxify some areas or water and to restore them to normal uses. Besides plant and animal breeding, biotechnology can also help in improving microorganisms for industrial purposes. Microorganisms are already being used to help clean up oil spills and to cleanse water from sugarcane processing residues. Genetically modified yeast is used for cheese and wine making procedures and for other fermentation products.

The breeding of interspecific hybrids predates molecular biotechnology as a method of bringing together desirable traits from different species. Although interspecific hybrids are difficult to produce, often plants of a species receive pollen from plants of other species and in rare cases fertilization can take place. The resulting hybrids are usually not fully viable or sterile, but they may survive and backcross with one or both of their parents, producing an offspring which will be more viable than the hybrid itself. In some cases the hybrid plants are able to explore new specialized niches or to adapt to a changed environment. The phenomenon of an occasional interspecific or even intergeneric cross, followed by hybrid survival and subsequent backcross to one or both parental species for several generations, results in the introduction in one species of some traits of the other and is called introgression. Hybrids that exploit new niches may also end up by evolving into new species (Futuyma, 1979). Genetic engineering is the method for the precise transfer of known genes from an organism to another beyond the limits imposed by the reproductive isolation of the species and can be considered a method of producing targeted introgression.

3. BIOSAFETY

Biosafety deals with the safe and environmentally sustainable use of all biological products and applications for human health, biodiversity, and environmental sustainability in support of improved global food security. Currently, for the general population, there are special issues of concern in regard to the biosafety of the products derived from biotechnology. These issues seem to be important only with regard to the products of genetic engineering (since molecular markers and marker-assisted breeding result usually in products that are equal to varieties created through the regular breeding procedures). It is called transgenic, an organism in which a foreign gene (a transgene) is incorporated into its genome. The transgene is present in both somatic and germ cells, is expressed in one or more tissues, and is

inherited by the offspring in a Mendelian fashion. Transgenics are also called "genetically modified organisms" or GMOs.
Some of the potential health risks associated with transgenics are:

- Allergenicity of the product of the introduced gene or genes. The nature of the allergenicity, in the few cases where it appeared, was the same as the allergenicity caused by the nontransgenic product coded by the transferred genes in the original species. The best known case involves a soybean variety that was transformed by a Brazilian nut gene, with the intention of enriching the soybean protein with methionin, an amino acid for which soybeans are deficient. Brazil nuts are allergenic, and the transformed soybeans showed the same allergenic compound that is present in Brazil nuts. A test against that allergenic compound allowed the identification of the problem and avoided the release of the modified soybean (FAO, 2001).
- Unexpected production of a toxic component due to the insertion of the new sequence (disruption or activation of gene sequences or of regulators). Analysis of the whole protein profile of the new organism and "in vivo" tests with experimental animals could identify the problem when it arises. In most cases, the utilization of proper insertion sequences through site-specific recombination will possibly prevent this problem.
- Expression of a gene product in undesired locations. An example of this is the production of the Bt toxin of *Bacillus thuringiensis* in the pollen grains of corn. This expression occurs because promoters that are not tissue specific are used (e.g., the 35S promoter). The utilization of tissue-specific promoters, which will be available for the "next generation" of genetically modified organisms should totally overcome the problem.
- Antibiotic resistance transferred from GMOs to bacteria in the human gut. Antibiotic resistance genes have been used in one of the steps for the production of the "first generation" GMOs. While the probability of such transfer is extremely low, it is theoretically possible (although not proven in practice). Because of this, the utilization of these genes is discouraged, and the new transgenics that are being currently produced are being made without the use of such markers.

A scientific panel commissioned by the British Ministry of Agriculture concluded that the risks to human health from GMO crops on the market are very low at the time of the analysis (GM Science Review Panel, 2003). However, it also concluded that on the crops developed GMOs may present greater challenges in risk management in the future.

Besides the health effects mentioned above, when making risk assessment of transgenics, all methods by which they have been produced should also be considered. There are many different procedures for DNA and/or RNA isolation, cloning, multiplication, introduction into a transmission vector, and insertion into other host organisms. The safety of the resulting organism should be evaluated according to its proposed use and in relation to the implications for health and for the environment. Additionally, questions to be considered should cover the environment in which it should be tested and released. The procedures must be examined carefully, and the properties of the organism in the environment analyzed. During this examination and analysis, one should try to anticipate any possible routes of escape (animal movement, pollen exchange, breakage, and transport of plant parts), any potentially harmful consequence for humans that may ingest or have contact with it (food safety, irritants, allergenics) and any harmful consequence for other species in the ecosystem that may somehow interact with it. When a transgenic is considered safe, it should be safe for all intended regular uses and for the environment for which the risk analysis was made. The Codex Alimentarius Commission, which is a joint FAO/WHO Commission, has recently created a task force to help countries to create and establish food safety standards for food derived from GMOs. Among options for enforcing compliance with such standards are inspections and audits, imposition of administrative and monetary penalties, and trade sanctions. However, significant practical, technical, and economic limitations mean that ensuring compliance is no easy task (McLean et al., 2002).

In addition to potentially negative impacts on human health, transgenics may also have some adverse impacts on ecosystems in general. Some of the potentially negative environmental impacts of the adoption of transgenics are:

- Displacement of local biodiversity and elimination of existing forms as new improved genotypes are adopted and old landraces are abandoned. This displacement has been occurring ever since new genotypes were first produced through traditional breeding procedures and does not represent a particular problem caused by the use of transgenics. Displacement is also caused by the introduction of alien species that are highly aggressive or competitive in comparison with the existing ones.
- Transgenic escape (due to natural out-crossing) may result in the creation of new plagues, especially in the creation of new weeds in the case of plants, or increased aggressiveness of previously existing weeds. This increase can happen only when there are wild relatives of the transgenic crop in the same area, such as in the center of origin or of diversity of the crop species.
- Harmful effects on nontarget species and waste of biological resources refer to possible effects on the population of pollinators, predators, symbionts, natural enemies, etc.
- Perturbation of biotic communities and adverse effects over the ecosystem that may happen whenever a new life form or genotype is introduced into an environment new to it. Since a genetically modified crop (e.g., maize) tends to behave in a way that is similar to the nonmodified variety, the

introduction of an unmodified new species causes a more significant disturbance on the ecosystem than the introduction of a transgenic variety of a previously existing species in that environment.

The need for biosafety evaluations prior to the release of genetically modified genotypes (in particular, transgenics) significantly delays their utilization and significantly increases research costs. In some cases the added costs may be so high that the eventual advantage of using the transgenic may not be worthwhile. These facts should be considered before engaging in any research using these techniques.

4. PRESENT STATE OF BIOTECHNOLOGY APPLICATIONS TO AGRICULTURE

To date, a relatively small number of commercial agricultural products have been developed through the application of molecular biotechnology techniques. Some vaccines against animal diseases, strains of yeast, and other products for cheese and wine fermentation, as well as few transgenic crops have been developed. Marker-assisted selection is still in its infancy due to the fact that detailed linkage maps are not yet available for several species. Requiring large investments, most of the research products that are being developed using genetic engineering are coming from the private sector in developed countries, even though all the research builds on the basic research funded by the public sector in developed countries. This section provides a conceptual overview of the present state of biotechnology applications to agriculture; the first section of Chapter 13 provides a technical overview that includes examples of specific products. ERS (2003) also provides a general overview of agricultural biotechnology issues.

Although traditionally private sector focuses more on applied research leading to marketable products, with the science-based biotechnology well underway, the private sector has increasingly taken on basic research. On the other hand, being origins of most scientific advances, public sector has also conducted applied agricultural research and, since passage of the Bayh-Dole Act in 1980, has become increasingly active in seeking intellectual property protection for their research outputs. In this new biotech era, the relationships between basic and applied research and between public and private R&D become closer than ever before. The close relationship between the University of California, Berkeley, and Novartis provide one example (Nestle, 2002, pp. 120-122).

Genetic engineering has been used to develop a variety of crop plants for different purposes, most of them related to solving production problems. Crops carrying the gene coding for the Bt insecticidal protein derived from *Bacillus thuringiensis* require less or no insecticide applications. Crops expressing the gene for resistance to gluphosate or to gluphosinate are easier to manage, because weed control in the field is done by the application of a contact herbicide (Roundup). At present, these two genes are the most common in the commercially available transgenic plants.

While few products are currently commercially available, a plethora of transgenic products covering a wide range of species are being developed, and are close to, or at, the field-testing stage, such as rice with pro-vitamin A, rice with increased iron content, male sterile Brassica, carnations with modified flower color and vase life, tomatoes and melons with increased shelf life, potatoes with altered starch composition, sweet potatoes with resistance to nematodes, and many species with resistance to a large number of different viruses. In the near future, it is expected that another group of "new generation" transgenic products will be created carrying more interesting and more complex traits, that can be of special interest for developing country farmers, such as crops with resistance to drought, to salinity, and to cold.

The most recent developments in the biotechnology area also indicate that the possible group of "new generation" products of genetic engineering should raise fewer safety concerns. One reason is the fact that new tissue-specific promoters (of the process through which RNA is formed along with DNA) are being set up. If these are used instead of the less-specific types of promoters available today, the problem of expressing a gene where it is not needed (like Bt maize expressing the Bt toxin in pollen grains) will be avoided. Another safety concern, that of giving rise to some unexpected product or toxin because the gene insertion occurs at random, and a regulatory sequence being disrupted and producing some unexpected product, will also subside. In fact, techniques for promoting site-specific insertion are being developed. Like any other field of science, more mistakes are made early on with a greater probability of defective products being developed. We expect future advances in biotechnology to yield a safer and cleaner generation of products that can be truly useful for the poor farmers in developing countries.

In assessing the safety of transgenics (or more generally, GMOs), it should be considered that their reproductive and agronomical behavior will be similar, if not equal to the behavior of the nonmodified species that was used to produce the transgenics, save for the added gene. Biosafety evaluations to date appear effective: To the present there appears to be no single case of documented environmental damage or of even an allergy reported as being due to their usage.

Unfortunately, implementation of biotechnology techniques is expensive to date, especially for research systems of developing countries. Not only is it expensive, potential users require considerable training, and its adoption requires the establishment of an appropriate regulatory environment with biosafety measures.

Biotechnology to date represents a set of techniques generally not available to developing countries. This means that for the full potential of the technology to be realized in developing countries, in particular, further innovations that can streamline its adoption and application are necessary. Biotechnology is not a cure-all nor an end-of itself; biotechnology should be seen just as one more tool to be used by research programs when appropriate and necessary. Biotechnology applications can make possible a large range of new products as long as there are not strong barriers to its adoption or applications. Nobody can realistically expect that researchers from developed countries will develop all the products necessary for agriculture in developing countries—it is not by chance that all the currently available transgenic crops are in a few species that are internationally traded. Biotechnology will have to become less costly both to make it worthwhile for researchers of developing countries to produce products specifically tailored to developing country needs, as well as for developing countries to use it themselves.

REFERENCES

Allard, R. W., 1960, *Principles of Plant Breeding,* John Wiley and Sons, Inc., New York, 485p.

Champoux, M. C., Wang, G., Sarkarung, S., MacKill, D. J., O'Toole, J. C., Huang, N., and McCouch, S. R., 1995, Locating genes associated with root morphology and drought avoidance in rice via linkage to molecular markers, *Theor. Appl. Genet.* **90**:969-981.

Economic Research Service, United States Department of Agriculture, 2003, Briefing Room: Agricultural Biotechnology, www.ers.usda.gov/Briefing/Biotechnology/.

Falconer, D. S., 1960, *Introduction to Quantitative Genetics*, Ronald Press, New York, 570p.

FAO, 2001, Los organismos modificados genéticamente, los consumidores, la inocuidad de los alimentos y el medio ambiente, Estudio FAO, Cuestiones de Etica 2, 26p.

Farias, J. C., Anjos, J. R. N., Costa, A. F., Sperandio, C. A., and Costa, C. L., 1996, Doenças causadas por virus e seu controle, in: *Cultura do Feijoeiro Comum no Brasil*, R. S. Araujo, C. A. Rava, L. F. Stone, and M. J. de O. Zimmermann, POTAFOS, Piracicaba, S.P. Brazil, pp. 731-760.

Futuyma, D. J., 1979, *Evolutionary Biology*, Sinauer Associated, Inc., Sunderland, Massachussets, 565p.

GM Science Review Panel, 2003, GM Science Review, First Report: An open review of the science relevant to GM crops and food based on the interests and concerns of the public, Report submitted to the Secretary of State for the Environment, Food and Rural Affairs, United Kingdom (July, 2003); http://www.gmsciencedebate.org.uk/report/default.htm.

Harlan, J. R., 1975, *Crops and Man*, American Society of Agronomy and Crop Science Society of America, Madison, 294p.

McCouch, S. R., Kochert, G., Yu, Z. H., Khush, G. S., Coffman, W. R., and Tanksley, S. D., 1988, Molecular mapping of rice chromosomes, *Theor. Appl. Genet.* **76**:815-829.

McLean, M., Frederick, R., Traynor, P., Cohen, J., and Komen, J., 2002, A conceptual framework for implementing biosafety: Linking policy, capacity and regulation, ISNAR Briefing Paper 47, ISNAR, The Hague; http://icsudqbo.alias.domicile.fr/events/GMOs/PDF/ISNAR%20Briefing%20Paper%2047.pdf.

Nakano, M., and Mii, M., 1990, Somatic hybridization between *Dianthus chinensis* and *D. barbarus* through protoplast fusion, *Japanese J. of Breeding* **40**, Supplement 1:26-27.

Nestle, M., 2002, *Food Politics*, University of California Press, Berkeley.

Ohgawara, T., Kobayashi, S., Ishii, S., Yoshinaga, K., and Oiyama, I., 1989, Somatic hybridization in citrus navel orange (*C. sinensis* Osb.) and grapefruit (*C. paradisi* Macf.), *Theor. Appl. Genet.* **78**:609-612.

Ramalho, M. A. P., Dos Santos, J. B., and Zimmermann, M. J. de O., 1993, Genética quantitativa em plantas autógamas. *Aplicações ao Melhoramento do Feijoeiro*. ed. Universidade Federal de Goiás, Goiânia, GO, Brazil, 271p.

The Crucible Group, 1994, *People, Plants, and Patents: Impact of Intellectual Property on Trade, Plant Biodiversity, and Rural Society*, IDRC, Ottawa, Ontario, Canada, 117p.

UNEP/CBD/94/1, 1998, *Convention on Biological Diversity: Text and Annexes*, OACI, Canada, 34p.

Vavilov, N. I., 1951, The origin, variation, immunity and breeding of cultivated plants, translated from the Russian by K. S. Chester. *Chronica Botanica* Vr. 1/6.

Chapter 13

THE POTENTIAL OF BIOTECHNOLOGY TO PROMOTE AGRICULTURAL DEVELOPMENT AND FOOD SECURITY*

Hoan T. Le[1]

[1]*Plant Biotechnology Officer, Plant Production and Protection Division, Food and Agriculture Organization of the United Nations, Rome, Italy*

Abstract: To satisfy the food demand of—and provide purchasing power to—the additional two billion people to be born in the next 30 years, mostly in developing countries, technological innovation, be it biotechnology and/or genetic engineering, is required to improve agricultural productivity. In fact, some technological advanced developing countries are not only applying the diverse range of biotechnology tools in agricultural research but also allowing the cultivation of the controversial genetically modified (GM) crops. There is also increasing evidence that the first generation of GM crops, developed for industrial agriculture, is having positive impacts on smallholders in some developing countries. Most importantly, biotechnology tools allow tremendous progress in tackling major production constraints that conventional means have failed to solve, such as biotic stress caused by diseases and insect pests and abiotic stresses caused by high temperature, salinity, drought, flooding, and problem soils. Preliminary research also indicates that GM technology can help to raise yield potential and produce crops that are more efficient in nutrient uptake, thus reducing not only fertilizer requirements but also food production costs, hence lowering food prices and making food more affordable to more people and contributing to food security and agricultural development.

---

* Disclaimers:
The designation employed and the presentation of the material in this information product do not imply the expression of opinion whatsoever on the part of the Food and Agriculture Organization of the United Nations concerning the legal status of any country, city, or area or of its authorities, or concerning the delimitation of its frontiers or boundaries.
The designations "developed" and "developing" economies are intended for statistical convenience and do not necessarily express a judgment about the stage reached by a particular country, territory, or area in the development process.
The views expressed herein are those of the author and do not necessarily represent those of the Food and Agriculture Organization of the United Nations.

Key words: agricultural biotechnology; agricultural development; biotic and abiotic stresses; environmental conservation; food security; genetically modified crops; gene revolution; green revolution; innovation; nutritional quality improvement; yield stability.

1. INTRODUCTION

The Food and Agriculture Organization of the United Nations (FAO) estimates that production of grain, fishery, fuel wood, round wood, and meat in developing countries needs to increase at 70, 40, 27, 83, and 100%, respectively, to meet the demand of more than two billion people projected to be born in the next 30 years (FAO, 1999). The projected increase is to be made with limited land and water resources. This requires technological innovation, in addition to conventional technologies such as the technological packages of the green revolution (GR) that have allowed food production to meet world food demand. However, there are signs that the GR technology has reached its plateau due to the decline in yields of rice and wheat (the GR crops) in the last decade. In addition, the GR has not met the needs of resource-poor farmers working the marginal environment, and there are still 800 million people who are malnourished due to inadequate food distribution. To meet their needs and those of future generations, the modern biotechnology tools of recombinant DNA, including genetic engineering, can form part of such innovation.

Indeed, modern biotechnology is revolutionizing the way in which the necessities of life—food, feed, fiber, fuel, and medical drugs—are being produced. In the agricultural arena, biotechnology tools have been used for animal and plant disease diagnostics, for production of recombinant vaccines against animal diseases, and for the improvement of livestock and crops. While the use of genetically engineered (GE) drugs and vaccines has not stirred much controversy, the deployment of genetically modified (GM) crops has met with fierce resistance, particularly in Europe, on ethical grounds and on concerns of perceived negative impacts of GM crops on the environment and food safety. Ethical considerations revolve around topics such as: (1) the "unnatural" nature of gene transfers across species, (2) possible widening of the gap between the rich and poor farmers and countries, and (3) the increase in global food supply's dependency on a few multinational corporations that control agricultural biotechnology and the seed industry. There are concerns that the negative publicity of and the resistance to GM crops by consumers in Europe may have hindered the transfer of this new innovation to the developing countries where increasing crop productivity is most urgent.

The importance of biotechnology for improving agricultural productivity was emphasized during the World Food Summit: Five Years Later (WFS:fyl) when the Heads of State and their governments renewed the pledge made in the 1996 WFS to halve the number of hungry people by 2015. However, the current low rate of hunger reduction indicates that to reach the goal, more commitments accompanied by concerted actions are required, and technological innovation is an important component of this effort Indeed, governments have "*called on the FAO, in conjunction with the CGIAR and other international research institutes, to advance agricultural research and research into new technologies, including biotechnology. The introduction of tried and tested new technologies including biotechnology should be accomplished in a safe manner and adapted to local conditions to help improve agricultural productivity in developing countries. We are committed to study, share and facilitate the responsible use of biotechnology in addressing development needs*" (FAO, 2002).

This chapter explores the potential contribution of biotechnology to food security and poverty alleviation. While the focus is on crop biotechnology, attention will also be given to forest, animal, and fishery biotechnology, when appropriate. The chapter is an updated version of a case study entitled "The Potential of Agricultural Biotechnology" originally written for a 2001 joint FAO/World Bank publication entitled *Farming Systems and Poverty: Improving Livelihoods in a Changing World* (Le, 2001).

2. TECHNICAL OVERVIEW OF BIOTECHNOLOGY

The Convention on Biological Diversity provides a broad definition for biotechnology, i.e., "any technological application that uses biological systems, living organisms, or derivatives thereof, to make or modify products or processes for specific use." This definition covers traditional tools and techniques commonly used in food and agriculture such as fermentation technology to produce the Chinese *hoisin* sauce (a soya-based sauce), the Korean *kim chee* (vegetable pickle), the Japanese *sake*, the Vietnamese *nuoc mam* (fish sauce), to mention some examples in Asian culture, as well as the wines and cheeses of Western cultures such as the French Camembert and the Italian Parmigiano. Non-controversial techniques such as tissue culture and traditional plant breeding, including the use of chemicals and radiation to induce mutagenesis for crop improvement, are also covered by this definition.

The Cartagena protocol on biosafety, however, defines "modern biotechnology" in a more narrow sense. This definition covers applications of: (a) *in vitro* nucleic acid techniques, including recombinant deoxyribo-nucleic acid (DNA) and direct injection of nucleic acid into cells or organelles or (b) fusion of cells beyond the taxonomic family (that overcome natural physiological reproductive or recombination barriers and that are not techniques used in traditional breeding and selection). Modern biotechnology, hence, encompasses the noncontroversial use of: (1) recombinant vaccine; (2) DNA markers for gene mapping, disease diagnostics, genetic resource characterization, DNA marker-assisted section in plant breeding, and DNA fingerprinting; and (3) the currently controversial genetic engineering to produce genetically modified organisms (GMOs), including the so-called transgenic crops, also called GM crops or GE crops.

Operationally, biotechnology consists of two components: (i) tissue and cell culture and (ii) DNA technologies, including recombinant DNA and genetic engineering. Both components currently are essential for the production of GM plants, crops, and animals.

Plant tissue and cell culture are relatively low-cost technologies, simple to learn, easy to apply, and widely practiced in many developing countries. Plant tissue culture aids crop improvement via micropropagation of elite stocks through *in vitro* culture of meristem, providing virus-free planting stock; generating somaclonal variants with desirable traits; overcoming reproductive barriers; and bringing desirable traits from wild relatives to crops through embryo rescue or *in vitro* ovary culture and pollination. Gene transfers from distantly related species can also be facilitated through cell culture and plant protoplast fusions whereas anther culture with doubling of chromosomes to obtain homozygous lines can help to speed up time in a plant breeding program. Tissue culture is particularly useful not only for *in vitro* conservation of plant germplasm but also for the exchange of disease-free germplasm among countries.

Animal cell and tissue culture are widely practiced in developed and technologically advanced developing countries, in particular for medical research. For livestock improvement, simple artificial insemination is widely practiced in developing countries whereas developed countries apply a wide ranging of advanced technologies, including semen sexing, embryo sexing, embryo transfer, *in vitro* fertilization, embryo cloning, and somatic cloning. The latter was used to clone Dolly, the sheep.

The second component of biotechnology includes DNA-based technologies and genetic engineering which make use of DNA sequences as molecular markers, the knowledge of the genes and the genetic code (DNA) for improvement of crops, trees, livestock and fish. The uses of DNA-based markers, which are not controversial, assist in the characterization of genetic resources for conservation and crop improvement. DNA markers are particularly useful for gene map construction for gene isolation and for marker-assisted selection (MAS) in conventional plant breeding programs. MAS is widely practiced in the private sector (Ragot, 2003) and can help to accelerate corn, soybean, and wheat-breeding programs (Mazur, Krebbers, and Tingley, 1999; Orf, Diers, and Boerma, 2004; and Ward, 2003). MAS are particularly useful in breeding for disease and pest resistance because they eliminate the need to introduce pests and pathogens for screening purposes. DNA markers are also important for diagnostics of diseases and pests, including monitoring pest populations for their management.

Although currently controversial, the most important feature of genetic engineering, also called genetic or DNA transformation, is the ability to move genes even across kingdoms, helping to enlarge the gene pools for all organisms. Genetic engineering allows useful genes from any living organism to be transferred to crops or animals for improving their productivity. Genetically altered bacteria or trees can be used in soil remediation. Furthermore, biosynthetic pathways can also be manipulated to produce added nutritional compounds in crops for food and feed, high value pharmaceuticals and other polymers, using plant and animal as bioreactors. The few examples of technologies present today only vaguely portent the vast implications for potential importance of biotechnology on agriculture in the next two decades.

3. CURRENT BIOTECHNOLOGICAL RESEARCH RELEVANT TO AGRICULTURAL DEVELOPMENT: THEIR POTENTIAL AND REALIZED IMPACT

Although the current commercial GM crops target simple traits and single genes, technological advances now permit the transfer of as many as 12 genes into a plant genome (Chen et al., 1998). Importantly, the recent development of binary bacterial artificial chromosome (BIBAC) (Hamilton et al., 1996) and transformation-competent artificial chromosome (TAC) (Qu et al., 2003) vector systems which are capable of transferring large foreign DNA fragments up to 150 kilobase into a plant nuclear genome, are useful breakthroughs for map-based cloning of agronomical important genes. This should accelerate gene identification and genetic engineering of plants (Hamilton et al., 1996). Such development may facilitate the alteration of more complex traits such as yield and tolerance to drought, salinity, heat, chill and freezing, as well as tolerance to problem soils such as salinity and aluminum toxicity.

3.1 Input replacement

One of the main criticisms of the GR has been that it bypassed poor farmers living in marginal environments and those who cannot afford the cost of inputs such as pesticides, fertilizers, and infrastructure cost for irrigation. The *gene* revolution is actually providing some measures to address these concerns, with GM crops that produce their own pesticides (such as the current crops of GM crops with various Bt genes transferred from different strains of the soil bacterium, *Bacillus thuringiensis*) and are efficient in nutrient uptakes. Concerning phosphorus, Mexican researchers at Centro de Investigaciones y Estudios Avanzados (CINVESTA) have demonstrated that GM tobacco and tropical corn are highly productive under low phosphorus soil conditions. However, these lines have not been tested under field conditions since 1999 due to the pressure from anti-GM groups (Herrera-Estrella, 2002). In addition, a research group at Purdue University has cloned a phosphate transporter gene from *Arabidopsis*. These genes were also found in other crops such as tomato, potato, and alfalfa. This will allow the development of GM plants with more efficient uptake of phosphate (Muchhal and Raghothama, 1999; Mukatira et al., 2001). Scientists are conducting research on biological nitrogen fixation with the objective of making nonleguminous crops, such as rice, fix their own nitrogen, or expanding the host range of nitrogen-fixing bacteria so that more crops can have such symbiotic relationships. This would also help to protect the environment by saving fossil fuel needed to produce nitrogen fertilizer.

3.2 Utilization and rehabilitation of marginal and degraded lands

In many regions of the developing countries, considerable areas of land exist that are unusable for agriculture due to soil and related constraints; other areas are utilized but produce suboptimal yields. Evidence indicates that there is great potential for increasing productivity in marginal areas. The pioneering work by Mexican, followed by American, researchers in elucidating the molecular mechanism of aluminum tolerance and in developing GM plants resistant to this toxic ion would have great impact on developing countries (de la Fuente et al., 1997; Mesfin et al., 2001), particularly in opening up vast areas in the Brazilian Cerrados and West African moist savannah to more intensive cultivation. Since acid soils cover 43% of tropical areas, aluminum-tolerant crops would help to extend crop production in these otherwise low-productivity lands without incurring the costs of soil amelioration.

On the other hand, 30% of arable land is alkaline, making iron unavailable for optimum crop production. Japanese workers recently demonstrated that a GM rice, engineered with barley genes, showed an enhanced tolerance to low iron availability and yielded four times more than nontransformed plants in alkaline soil (Takahashi et al., 2001). Encouraging results are also being made in the area of salinity tolerance. In the presence of 200 mM NaCl, GM tomato and canola plants reached maturity with very good fruit set and oil quality, respectively (Apsc ct al., 1999; Zhang and Blumwald, 2001; Zhang et al., 2001). In addition, climatic variability such as sudden drought or frost may have severe consequences for resource-poor farmers living in marginal environments.

3.3 Stabilizing yield potentials under dehydration stress of drought, salinity, freezing, and chilling

Biotechnology applications of research on environmental stress tolerance may ensure poor farmers of a stable harvest. Research into the physiological and biochemical basis for abiotic tolerance has been greatly aided by advances in molecular biology. American researchers working on freezing resistance (Jaglo-Ottosen et al., 1998) and Japanese researchers working on drought tolerance (Kobayashi et al., 1999) have isolated the same transcription factor from *Arabidopsis thaliana*, commonly known as thale cress, a weedy relative of canola (rapeseed), that when overexpressed in GM plants resulted in significant tolerance to drought, salt, and freezing stresses. The transcription factor, named CBF1 by the Americans, and DREB1A by the Japanese, was responsible for controlling the expression of other regulatory genes when plants undergoing dehydration stress. The WeatherGard™ technology is based on the CBF family of transcription factors. Accordingly, transgenic canola with the *Arabidopsis* transcription factor gene, CBF1, also shows drought tolerance as compared with its nontransgenic control. The WeatherGard™ technology also confers freezing, salinity, and drought tolerance in tomatoes. At the Centro Internacional de Mejoramiento de Maíz y Trigo (CIMMYT), the WeatherGard™ transgenic wheat seedlings with a drought-inducible promoter also show better recovery after 15 days without water (Goure, 2002), whereas at the International Rice Research Institute (IRRI), transgenic rice with DREB1A gene driven by the stress inducible promoter, rd 29A, showed very high drought tolerance at vegetative phase after three weeks without water and at reproductive phase after one week without water (Datta, 2004). Recently, German researchers investigating the molecular mechanism of drought resistance in the resurrection plant *Craterostigma plantagineum* (a native of South Africa) uncovered novel ABA- and dehydration-inducible aldehyde dehydrogenase genes which also have their homologues in *Arabidopsis*. Transgenic *Arabidopsis* overexpressing the genes were found to survive longer periods of drought (16 days) as compared with 12 days for nontransgenic (Kirch et al., 2001). Researchers at Cornell University recently reported that they had developed transgenic rice overexpressing trehalose that was more drought-resistant than the nontransgenic control (Garg et al., 2002).

The above examples indicate that biotechnology tools may help to bridge the gap between potential and actual average yields in developing as well as developed countries. Furthermore, these tools may help to move the yield potential to a higher level. The demonstrated yield increases—of 10% to 35% for GM rice overexpressing corn's photosynthetic enzymes (Ku et al., 2000) and fourfold for GM rice with a barley gene for its tolerance to low soil iron in alkaline soils (Takahashi et al., 2001)—allow for such optimism.

3.4 Improved postharvest storage life

It is envisaged that food production would increase through such activities as well as through the prevention of postharvest losses by extending shelf life of produce such as tropical fruits. Such increases in food production will lower prices and benefit rural wage earners and the urban poor. The net return to producers, including the poor ones, should remain high due to reduction in production costs from biotechnology. In fact, evidence shows that the technology is scale-neutral, i.e., resource-poor farmers benefit as much and even to a greater degree than larger farmers, as shown in Bt cotton studies in China, Mexico, and South Africa (Pray et al., 2001; Traxler et al., 2001, Ismaël et al., 2001), respectively.

3.5 Improved nutritional and medicinal quality

With 800 million malnourished people in developing countries, malnutrition can be addressed with nutritional genomics that use metabolic engineering to manipulate plant micronutrients for human health (DellaPenna, 1999; Tian and DellaPenna, 2001; Lucca, Hurrell, and Potrykus, 2002; Mackey, 2002). Although the production of the so-called functional foods may initially focus on wealthy consumers in the developed world, genes can be engineered into crops cultivated and consumed by poor farmers to improve their dietary requirements. Efforts are also being made to enhance nutritional values and/or reduced toxic or allergenic properties in food. These may be especially beneficial to poor farmers and people who do not have a balanced diet composed of diverse food sources. The example of GM rice with enhanced beta-carotene and iron is just the beginning of efforts of what has been coined "nutraceuticals." This would benefit people whether rich or poor in developed or developing countries. Indeed, the rice-consuming nations may benefit from the vitamin A-enriched Golden Rice. This GM rice can provide up to 40% of the daily allowance of vitamin A, based on a diet of 300 grams of rice per day, to prevent severe problems of vitamin A deficiency (Potrykus, 2001); whereas GM cassava with a reduction in cyanogen glycosides can prevent food poisoning (Sayre, 2000). Work is being carried out to produce GM Golden Mustard to provide the daily allowance of vitamin A in one teaspoon of oil (Dahwan, 2002). A GM rice with a high iron content and high in phytase (an enzyme that degrades phytic acid, an inhibitor of iron absorption) has been obtained that has the potential to alleviate iron deficiency anemia in rice-consuming populations (Lucca et al., 2002). A GM potato with an engineered gene from *Amaranthus hypochondriacus* (Chakraborty, Chakraborty, and Datta, 2000), called "protato" due to its high protein, will soon be available (Coghlan, 2003). Antioxidant compounds, such as lycopene and vitamin E, are being enhanced in GM tomato and canola, respectively. GM soybean and canola with modified oil are also in the pipeline.

In addition, substantial progress is being made in using GM crops to produce vaccines at low cost and suitable for storage conditions in developing countries (Arntzen, 1995, 1996; Langridge, 2000; Kong et al., 2001). Lower cost production of drugs using transgenic crops has the potential to improve the health of the poor who may not have access to currently expensive drugs. Furthermore, the availability of inexpensive, plant-derived vaccines against diseases endemic in developing countries such as hepatitis B, cholera, and malaria would offer poor people a chance to lead healthy and productive lives. It is hoped that an inexpensive plant-derived vaccine against AIDS may one day be developed. Currently, there is evidence that GM Bt-crops play an important role in providing safer food than that of traditionally bred crops through the reduction in mycotoxins produced by fungi infection through insect attack (Bakan et al., 2002).

While there are concerns of transgenic flow of crops that are used as bioreactors to produce drugs, genetic-use restriction technologies (GURTs) may be very useful for preventing contamination of the environment with certain drugs and/or vaccines (through gene leakages) by restricting unwanted transgene flow. On the other hand, in forestry, researchers have also been manipulating genes involved in floral development to produce nonflowering trees, thus improving not only wood productivity but also preventing unwanted gene flow through pollens and/or seeds (Meilan, 2001).

Biotechnology tools are being used to investigate the mechanism of apomixis in plants for its potential applications in agriculture. This important trait could have enormous potential impact if the technology can be made available to resource-poor farmers who could replant hybrid seeds which retain permanent hybrid vigor in apomictic hybrid varieties.

3.6 Reduced environmental pollution and improved forest plantations' productivity

In forestry, researchers at Michigan Tech University made a breakthrough in tree engineering with potentials for reductions in energy, costs, and environmental pollution from pulp mills. Using antisense technology to suppress an enzyme *Pt4CL1* encoding 4-coumarate: coenzyme A ligase (4CL) in the lignin biosynthetic pathway, these workers have produced GM aspen with a 45% reduction of lignin and a 15% increase in cellulose (Hu et al., 1999). Recently, the same researcher (now at North Carolina State University) and his colleagues have modified the expression of both 4CL and a second gene, CAld5H, and reduced lignin content in transgenics by 45% to 50%, while increasing cellulose by 30%. The transgenic trees also grew faster, thus demonstrating the first successful dual-gene alteration achieved through genetic transformation in forestry (Li et al., 2003).

GM technology can be valuable for tree domestication by making large improvements in tree productivity (Strauss, DiFazio, and Meilan, 2001). Although forest plant plantations account for only 0.2% to 17.1% of forest areas in several southern hemisphere countries, these plantations produce 50% to 95% of those countries' wood production (Nambiar, 1999). Coupled with GM technologies, these plantations can be more productive, thus further reducing the area needed for plantations and the pressure on natural forests, allowing their restoration and conservation.

3.7 Restored environmental degraded soils

Biotechnology applications will have positive impacts on environments degraded through conventional practices, e.g., restoration of degraded soil using phytoremediation with engineered crops and/or microorganisms. In fact, French et al. (2001) and Hannink et al. (2001) demonstrated detoxification of explosives by transgenic plants expressing a bacterial nitroredu-ctase. Recently, transgenic cottonwood with mercuric ion reductase gene has been field tested for remediation of soil contaminated with mercury (Che et al., 2003). The GM tomato and canola engineered with the anti-port protein described previously (Apse et al., 1999; Zhang and Blumwald, 2001; Zhang et al., 2001) also have potential for phytoremediation of saline soils as the uptake salt was sequestered into leaves' vacuoles that can be removed while the fruits and oil seeds can be harvested for food.

3.8 Bio-fuels to reduce carbon dioxide emissions to the environment

As a source of renewable energy, GM crops can be engineered to produce fuel directly or indirectly through the processing of their biomass. Production of biomass for fuels such as alcohol would not necessarily contribute to additional carbon dioxide emissions into the atmosphere and could be especially beneficial if such fuels were used instead of petroleum-based fuels to meet the growing needs of the Third World (Guy et al., 2000).

3.9 Animal husbandry

In the livestock and poultry sector, while transgenic research has focused mainly on the production of therapeutic compounds, agricultural applications include the production of transgenic livestock and poultry for enhanced growth and feed efficiency, enhanced product quality, and reduced environmental burden. While the earlier transgenic pigs suffered from deleterious effects, recently-produced transgenic pigs—using different constructs such as (i) a bovine growth hormone transgene driven by the zinc-inducible methallothionine promoter (Pursel et al.,

1997); or (ii) the human insulin-like growth factor 1 driven by chicken regulatory sequences (Pursel, Coleman, and Wall, 1996)—were leaner than nontransgenic pigs (although in the latter, leanness were expressed only in female transgenic pigs).

While increasing lean meat yield helps to improve product quality, altering milk composition in transgenic cows to acquire traits such as reduced lactose content would suit the needs of the lactose-intolerant population, while the removal in transgenic cow's milk of beta-lactoglobulin, an allergen to 10% of the consumer population, would help those allergy sufferers. Modifying milk composition would also affect cheese processing and quality. Recently, New Zealand's researchers have enhanced milk composition and milk processing efficiency by increasing the casein concentration in milk. The transgenic cows were engineered with additional copies of the genes encoding bovine beta- and kappa-casein and the resulting nine cows, representing two high-expressing lines, produced milk with an 8% to 20% increase in beta-casein, a twofold increase in kappa-casein levels, and a markedly altered kappa-casein to total casein ratio. High-protein milk is desired by cheese makers although there was no information on the cheese derived from this altered milk (Brophy et al., 2003). It should be noted that modern day cheese making has been fundamentally changed due to the use of chymosin enzyme, also called rennin, produced by GE bacteria. Traditionally, rennet (which contains chymosin) was processed from the fourth stomach of slaughtered, newly born calves for use in cheese making.

Highly intensive industrial livestock production has caused environmental pollution. Canadian researchers have recently developed transgenic pigs with a bacterial phytase gene. The transgenic pigs require almost no inorganic phosphate supplementation and excrete up to 75% less phosphorus than nontransgenic pigs, thus lessening pollution caused by phosphorus. The development of disease-resistant GM livestock, poultry, and aquatic species can benefit human health and the environment through decreased use of antibiotics, thus minimizing the chance of antibiotic-resistant "super bugs" and reducing production costs (Sang, 2003). While the benefits of transgenesis in livestock is clear—for protecting animals against diseases, reducing pollutants, optimizing digestion, improving growth and fertility, optimizing meat and milk composition, etc.—the limitations to its success during the last 15 years include the technical difficulty and enormous costs in generating transgenic farm animals (Houdebine, 2002).

On the GM aquatic species front, a handful of countries are working to increase their productivity, using mainly the growth hormone genes. Nevertheless, there has not been any commercial release of transgenic fish although taste testings were conducted for GM trout in Canada (Entis, 1998) and GM tilapia in Cuba (Guillen et al., 1999).

In parallel with crop biotechnology, poor farmers can also benefit from advances in animal biotechnology due to a livestock revolution occurring now in most developing countries. Research has shown that the rural poor and landless get a higher share of their income from livestock than better-off rural people. Hence, an increase in the consumption of animal products can actually help to increase the food purchasing power of the poor. This livestock revolution could become a key means of alleviating poverty in the next 20 years if proper policies and investments are in place (IFPRI, 1999). Animal biotechnology can supply abundant and healthier animal protein at lower cost which may also be achieved due to faster growth on less feed. The investments of poor farmers who own few livestock will be better protected through improvements in animal health, which may be achieved through better and cheaper vaccines. The vaccines may be produced by recombinant DNA, including plant-derived recombinant vaccines for livestock, suitable for storage conditions in developing countries. The detection of diseases via molecular-based diagnostics will also help control the spread of disease among village herds and will upgrade livestock health. This will help rural communities in general and provide household food security at the individual family level.

Agricultural biotechnology, in particular plant biotechnology, is benefiting greatly from the *Arabidopsis* and the rice genome sequencing projects and post-genomics research in functional genomics, proteomics, and metabolomics. The knowledge of *Arabidopsis*—its genes, clustering of genes with similar expression patterns, and their order—may be used to isolate, characterize the corresponding genes, and understand gene order and expression patterns in other crop plants (Somerville and Somerville, 1999). This knowledge, coupled with opportunities to move genes across species barriers, broadens crop gene pools, which has not been possible (or possible only with tremendous difficulties) using conventional approaches. Similarly, animal biotechnology should benefit from the completion of the human and mouse genome sequencing projects (Mouse Genome Sequencing Consortium, 2002; The International Genome Sequencing Consortium, 2001) whereas fish biotechnology is the beneficiary of the DNA sequencing of the Japanese pufferfish, a specialty in sushi restaurants. In fact, this fish has the smallest known genome among vertebrates, eight times smaller than its human counterpart, but has about the same number of genes. Hence, the pufferfish genome, with very little repetitive sequence, has become a tool for discovering genes in the human genome (Chapman et al., 2002). It should be noted that the universality of the genetic code and the common ancestor shared by all organisms during their evolutionary pathways have made comparative genomics possible as well as the expression of genes from bacteria in plants or vice versa, with some modifications.

4. GM CROP ADOPTION

4.1 Global trends

Since 1996, GM crops have been adopted at an exponential rate in developed and technologically advanced developing countries. In only eight years, GM crops' hectarage has extended from 1.7 to 67.7 million hectares (mha), a more than 40-fold increase. This indicates that in spite of the GM crops' controversy, there has been no sign of a slowdown in its adoption by farmers.

The 2003 GM crops' hectarage was cultivated by 7 million GM farmers in 18 countries, including Brazil and the Philippines that officially approved GM crops for the first time, an increase from 6 million farmers in 16 GM growing countries in 2002. Furthermore, the year 2003 saw that six countries, including Brazil and South Africa, grew 99% of global GM crops areas. This has broken the three consecutive years (2000, 2001, and 2002) that the four countries, namely, the United States, Argentina, China, and Canada, held such record but also indicated an increase in GM growing areas by other countries. In 2003, the United States grew 63% of global total, followed by Argentina (21%), Canada (6%), Brazil (4%), China (4%), and South Africa (1%). While the dominant GM crops have been commodity crops, i.e., soybean, cotton, canola, and corn, the dominant traits have been herbicide resistance and insect resistance (Fig. 13-1).

4.2 Developing country trends

In the developing countries, the adoption of GM crops has increased steadily—from 14% in 1997, to 16% in 1998, 18% in 1999, 24% in 2000, 26% in 2001, 27% in 2002, and 30% in 2003 (Fig. 13-2). In fact, the growth rates in recent years are faster in developing countries as compared with industrialized countries: 26% versus 17% between 2000 and 2001, 19% versus 9% between 2001 and 2002, and 28% versus 11% between 2002 and 2003 (James, 2001, 2002, 2003). However, the bulk of GM area in developing countries has been mainly in Argentina's industrial agriculture (which grew 22%, 23%, and 21% of world GM area in 2001, 2002, and 2003, respectively) and mainly for the industrial crops soybean, cotton, and corn. If Argentina's statistics were excluded, developing countries would have about 9% global GM area, contributed largely by China and South Africa cultivating mainly Bt cotton. Significantly, more than 85% of the 7 million farmers benefiting from GM crops in 2003 were resource-poor farmers planting Bt cotton in nine Chinese provinces and in South African's province of KwaZulu Natal.

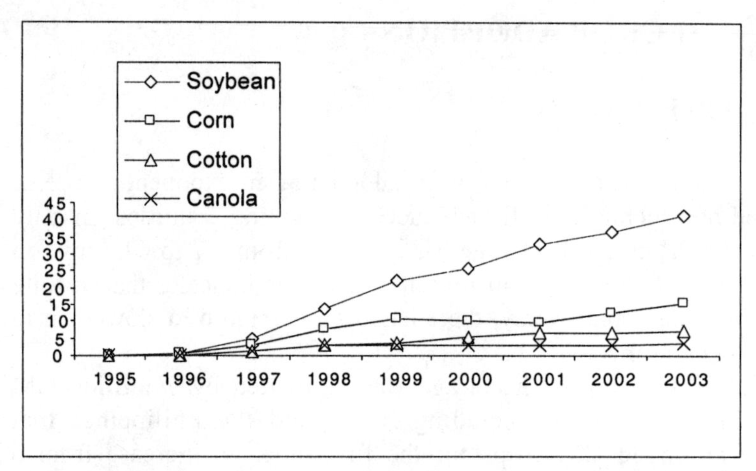

Figure 13-1. Global area of major transgenic crops from 1996 to 2003 (million hectares)
Source: James (2003).

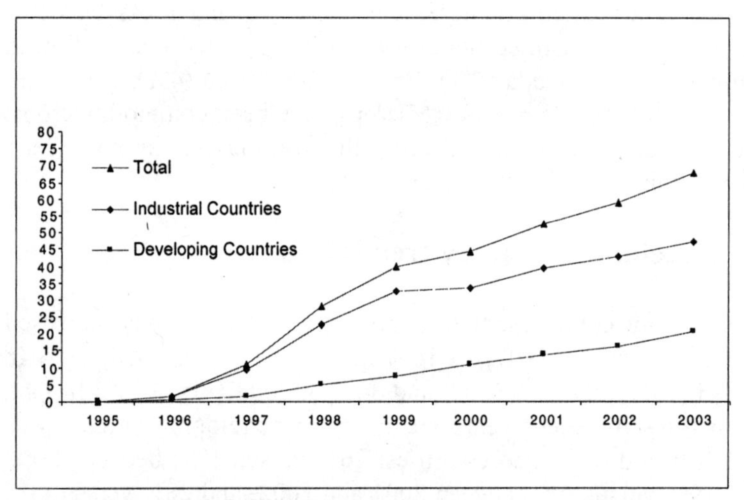

Figure 13-2. Global trends of transgenic crops in developing and industrial countries
Source: James (2003).

In 2003, the three most populous countries in Asia (China, India, and Indonesia) and the three largest economies in Latin America (Argentina, Brazil, and Mexico) are all officially growing GM crops. However, there remain a large number of developing countries with low technical capacity and their vast resource-poor farmers whose lives remain untouched by innovation. The call by the seven national academies of sciences for the development of GM crops to benefit poor farmers in developing countries remains to be fulfilled (NAS, 2000).

5. RESEARCH TRENDS: CROPS AND TRAITS

In January 2003, FAO launched an online database (FAO-BioDeC, 2003) to monitor the trends in the status of development, adoption, and application of crop biotechnologies, including GM crops in developing countries. The database encompassed data collected and kindly provided by ISNAR Biotechnology Service (IBS), part of the International Service for National Agricultural Research (ISNAR). Both FAO-BioDec (2003) and IBS/ISNAR's Next Harvest© Databases (Cohen et al., 2003) show that not only a diverse range of crops important for resource-poor farmers, including food security crops (e.g., rice and corn), are being studied, but also *traits* important to resource-poor farmers in developing countries (e.g., tolerance to abiotic stresses and quality traits). The number of transformation events was recorded. A transformation event is a unique insertion of the gene into the plant genome. Each transformation event may contain a unique gene, gene promoter, gene marker, and gene location within the plant genome. Cohen et al. (2003) compiled distributions by crop groups showing that the percentages of transformation events are 35% for cereals, 15% for vegetables, 12% for fruits, 10% for roots and tubers, 7% for fibers, 9% for oil crops, and 12% for others. The most researched crops are: corn (21% of all projects), rice (16%), potato (9%), cotton (7%), soybean (6%), tomato (6%), other fruits (7%), other vegetables (11%), and other crops (22%). Rice is the most common research subject in Asia while potatoes and corn are most common in Africa. In Latin America, potatoes, corn, soybean, and sugarcane are most common (Fig. 13-3 and Cohen et al., 2003).

The FAOBioDec (Fig. 13-4) and Cohen et al. (2003) show 15 and 8 commercial events, i.e., transformation events that have gone through all required regulatory tests and released for commercialization, respectively, in developing countries. As the ISNAR work focused exclusively on 16 selected countries, there are differences between its data and that of FAO. Even with minor differences in numbers of events in commercial, field, and experimental phases, the trends are clear that in all phases, Latin America and the Caribbean have the highest number of activities in GM crop research, followed by Asia, whereas Africa ranks third. European countries in transition have high numbers in experimental phase and fewer in field trials, whereas the reverse is for the Near East, with high field trials and low numbers of GM crops in experimental phases. The last two groups, i.e., countries in transition in Europe and countries in the Near East, have no commercial GM crops (Fig. 13-4).

Both databases are consistent in showing that in spite of the "molecular divide" (Fresco, 2003), public sector researchers in technologically advanced developing countries are forging ahead to develop a diverse range of GM crops for traits important for their own needs (Fig. 13-4; FAO-BioDec, 2003; Cohen et al., 2003). Furthermore, the so-called "molecular divide" appears to exist even among developing regions: The development of GM crops in Asia and Latin America are running neck to neck, with 181 and 199 GM crops, respectively, while Africa has 33, and the Near East and Eastern Europe with 15 and 16, respectively (Fig. 13-5). In developing countries, disease resistance (bacterial, fungal, and viral diseases) GM research is highest at 147, with insect resistance (IR) at 91, product quality (PQ) traits at 77, herbicide resistance (HR) at 72, tolerance to abiotic stresses at 36, and multiple traits at 21 (Table 13-1). While GM research on herbicide resistance is

Table 13-1. General trend of GM research in developing countries

Traits	Number of events
Disease resistance (bacterial, fungal, viral diseases)	147
Insect resistance	91
Product quality	77
Herbicide resistance	72
Abiotic stress tolerance	36
Multiple traits	21

Source: FAO-BioDec (2003).

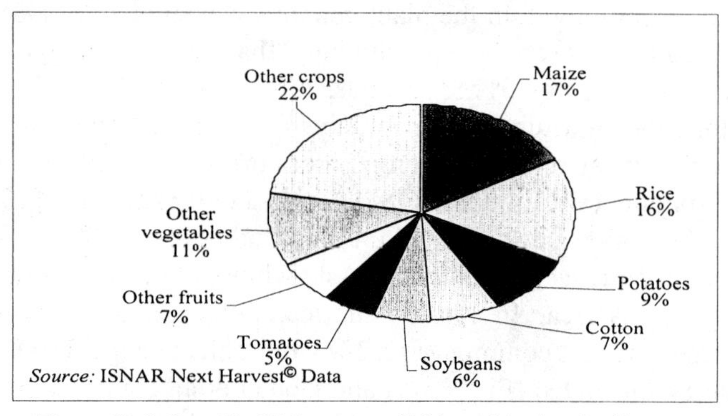

Figure 13-3. Specific GM crops available and in the pipeline

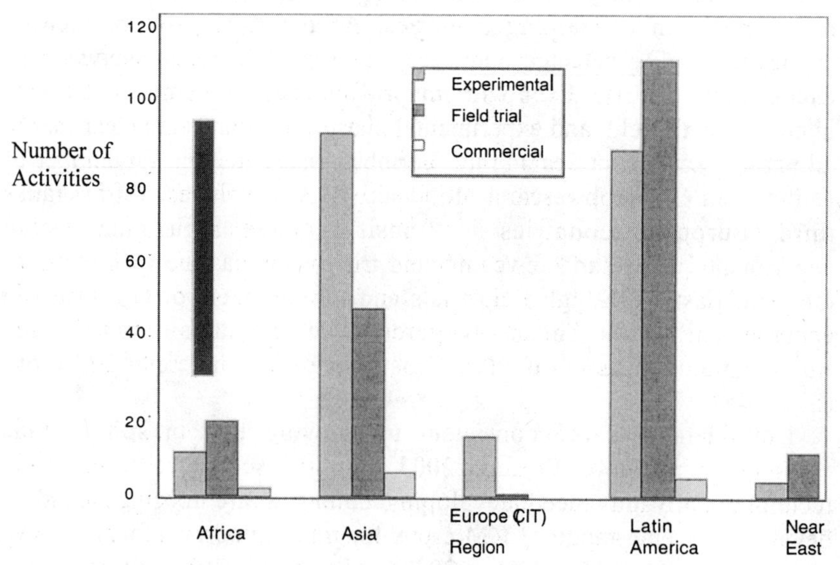

Figure 13-4. Regional status of GM crops in developing countries
Source: FAO-BioDec (2003).

high in Latin America, its magnitude is lower in Africa, Asia, and Eastern Europe. Not surprisingly, insect pest and pathogen resistance is high on Asia and Latin America's research agendas as well as product quality traits, with Asia leading in numbers. Asia is also leading in the amount of GM crop research for abiotic stresses. In Eastern European countries in transition, research on product quality is highest; whereas tolerance to abiotic stress and herbicide research is comparatively lower (Fig. 13-5).

Cohen et al. (2003) reported that the genes used are mainly "off the shelf," i.e., genes or genetic elements that are already available in commercial products, or that are the property of public research institutes and universities. These are not locally isolated genes, some already being developed by developing countries public sector institutes. Surprisingly, the number of successful projects involving public and private sectors is very low. National work includes many exciting developments, e.g., Chinese public sector researchers isolated 20 new Bt genes, Malaysian researchers isolated a tissue-specific promoter for rubber, Egypgian researchers at the Agricultural Genetic Engineering Research Institute (AGERI) collaborated with Pioneer Hybrid to isolate four corn promoters, and the Egyptian researchers also developed new transformation protocols and regeneration systems for wheat (Cohen et al., 2003). Except for China, no public sector product is close to commercialization. Closest is Egypt with squash and potato, whereas South Africa's GM sugarcane and potato are three to four years away (Cohen et al., 2003).

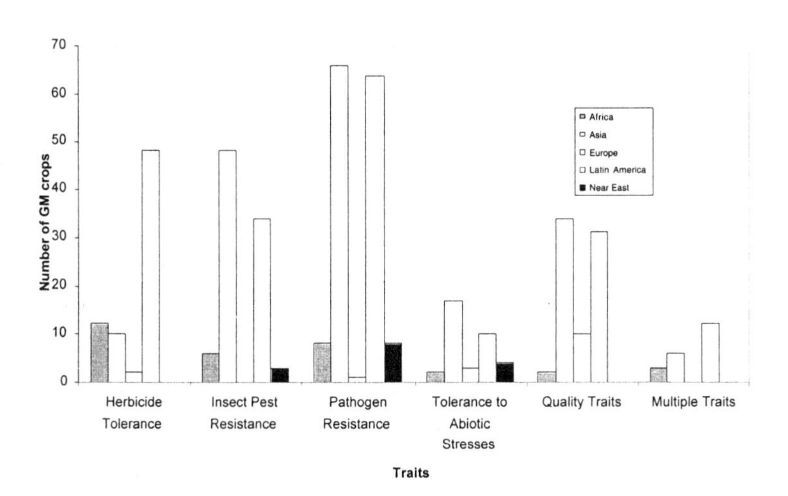

Figure 13-5. Regional distribution of trait groups being researched for GM crops in the pipeline
Source: FAO-BioDec (2003).

6. PRODUCTS IN THE PIPELINE

6.1 Crops

As described above, a diverse range of traits and crops are being studied in developing countries. These include staple food such as rice, banana and plantain, fruit crops such as mango and papaya, and tropical industrial crops such as palm oil and coconut. The traits engineered are abiotic stress tolerance such as for drought, salinity, freezing, and aluminum toxicity, important for marginal and small holder farmers (Table 13-2).

Table 13-2. Some examples of GM crops in the pipeline with traits important for developing country/resource-poor farmers

Country	Crop(s)	Traits
Argentina	Wheat	Baking quality (high molecular weight gluten)
	Corn	Oil composition
Brazil	Bean	Golden mosaic virus resistance
Bulgaria	Grape	Freezing tolerance
China	Corn	High lysine
	Rice	Salt tolerance
China, India, Philippines	Rice	Resistance to bacterial blight
Egypt	Wheat	Drought tolerance
	Wheat	Salinity tolerance
	Squash and melon	Zucchini yellow mosaic virus resistance
India	Cabbage	Insect resistance (black diamond moth)
	Potato	High protein with gene Ama1 from *Amaranthus*
	Mustard	High vitamin A
Indonesia	Ground nut	Virus resistance
Indonesia, Malaysia, Philippines, Thailand, Vietnam	Papaya	Papaya ringspot virus resistance
Indonesia, Kenya, Zimbabwe	Sweet potato	Virus resistance
Malaysia	Rice	Tungro virus resistance
Malaysia and Indonesia	Palm oil	Modified oil composition
Mexico	Corn	Aluminum tolerance
Philippines	Bananas and plantains	Virus resistance (banana bunchy top virus, banana bract mosaic virus)
	Coconut	Improved fatty acid content
	Mango	Delay ripening
	Papaya	Resistance to papaya ringspot virus
	Rice	Drought tolerance
	Rice	Salinity tolerance
South Africa	Potato	Drought and heat tolerance
Zimbabwe	Cowpea	Cowpea mosaic virus
	Corn	Insect resistance

Sources: FAO-BioDeC (2003) and Next Harvest© databases.

Table 13-3. GM trees in research and/or in field trials

Country	Species	Traits
Canada	Black Spruce, tamarack, white spruce, European larch	Disease tolerance, insect resistance
China	Poplar	Insect resistance
Malaysia	Rubber	Pharmaceutical compounds
Belgium, Netherlands, Sweden	Apple	Fungal resistance, markers
United Kingdom (UK), Netherlands	Paradise apple	
Spain	Eucalytus	Markers
UK	Eucalytus grandis	
Portugal	*E. globules*	
Denmark, Norway, Germany	European Aspen	Markers, rol genes, herbicide resistance, phytochrome A synthesis
France, Spain, UK	Hybrid aspen	Altered lignin, markers, and growth tests
Denmark, France, Germany, UK	Cottonwood poplar	Insect resistance, herbicide resistance, phytoremediation and male sterility
France	Quaking aspen	
Spain	Plum	Plum pox poty virus resistance
Finland	Norway Spruce, Scots pine, Silver birch	Markers, nitrate reductase
Italy	Olive	Water stress and disease resistance, improved rooting
	Cherry	Improved rooting
Spain	Orange	Markers
United States of America	American chestnut	Fungal resistance
	American black walnut	Insect resistance and reduced aflatoxin contamination
	Poplar	Altered lignin content
	Apple	Insect resistance
	Plum	Plum pox poty virus resistance
	Black spruce and white spruce	Insect resistance
	Teak	
	Pine	
	Larch	
	Coffee	Caffeine-free coffee beans, delay, and uniformed ripening
France	Coffee	Insect resistance
Japan and UK	Coffee	Caffeine-free coffee beans

Sources: Huang and Wang (2002); Gartland, Dellison, and Fenning (2002); Fenning and Gershenzon (2002); Lubick (2002); Natural Resource Canada (2003).

6.2 GM trees/forests

While the widespread adoption of GM crops worldwide has made headlines, the only commercialized GM tree is GM papaya, which is resistant to papaya ringspot virus, being planted in Hawaii. In the pipeline are transgenic fruit and industrial crops (Table 13-3). Substantial research has been made on GM trees, and there are now a total of 117 experimental plantations with GM trees belonging to 24 species around the world (Carnus et al., 2003). Table 13-3 shows examples (not an exhaustive list) of GM trees being researched and/or in field trials in a number of countries.

6.3 GM livestock, poultry, and aquatic species

As of now, there is no commercial production of transgenic livestock, poultry, and aquatic species for food. Table 13-4 shows examples of GM livestock and aquatic species in research phase and/or in tasting trials.

There is no commercial release of GM aquatic species but that may change since AquaBounty, a private company, has been seeking regulatory approval for commercialization of transgenic Atlantic salmon with a growth hormone gene. This GM salmon has 4-6 times the growth rate and a 10%-20% improvement in feed conversion efficiency relative to nontransgenic salmon, and thus can be grown with less environmental pollution. If commercially successful, this transgenic salmon can improve profitability for fish farmers due to its shorter production times and reduced feed costs. The same company also has transgenic lines of rainbow trout and tilapia (Entis, 2003). Worldwide, 11 countries are engaging in genetic engineering of 23 aquatic species, using nine genes (Hallerman, 2003; Table 13-4). The majority of genetic modification involves growth hormone genes, although transgenic aquatic species for environmental tolerance and disease resistance are also being developed. The United States, China, Canada, UK, Republic of Korea, and Cuba are the main players in this area. There have been environmental and food safety concerns about GM aquatic species that are being addressed to harness the benefits, such as only using sterile, triploid female fish to prevent gene escape.

Table 13-4. Some examples of GM livestock, poultry, and aquatic species in the pipeline

Country	Livestock	Traits
Australia	Sheep	Increased wool growth (sheep)
		Increased milk
	Cattle	Improved meat quality
Canada	Goat	Biosteel in milk
	Pig	Phytase gene (reduced phosphorus in excretion)

Table 13-4. (continued).

Country	Livestock	Traits
China	Pig	
New Zealand	Cattle, sheep	Altered milk composition (cow)
United Kingdom	Chicken, cattle, sheep	Pharmaceutical compounds
United States	Cattle, pig, sheep	Faster growth, leaner meat
	Chicken	Disease resistance
	Rabbit	
Canada, United States	Rainbow trout, Coho salmon, Atlantic salmon, Pacific salmon	Faster growth, disease resistance
Canada, China	Goldfish	Increased cold tolerance
China	Carp, goldfish, Wuchang fish	Faster growth and disease resistance
	Loach	Human growth hormone
Cuba	Tilapia	Faster growth
Korea	Mud loach	Faster growth
France	Oyster	Reporter gene
India	Indian catfish	
Israel	Common carp, tilapia, gilthead seasbream	Faster growth and cold tolerance
Japan	Japanese abalone, rainbow trout	
Japan	Medaka	Reporter genes
Norway	Zebra fish	Sterility, vaccines
Singapore	Zebra fish, Killi fish, walleye, riceland shrimp, the green mussel	Reporter genes for environmental monitoring
United Kingdom, United States	Tilapia	Pharmaceutical compounds, vaccines
United States	Shellfish: brine shrimp, white shrimp, fresh water prawn, crayfish	
	Medaka	Reporter genes for contaminants
	Striped bass, largemouth bass	
	Japanese abalone, red abalone, blue mussel, Pacific oyster, Eastern oyster,	Improve growth rate
	Northern pike	Disease resistance
	Channel catfish	Pharmaceutical compounds

Sources: Hallerman (2003); McLean (2002); Sang (2003); Hulata (2001); and NAS (2002).

7. IMPACTS

Although there have been arguments that the first generation of GM crops concentrated on input and simple traits designed for industrial agriculture in developed countries, which may not benefit small farmers in developing countries, there is increasing evidence to the contrary. The cultivation of GM crops in some developing countries with high research and extension capacity in biotechnology demonstrates that GM is already making an impact through reduced pesticide costs, reduced risks of poisoning, environmental benefits, and productivity gains. The number of farmers that benefited from GM crops increased from 3.5 million farmers in 2000 to 5.5 million in 2001 to 7 million in 2003. More than 85% of the farmers that benefited from GM crops in 2003 were resource-poor farmers planting Bt cotton, mainly in China and also in South Africa (James, 2003).

In China, Pray et al. (2001) presented evidence of higher economic returns to small farmers who planted Bt cotton, who also required less hospitalization due to pesticide poisoning, than those cultivating non-Bt cotton. The use of Bt cotton has reduced pesticide use by 80% in Hebei Province in China. Since pesticide use in cotton accounts for 25% of global pesticide consumption in crops, this has great potential environmental and health benefits. South Africa's experience has shown that small farmers can also benefit from Bt cotton. The number of small farmers participating in the cultivation of Bt cotton in KwaZulu Natal province increased from four in 1997, to 400 in 2000, and 644 in 2001 (Thompson, 2001; Webster, 2000), i.e., from only 0.1% of farmers in 1997/98 to over 90% of farmers by 2001/02 (Bhattacharya, 2003). The farmers, 60% of whom are women, typically farm between 1 and 3 hectares. The GM cotton boosted the yields between 50% and 89% compared to its conventional counterpart (Bhattacharya, 2003). Besides increasing yields, the GM cotton also reduced the need for pesticide spraying, which had the additional benefit of saving labor, important in a region ravaged by HIV/AIDS. This indicates that small farmers are realizing the benefits in growing GM crops. In Kenya, it has been projected that two sweet potato biotechnologies, GM virus and weevil resistance, will generate annual gross benefit of US$5.4 million and US$ 9.9 million, respectively (Qaim, 2000). Due to the semi-subsistence nature of sweet potato, the producing households will be the main beneficiaries. The high efficiency of the research projects is confirmed by significantly positive returns on their investments (Qaim, 2000). In Kenya, disease-free banana plantlets derived from plant tissue culture have greatly increased yields from 8-10 to 30-40 t/h (Africa News Service, 2000; Thompson, 2001).

In Argentina, the high adoption rate of GM crops shows that they have already had an impact there. The illegal smuggling of GM seeds from Argentina to Brazil for cultivation indicates that Brazilian farmers, largely commercial growers, appreciate the benefits of GM crops over conventional ones (Smith, 2003) In Mexico, cultivation of Bt cotton for two years has resulted in an estimated US$5.5 million in economic surplus, of which about 84% accrued to farmers and 16% to seed suppliers (Traxler et al., 2001). In Cuba, the country's biotechnology strategy is already giving high payoffs in terms of royalties from proprietary technologies, particularly in biomedicine. For agriculture, Cuba is developing toolkits for plant disease diagnosis. Cuba's GE vaccines against cattle ticks and against an enterotoxic *Escherichia coli*, has already been sold on international markets (Borroto, 2000). The use of recombinant vaccine against cattle tick has reduced Cuba's pesticide imports from US$2.5 million to only US$0.5 million annually (Borroto, 2000). Its production of a patented bionematicide will allow the reduction of toxic nematicides used in banana plantation (Lehman, 2000).

7.1 Immediate impacts of tissue culture and micro-propagation

Although DNA technologies are beginning to benefit small farmers in developing countries, the immediate impact for many countries, particularly those with low technical capacity, will be in the production and distribution of disease-free and high-quality planting material of the native clones of vegetatively propagated plants. These include banana, plantain, cassava, yams, potato, sweet potato, pineapple, sugarcane; many fruit trees such as apple, pear, plum, date palm, mango, and litchi; and many ornamental shrubs and flowers. The benefits of micro-propagation are immediate, and the availability of cheap labor in the developing countries provides a competitive edge in the use of this technology. Micro-propagation of banana and sugarcane has created rural jobs in Cuba and promoted exports of propagules of ornamental plants from India to Europe. Within the last five years, nearly 100 plant micro-propagation companies have been established in India by the private sector. In Cuba, if micro-propagation capacity can be scaled up to satisfy domestic demand, the country can save $15 million (U. S.) annually for expenditure on imported potato seed stock. Cuban cottage industry, based on tissue culture, is providing part-time employment opportunities for rural housewives. In China, micro-propagation of virus-free sweet potato seed in Shandong, which resulted in an average yield increase of at least 30%, gave an internal rate of return at 202% and a net value of $550 million (U. S.) (Fuglie et al., 2001).

Applications of agricultural biotechnology will continue in a dynamic manner in both developed and developing countries, albeit at different paces. It is difficult to foresee the full impact of the technology with regard to agricultural growth and poverty alleviation, as it is contingent both on technology development and on how the technology is integrated into national programs. However, economic studies conducted by FAO indicate that the current trend in biotechnology will only reinforce existing trade patterns in cereal and oil crops unless developing countries take measures to strengthen their technical capability (FAO, 1999). If sustainability is factored in, oil-producing perennial crops such as oil palm may be more advantageous in the long run than annual crops, although there are short-term disadvantages due to the longer time it takes to establish a perennial crop stand. However, it is expected that biotechnology advances will reduce production costs and raise yields, resulting in lower food prices. If developing countries continue to use conventional technology, food prices will remain high in those countries due to higher production costs, reducing their competitiveness in world markets not only with developed countries but also with advanced developing countries that use biotechnology to improve agricultural productivity.

Even if access and use of proprietary technologies are not major constraints, the dissemination of biotechnology products—such as improved seeds to small farmers in developing countries—may be problematic. In many countries, there is a general lack of infrastructure for seed delivery and functional extension services to serve poor farmers. Thought will have to be given not only to technologies but also to proper channels for their delivery to small farmers and to sufficient extension services, market access, and rural infrastructure for proper crop system development.

There is always inherent risk in any technology, old and new. As is the case with conventional insecticides, there are concerns about increases in pest resistance as a result of Bt crops that may result in the loss of Bt as an important pesticide. Such risks can be addressed through scientific-based risk analysis and risk management, including post-commercial monitoring, coupled with proper management of cropping systems. Recent experience with large-scale Bt GM crops in the United States supports this approach. Tabashnik et al. (2000) reported that, contrary to expectations, insect resistance has not been observed in the Bt cotton-growing region in Arizona. Furthermore, results of a seven-year study by the same team of researchers showed that Bt cotton caused long-term suppression of the pink bollworm, *Pectinophora gossypiella,* a major pest (Tabashnik et al., 2003). While it is critical to monitor post-commercialization and research for effective strategies to delay the buildup of insect resistance to Bt toxins, additional genes for insect resistance beyond Bt genes are being tested and more will be discovered to provide an array of arsenals for protecting crop plants. By combining several different types of insect-resistant genes into one crop, it should be possible to develop crops with more durable insect resistance even in the absence of special management practices.

8. CONCLUDING REMARKS

The impact of biotechnology in the next 30 years will depend largely on the strategies that countries adopt to improve their technical capital and, thus, capture the benefits of biotechnology. Although biotechnology cannot by itself stimulate economic growth and alleviate poverty, the new innovation certainly provides an additional tool in the fight against hunger. Theodore Schultz showed more than 30 years ago that poor farmers are effective business people who use resources and technology at their disposal to obtain maximum return to their investments. The problem is that they reach equilibrium at a very low level. To bring this equilibrium to higher levels, new innovations are needed.

In the GR, many small producers were left behind due to lack of access to the inputs required, as well as inappropriate policies. The "gene revolution" may finally provide the opportunity for them to share in the benefits of technology, provided appropriate enabling policies and investments are in place. Indeed, since the last version of this paper published in the joint FAO-World Bank publication, there have been initiatives to help developing countries to access proprietary technologies such as the African Agriculture Technology Foundation (AATF) to facilitate access to proprietary technologies on behalf of resource poor farmers in Africa (Terry, Monyo, and Matlon, 2002). In addition, the recently launched Public-Sector Intellectual Property Resource for Agriculture (PIPRA), supported by the Rockefeller and McKnight Foundations, is also a positive development to bring proprietary technology packages royalty free or at low costs to poor countries (Atkinson et al., 2003).

It is important to note that except for Bt corn approved for commercial cultivation in the Philippines and in South Africa, the current GM crops have not targeted staple crops consumed in many developing countries such as cassava, millet, and tef, to name a few. Furthermore, the size of the commodity seed market (except for China, Brazil, and India) and the ability to pay for the seeds by developing countries do not attract developed countries' private sector's investment whereas the private sector is almost nonexistent in many developing countries. This leaves the public sector and public-private collaboration critical for the development of GM crops for resource poor farmers.

While investment in biotechnology is considered as out of the reach of many developing countries, the reality is that developing countries can benefit from biotechnology innovation, particularly GM crops, if they have developed some capacity in traditional breeding and if they have regulatory frameworks such as biosafety and intellectual property rights (IPRs) in place. A simple plant-breeding program would allow the transfer of engineered genes from GM crops developed elsewhere into their local varieties. Furthermore, the seemingly insurmountable barriers of IPRs and biosafety can be addressed properly by collective efforts by all concerned, particularly developing countries themselves through a regional approach. This will help to reduce the cost of testing and maximizing the use of regional experts while waiting for the developing of critical mass in individual nations.

The prospect of not meeting the millennium goal of halving the number of malnourished people by 2015, i.e., only 11 years from now, is looming. The global community needs to use all the means at our disposal, including GM crops, to combat hunger and poverty. Although John Maynard Keynes pointed out that "In The Long Run We're All Dead", it is not acceptable that in our contemporary time with great advances in biotechnology, including biomedicine, the developed countries' citizens have a "longer run" than that of the developing countries' poor and hungry people. Strategic applications of biotechnology, including GM crops, will help to improve crop productivity and food quality while conserving the environment. Import-antly, this will help to bridge the longevity gap between developed and developing countries and the existing genomics divide, leading towards a more equitable world.

ACKNOWLEDGMENTS

The contribution of Zephaniah Dhlamini, FAO Associate Professional Officer, who gathered information from the FAO-BioDec database and produced the associated charts, is gratefully acknowledged. Joel Cohen, John Komen, Jose-Falck Zepeda, and Patricia Zambrano shared information and provided charts from the Next Harvest© Database prior to publication, for which the author is grateful. The critical reviews offered by Dr. Wayne Parrott of the University of Georgia, Athens, USA, and Henry Nguyen of the University of Missouri, Columbia, USA are greatly appreciated.

REFERENCES

Africa News Service, 2000, *Central Kenya Farmers Embrace Biotech Farming,* (November 1, 2000).

Apse, M. P., Aharon, G. S., Snedden, W. A., and Blumwald, E., 1999, Salt tolerance conferred by overexpression of a vacuolar Na+/H+ antiport in Arabidopsis, *Sci.* **285:**1256-1258.

Arntzen, C. J., 1995, Oral immunization with a recombinant bacterial antigen produced in transgenic plants, *Sci.* **268:**714-716.

Arntzen, C. J., 1996, *Crop Biotechnology in the Service of Medical and Veterinary Science*, NABC, Ithaca, New York.

Atkinson, R. C., Beachy, R. N., Conway, G., Cordova, F. A., Fox, M. A., Holbrook, K. A., Klessig, D. F., McCormick, R. L., McPherson, P. M., Rawlings III, H. R., Rapson, R., Vanderhoef, L. N., Wiley, J. D., and Young, C. E., 2003, Public sector collaboration for agricultural IP management, *Sci.* **301:**174-175.

Bakan, B., Melcion, D., Richard-Molard, D., and Cahagnier, B., 2002, Fungal growth and fusarium mycotoxin content in isogenic traditional maize and genetically modified maize grown in France and Spain, *J. Agric. Food Chem.* **50:**728-731.

Bhattacharya, S., 2003, KwaZulu farmers boosted by GM cotton, *New Sci. Online News* **16:**31; http://www.newscientist.com/hottopics/gm/gm.jsp?id=ns99993473.

Borroto, C., 2000, Biotechnology seminar: Cuban national program on agricultural biotechnology: Achievements, present and future, seminar held at the Food and Agriculture Organization of the United Nations, Rome, Italy (October 11, 2000).

Brophy, B., Smolenski, G., Wheeler, T., Wells, D., L'Huillier, P., and Laible, G., 2003, Cloned transgenic cattle produce milk with higher levels of beta-casein and kappa-casein, *Nat. Biotech.* **21**(2)**:**157-162.

Carnus, J. M., Parrotta, J., Brockerholl, E. G., Arbez, M., Jactel, H., Kremer, A., Lamb, D., O'Hara, K., and Walters, B., 2003, Planted forests and biodiversity, paper presented at the UNFF Intersessional Experts Meeting on the Role of Planted Forests in Sustainable Forest Management, New Zealand (March, 24-30, 2003).

Chakraborty, S., Chakraborty, N., and Datta, A., 2000, Increased nutritive value of transgenic potato by expressing a nonallergenic seed albumin gene from *Amaranthus hypochondriacus, PNAS* **97:**3724-3729.

Chapman, A. J., Stupka, E., Putnam, N., Chia, J. M., Dehal, P., Christoffels, A., Rash, S., Hoon, S., Smit, A., Gelpke, M. D., Roach, J., Oh, T., Ho, I. Y., Wong, M., Detter, C., Verhoef, F., Predki, P., Tay, A., Lucas, S., Richardson, P., Smith, S. F., Clark, M. S., Edwards, Y. J., Doggett, N., Zharkikh, A., Tavtigian, S. V., Pruss, D., Barnstead, M., Evans, C., Baden, H., Powell, J., Glusman, G., Rowen, L., Hood, L., Tan, Y. H., Elgar, G., Hawkins, T., Venkatesh, B., Rokhsar, D., and Brenner, S., 2002, Whole-genome shotgun assembly and analysis of the genome of *Fugu rubripes, Sci.* **297**(5585)**:**1283-1285.

Che, D., Meagher, R. B., Heaton, A. C. P., Lima, A., Rugh, C. L., and Merkle, S. A., 2003, Expression of mercuric ion reductase in Eastern cottonwood (*Populus deltoides*) confers mercuric ion reduction and resistance, *Plant Biotech. J.* **1**(4)**:**311-317.

Chen, L., Marmey, P., Taylor, N. J., Brizard, J. P., Espinoza, C., D'Cruz, P., Huet, H., Zhang, S., de Kochko, A., Beachy, R. N., and Fauquet, C. M., 1998, Expression and inheritance of multiple transgenes in rice, *Nature Biotech.* **16:**1060-1064.

Coghlan, A., 2003, Genetically modified 'protato' to feed India's poor, *New Sci.* (January, 2003); http://www.newscientist.com/news/news.jsp?id=ns99993219.

Cohen, J., Komen, I., Falck-Zepeda, J., and Zambrano, P., 2003, ISNAR-IBS Next Harvest©, data presented at an Expert Consultation on Biotechnology Next Harvest© – Advancing Biotechnology's Public Good: Technology Assessment, Regulation and Dissemination (October, 7-9, 2002), ISNAR Headquarters, The Hague.

Dahwan, V., 2002, Personal communication.

Datta, S., 2002, Personal communication.

De la Fuente, J., Ramirez-Rodriguez, M., Cabrera Ponce, J. J., and Herrera-Estrella, L., 1997, Aluminum tolerance in GM plants by alteration of citrate synthesis, *Sci.* **276:**1566-1568.

DellaPenna, D., 1999, Nutritional genomics: Manipulating plant micronutrients to improved human health, *Sci.* **285:**375-379.

Entis, E., 1998, Taste testing at a top Canadian restaurant, *Aqua Bounty Farms*, **1:**1-4.

Entis, E., 2003, Biotech at sea: Innovation required; http://pewagbiotech.org/events/0131/.

FAO, 1999, Committee on Agriculture, 15th Session, Biotechnology; http://www.fao.org/unfao/bodies/COAG/COAG15/X0074E.htm.

FAO, 2002, World Food Summit: Five years later (Draft declaration), Article 25; http://www.fao.org/DOCREP/MEETING/004/Y6948E.HTM.

FAO-BioDec, 2003; http://www.fao.org/biotech/inventory_admin/dep/default.asp (accessed July 11, 2003).

Fenning, T. M., and Gershenzon, J., 2002, Where will the wood come from? Plantation forest and the role of biotechnology, *Trends in Biotech.* **20:**291-296.

French, C. E., Rosser, S. J., Davies, G. J., Nicklin, S., and Bruce, N. C., 2001, Biodegradation of explosives by transgenic plants expressing pentaerythritol tetranitrate reductase, *Nature Biotech.* **19:**1168-1172.

Fresco, L. O., 2003, A new social contract on biotechnology; http://www.fao.org/ag/magazine/0305sp1.htm.

Fuglie, K., Zhang, L., Salazar, L. F., and Walker, T., 2001, *Economic Impact of Virus-Free Sweet Potato Planting Material in Shandong Province, China,* Future Harvest, Washington, D. C.

Garg, A. K., Kim, J. K., Owens, T. G., Ranwala, A. P., Do Choi, Y., Kochian, L. V., and Wu, R. J., 2002, Trehalose accumulation in rice plants confers high tolerance levels to different abiotic stresses, *PNAS* **99:**15898-15903.

Gartland, K. M. A., Kellison, R. C., and Fenning, T. M., 2002, Forest biotechnology and Europe's forests of the future, paper presented at the Forest Botechnology in Europe: Impending Barriers, Policy, and Implications, Edinburgh.

Goure, W., 2002, Mendel biotechnology's transcription factors genes and food security in developing countries, paper presented at the Symposium on Plant Biotechnology: Perspective from Developing Countries (November 12-14, 2002), Annual Meetings of the American Society of Agronomy, the Crop Science Society of America and the Soil Science Society of America, Indianapolis, Indiana.

Guillen, I., Berlanga, J., Valenzuela, C. M., Morales, A., Toledo, J., Estrada, M. P., Puentes, P., Hayes, O., and de la Fuente, J., 1999, Safety evaluation of transgenic tilapia with accelerated growth, *Marine Biotech.* **1:**2-14.

Guy, C. L., Irani, T., Gabriel, D., and Fehr, W., 2000, Workshop reports: Workshop C: Food and environmental issues associated with the bio-based economy of the 21st century, *NABC News* **19:**8-11.

Hallerman, E., 2003, Status of development of transgenic aquatic animals, *ISB News Report*; http://www.isb.vt.edu/news/2003/news03.apr.html#apr0304.

Hamilton, C. M., Frary, A., Lewis, C., and Tanksley, S. D., 1996, Stable transfer of intact high molecular weight DNA into plant chromosomes, *PNAS* **93:**9975-9979.

Hannink, N., Rosser, S. J., French, C. E., Basran, A. J., Murray, A. H., Nicklin, S., and Bruce, N. C., 2001, Phytodetoxification of TNT by transgenic plants expressing a bacterial nitroreductase, *Nature Biotech.* **10:**1038-1168.

Herrera-Estrella, L., 2002, Plant biotechnology in the postgenomic era: Can it benefit developing countries? Paper presented at ISNAR-FAO Expert Workshop on Policy Planning and Decision Support: The Case of Biosafety (May 14-16, 2002), Rome.

Houdebine, L. M., 2002, Transgenesis to improve animal production, *Livestock Prod. Sci.* **74**(3)**:**255-268.

Hu, W. J., Harding, S. A., Lung, J., Popko, J. L., Ralph, J., Stokke, D. D., Tsai, C. J., and Chiang, V. L., 1999, Repression of lignin biosynthesis promotes cellulose accumulation and growth in transgenic trees, *Nature Biotech.* **17**(8)**:**808-812.

Huang, J., and Wang, Q., 2002, Agricultural biotechnology development and policy in China, paper presented at the Symposium on Plant Biotechnology: Perspective from

Developing Countries, Annual Meetings of the American Society of Agronomy, the Crop Science Society of America and the Soil Science Society of America, Indianapolis, Indiana (November 12-14, 2002).

Hulata, G., 2001, Israeli aquaculture genetic improvement programs, in: *Fish Genetics Research in MemberCountries and Institutions of the International Network on Genetics in Aquaculture*, M. V. Gupta and B. O. Acosta, eds., ICLARM Conf. Proc. 64, Penang, Malaysia, pp. 103-108.

IFPRI, 1999, Are we ready for a meat revolution? *20/20 Vision News & Views*; http://www.ifpri.org/2020/newslet/nv_0399/nv0399a.htm.

Ismaël, Y., Beyers, L., Lin, L., and Thirtle, C., 2001, Smallholder adoption and economic impacts of Bt cotton in the Makhathini Flats, South Africa, paper presented at the 5th International Conference, *Biotechnology, Science and Modern Agriculture: A New Industry at the Dawn of the Century*, Ravello (June 15-18, 2001).

Jaglo-Ottosen, K. R., Gilmour, S. J., Zarka, D. G., Schabenberger, O., and Thomashow, M. F., 1998, *Arabidopsis* CBF1 overexpression induces COR genes and enhances freezing tolerance, *Sci.* **280:**104-106.

James, C., 2001, *Global Status of Commercialized Transgenic Crops 2001,* ISAAA Briefs No. 24. Preview, ISAAA, Ithaca, New York; http://www.isaaa.org/publications/briefs/Brief_24.htm.

James, C., 2002, *Global Status of Commercialized Transgenic Crops 2002,* ISAAA Briefs No. 27, ISAAA, Ithaca, New York.

James, C., 2003, *Global Status of Commercialized Transgenic Crops 2003,* Preview, ISAAA, Ithaca, New York.

Kirch, H. H., Nair, A., and Bartels, D., 2001, Novel ABA- and dehydration-inducible aldehyde dehydrogenase genes isolated from the resurrection plant *Craterostigma plantagineum* and *Arabidopsis thaliana, Plant J.* **28**(5):555-567.

Kobayashi, N. T., Yoshiba, M., Sanada, Y., Wada, Y., Tsukaya, K., Kakubari, H., Yamaguchi-Shinozaki, K., and Shinozaki, K., 1999, Improving plant drought, salt, and freezing tolerance by gene transfer of a single stress-inducible transcription factor, *Nature Biotech.* **17:**287-291.

Kong, Q. K., Richter, L., Yang, Y. F., Arntzen, C., Mason, H. S., and Thanavala, Y., 2001, Oral immunization with hepatitis B surface antigen expressed in transgenic plants, *Proc. Natl. Acad. Sci.* **98:**11539-11544.

Ku, M. S. B., Cho, D., Ranade, U., Hsu, T.-P., Li, X., Jiao, D.-M., Ehleringer, J., Miyao, M., and Matsuoka, M., 2000, Photosynthetic performance of transgenic rice plants overexpressing the maize C4 photosynthesis enzymes, in: *Redesigning Rice Photosynthesis to Increase Yield*, J. Sheehy, P. Mitchell, and B. Hardy, eds., IRRI, Los Baños, Philippines.

Langridge, W. H. R., 2000, Edible vaccines, *Scientific Amer.* **283:**66-71; www.scientific american.com/2000 /0900issue/0900langridge.html.

Le, H., 2001, The potential of agricultural biotechnology, in: *Farming Systems and Poverty: Improving Farmers' Livelihoods in a Changing World*, J. Dixon, and A. Dixon, with D. Gibbon, eds., FAO and WB, Rome and Washington, D. C., pp. 366-374; http://www.fao.org/docrep/003/y1860e/y1860e12.htm#P147_26118.

Lehman, V., 2000, Cuban agrobiotechnology: Diverse agenda in times of limited food production, *Biotech. and Dev. Monitor,* **42:**18-21.

Li, L., Zhou, Y., Cheng, X., Sun, J., Marita, J. M., Ralph, J., and Chiang, V. L., 2003, Combinatorial modification of multiple lignin traits in trees through multigene cotransformation, *PNAS,* **100**(8):4939-4944.

Lubick, N., 2002, Designing trees, *Scientific Amer.*; http://www.sciam.com/article.cfm?articleID=0000BA96-AE60-1CDA-B4A8809EC588EEDF&pageNumber=1.

Lucca, P., Hurrell, R., and Potrykus, I., 2002, Fighting iron deficiency anemia with iron-rich rice, *J. of the Amer. College of Nutr.* **21:**184S-190S.

Mackey, M., 2002, The applications of biotechnology to nutrition: An overview, *J. of the Amer. College of Nutr.* **21:**157S-160S.

Mazur, B., Krebbers, E., and Tingley, S., 1999, Gene discovery and product development for grain quality traits, *Sci.* **285:**372-375.

McLean, M. A., Frederick, R. J., Traynor, P., Cohen, J. I., and Komen, J., 2002, *A Conceptual Framework for Implementing Biosafety: Linking Policy, Capacity and Regulation,* Briefing Paper No. 47, 12 pp.

Meilan, R., 2001, Personal communication.

Mesfin, T. S., Temple, J., Allan, D. L., Vance, C. P., and Samac, D. A., 2001, Overexpression of malate dehydrogenase in transgenic alfalfa enhances organic acid synthesis and confers tolerance to aluminum, *Plant Physiol.* **127:**1836-1844.

Mouse Genome Sequencing Consortium, 2002, Initial sequencing and comparative analysis of the mouse genome, *Nature,* **420:**520-562.

Muchhal, U. S., and Raghothama, K. G., 1999, Transcriptional regulation of plant phosphate transporters, *PNAS,* **96**(10):5868-5872.

Mukatira, U. T., Liu, C., Varadarajan, D. K., and Raghothama, K. G., 2001, Negative regulation of phosphate starvation-induced genes, *Plant Physiol.* **127:**1854-1862.

Nambiar, E. K. S., 1999, Pursuit of sustainable plantation forestry, *S. Afr. For. J.* **184:**45-62.

NAS (National Academy of Sciences), 2002, *Animal Biotechnology: Science-Based Concerns*, NAS, USA.

NAS, 2000, *Transgenic Plants and World Agriculture*; http://www.nap.edu/html/transgenic/.

Natural Resource Canada, 2003, http://www.nrcan-rncan.gc.ca/cfs-scf/science/biotechnology/treeim_e.html).

Orf, J. H., Diers, B. W., and Boerma, H. R., 2004, Genetic improvement: Conventional and molecular-based strategies, in: *Soybeans: Improvement, Production, and Uses*, H. R. Boerma, and J. E. Specht, eds., ASA-CSSA-SSSA, 3rd ed., Madison, Wisconsin, pp. 417-450.

Potrykus, I., 2001, Potrykus responds to Greenpeace criticism of 'Golden Rice', *Agbioworld,* 9 February; http://www.agbioworld.org/biotech_info/topics/goldenrice/criticism.html.

Pray, C. E., Ma, D., Huang, J., and Qiao, F., 2001, Impact of Bt-cotton in China, *World Dev.* **29:**813-825.

Pursel, V. G., Wall, R. J., Solomon, M. B., Bolt, D. J., Murray, J. D., and Ward, K. A., 1997, Transfer of an ovine metallothionein-oveine growth hormone fusion gene into swine, *J. of Animal Sci.* **75:**2208-2214.

Pursel, V. G., Coleman, M. E., and Wall, R. J., 1996, Regulatory avian skeleton-actin directs expression of insuline-like growth factor-1 to skeletal muscle of transgenic pigs, *Theriogen.* **35:**348.

Qaim, M., 2000, A prospective evaluation of biotechnology in semi-subsistence agriculture, paper presented at the XXIV Conference of the International Association of Agricultural Economists, Berlin.

Qu, S., Coaker, G., Francis, D., Zhou, B., and Wang, G. L., 2003, Development of a new transformation-competent artificial (TAC) vector and construction of tomato and rice TAC libraries, *Mol. Breed. New Strat. in Plant Improv.* **12**(4):297-308.

Ragot, M., 2003, Personal communication.

Sang, H., 2003, Genetically modified livestock and poultry and their potential effects on human health and nutrition, *Trends in Food Sci. and Tech.* **14**(5-8):253-263.

Sayre, R., 2000, Cyanogen reduction in GM cassava: Generation of a safer product for subsistence farmers, *ISB News Report* (August, 2000).

Smith, T., 2003, Brazil to lift ban on crops with genetic modification, *The New York Times*, September 27, 2003.

Somerville, C., and Somerville, S., 1999, Plant functional genomics, *Sci.* **285:**380-383.

Strauss, S. H., DiFazio, S. P., and Meilan, R., 2001, Genetically modified poplars in context, *For. Chron.* **77**(2):271-279.

Tabashnik, B. E, Carrière, Y., Dennehy, T. J., Morina, S., Sisterson, M. S., Roush, R. T., Shelton, A. M., and Zhao, J. Z., 2003, Insect resistance to Bt Crops: Lessons from the first seven years, ISB News Report, Virginia Tech., Blacksburg, VA (November, 2003); http://www.isb.vt.edu/news/2003/nov03.pdf.

Tabashnik, B., Patin, A. L., Dennehy, T. J., Liu, Y. B., Carrière, Y., Sims, M. A., and Antilla, L., 2000, Frequency of resistance to Bacillus thuringiensis in field populations of pink bollworm, *Proc. Natl Acad. Sci.* **97**(24):12980-12984.

Takahashi, M., Nakanishi, H., Kawasaki, S., Nishizawa, N. K., and Mori, S., 2001, Enhanced tolerance of rice to low iron availability in alkaline soils using barley nicotianamine aminotransferase genes, *Nature Biotech.* **19**(5):466-469.

Terry, E., Monyo, J., and Matlon, P., 2002, A technology transfer model for smallholder farmers in sub-Saharan Africa, Paper presented at the Symposium on Plant Biotechnology: Perspective from Developing Countries, Annual Meetings of the American Society of Agronomy, the Crop Science Society of America and the Soil Science Society of America, Indianapolis, Indiana (November, 12-14, 2002).

The International Genome Sequencing Consortium, 2001, Initial sequencing and analysis of the human genome, *Nature*, **409**:860-921.

Thompson, J., 2001, Appropriate technology for sustainable food security, *Modern Technology for African Agriculture,* IFPRI 20/20 Focus 7; http://www.ifpri.org/2020/focus/ focus07/focus07_05.htm.

Tian, L., and DellaPenna, D., 2001, *The Promise of Agricultural Biotechnology for Human Health*, Meeting report on the Keystone Symposium "Plant Foods for Human Health: Manipulating Plant Metabolism to Enhance Nutritional Quality," Breckenridge, CO (April 6-11, 2001); http://scope.educ.washington.edu/gmfood/member/commentary/show. php?author=Tian.

Traxler, G., Godoy-Avila, S., Falck-Zepeda, J., and Espinoza-Arellano, D. J., 2001, Transgenic cotton in Mexico: Economic and environmental impacts, Paper presented at the 5th ICABR International Conference, Biotechnology, Science and Modern Agriculture: A New Industry at the Dawn of the Century, Ravello (June 15-19, 2001).

Ward, R., 2003, Personal communication.

Webster, J., 2000, Enabling biotechnology in Africa: Current situation and future needs, Seminar presented December 8, 2000, FAO, Rome.

Zhang, H. X., and Blumwald, E., 2001, Transgenic salt-tolerant tomato plants accumulate salt in foliage but not in fruit, *Nature Biotech.* **19**(8):765-768.

Zhang, H. X., Hodson, J. N., Williams, J. P., and Blumwald, E., 2001, Engineering salt-tolerant *Brassica* plants: Characterization of yield and seed oil quality in transgenic plants with increased vacuolar sodium accumulation, *Proc. Natl. Acad. Sci.* **98**:12832-12836.

Chapter 14

IMPACT OF BIOTECHNOLOGY ON CROP GENETIC DIVERSITY

Matin Qaim,[1] Cherisa Yarkin,[2] and David Zilberman[3]
[1]Professor, Department of Agricultural Economics and Social Sciences, University of Hohenheim, 70593 Stuttgart Germany; [2] Director, Economic Research and Assessment, University of California Industry-University Cooperative Research Program, University of California, Berkeley, CA 94720; [3]Professor, Department of Agricultural and Resource Economics, 207 Giannini Hall, University of California, Berkeley, CA 94720

Abstract: While there are widespread concerns that agricultural biotechnology might contribute to a further erosion of crop genetic diversity, in this chapter it is argued that the opposite could actually be true. Biotechnology allows for separation between the act of developing novel crop traits and the process of breeding plant varieties. As a result, a given biotechnology innovation may be incorporated into a large number of locally adapted plant varieties. This is confirmed by first empirical evidence from different countries. However, a theoretical model is developed which shows that the outcome is situation specific and depends on various institutional factors. Local research capacities, intellectual property policies, and biosafety regulation schemes are identified as important determinants for the actual impact of biotechnology on crop genetic diversity. Policy implications are discussed with a particular emphasis on developing countries.

Key words: biodiversity; biosafety; biotechnology; intellectual property rights; plant breeding; transaction costs.

1. INTRODUCTION

Agricultural productivity enhancements have often been based upon development and dissemination of a small number of highly competitive plant varieties or animal species and, thus, have been associated with a decrease in crop genetic diversity (CGD) (FAO, 1998). Mechanical innovations such as tractors and combines, supported by breakthroughs in chemical fertilizers and pesticides, have facilitated the mono-cropping of vast areas of land, so it is not surprising there is concern that the advent of agricultural biotechnology may exacerbate these trends.[1] In this chapter, we argue that agricultural biotechnology may instead offer unique opportunities to preserve CGD, but the speed and extent to which this potential is realized depend upon institutional factors, including the distribution and level of protection afforded to intellectual property rights (IPRs), transaction costs associated with licensing, technology transfer, and biosafety regulations.

As is common in the literature (FAO, 1998), CGD is used to describe the genetic diversity of agricultural crops. The number of different varieties or landraces being used by farmers is an important indicator of the *in situ* CGD of a particular crop species.

Biotechnology introduces a fundamental change in the way that seeds and other genetic materials can be produced. With traditional breeding techniques, existing varieties are selectively combined to develop new varieties or hybrids. This is a lengthy process and involves a significant degree of randomness. The outcome is usually a novel variety that has a number of new traits and characteristics, not all of which are desirable. Biotechnology, however, allows the targeted introduction of selected genetic materials into existing crop varieties. Once the genetic sequence coding for a desirable trait such as insect resistance has been identified, a "transformation event" is created by transferring this genetic sequence to a particular receptor variety. Additional genetically modified varieties (GMVs) are then developed by crossing existing conventional varieties with this transgenic receptor variety.[2] Although several backcrossing generations are necessary to eliminate unwanted characteristics, this process is far quicker, easier, and cheaper than developing a new conventional variety through cross-breeding (Traxler, Falck-Zepeda, and Sain, 1999). It usually results in GMVs that are virtually identical to their conventional counterparts except for the new desirable trait. In other words, biotechnology permits a separation between the act of developing a specific agronomic trait and the breeding of a particular, locally adjusted variety. As such, it has important institutional and economic implications that will affect the diversity of crop plants in agricultural production.

In this chapter, we evaluate the potential for adoption of seeds and genetic materials developed using biotechnology and assess the impact on the diversity of crop varieties produced under alternative industrial and policy structures. We examine the roles of IPRs and the research capacity and efforts of private and public sectors in determining the utilization of

---

[1] We use the term "biotechnology" to refer to the subset of techniques associated with modern molecular biology that allows the selective introduction of specific genes into crop plants, in a manner that leads to the transmission of the input gene (transgene) to successive generations (FAO, 1999).

[2] Exceptions are clonally propagated, complex heterozygous species, such as potato and cassava. Because cross-breeding in these species is difficult, each GMV is usually created through a separate transformation event.

agricultural biotechnology to obtain policy implications for agricultural research efforts.

The next section provides an overview of our framework and main findings. It is followed by a detailed mathematical derivation of the main results. The final section presents a summary of our results, and it demonstrates their validity using available information on a number of GMVs of various crops grown in different countries.

2. SUMMARY OF MODEL AND MAJOR FINDINGS

The legal framework in the United States, Europe, and other developed countries has gradually evolved so that those who decipher genetic structures, discover the functions of genes, or identify mechanisms to alter genes can register patents and own the IPRs for the utilization of these discoveries. Private parties have an incentive to conduct research leading to new discoveries because they expect to gain financially from selling the rights to utilize the IPRs, or to utilize them directly in their own commercialization efforts. However, the extent to which IPRs are protected and traded varies among nations, and these variations may affect the way the products and processes of biotechnology are managed and utilized.

In our conceptual analysis, we consider the case of an agricultural industry sector that produces a single crop where producers are characterized by high degrees of heterogeneity. Variation in land quality, topography, and climatic conditions, even within a region, may result in growers adopting different varieties of the same crop. We assume that, technically speaking, all the existing varieties can be modified using biotechnology. This modification, for example, may reduce susceptibility to pests and diseases or increase the efficiency of nutrient uptake. For simplicity, we assume that the effect at the farm level is an increase in yields.

It is assumed that utilization of a biotechnology-based innovation, such as a *Bacillus thuringiensis* (Bt) gene that codes for the expression of an insecticidal protein in plant tissue, requires a large fixed cost in infrastructure for technology development, a modest fixed cost to obtain the capacity to incorporate the technology into each specific variety, and a relatively small variable cost of seed production.

We argue that biodiversity and the impact of a new biotechnology-based innovation on grower and consumer welfare depend on the manner in which its introduction and pricing strategies relate to the existing cropping system. In particular, we distinguish between situations where private sector companies introduce the technology and situations where it is introduced by the public sector. We also distinguish between situations where traditional local varieties are replaced by "generic" GMVs and situations where specific genes are introduced to local varieties, which continue to be planted in the new, modified form. The generic GMV may be an imported variety that, on average, performs well, but since it may not be highly suited to each location's conditions, will likely not perform as well as GMV versions of local varieties. These features will determine the outcomes of farmer and consumer welfare and biodiversity preservation.

When a private company introduces GMVs, we assume that it charges a monopoly price. This price is set at a level where marginal revenues are equal to marginal costs of producing the modified seeds. When farmers are heterogeneous in their conditions, the impacts of the technology may vary. GMVs will be adopted only where the gain from adoption is sufficient to cover the technology fee. The technology fee will increase with the variable cost of modification (and when the cost of modification is assumed to be fixed, only the cost of seed production is variable). Therefore, low variable costs increase the adoption of GMVs.

When a local variety is genetically modified (GM), the effect of environmental heterogeneity is likely to be an increase in the yield effect and cost savings compared to the case in which the local variety is replaced by a generic GMV. However, the modification of each variety entails a fixed cost. In deciding whether to genetically modify a specific variety or to offer growers a generic GMV, the private company will compare the extra cost of modification with the extra revenues earned from a modified variety relative to a generic variety. Two factors that will affect this decision are the cost of modification and the size of the market. When modification costs are high or when the market is small, it will be more profitable to sell a generic GMV. Thus, regions where the crop-breeding industry has low capacity and the cost of modification is relatively high are more likely to adopt generic GMVs, and their introduction may lead to a reduction of crop biodiversity.

Thus far, private sector companies have developed and introduced most of the GMVs globally. The history of the Green Revolution (Evenson and Gollin, 2003) suggests that eventually public sector institutions will also develop and introduce GMVs, and these varieties will be distributed through small seed companies. We analyze outcomes where the public sector makes the choice of whether or not to introduce a GMV to a region, and whether there will be a genetic modification of the local variety or a generic GMV imported from elsewhere. The public sector organization is assumed to maximize the domestic economic surplus, including seed producers, farmers, and consumers. Under these scenarios, the seed companies will charge competitive prices that are less than the price charged by a monopoly seller. As a result, adoption levels are likely to be higher than under the monopolistic scenario if all the other parameters are the same. Furthermore, since the total domestic surplus is larger than the profit of monopolistic seed companies, the public sector is more likely to invest in development of GMVs than a private sector monopolist. The public sector is more likely to develop local GMVs rather than import generic GMVs, since the development of local GMVs is likely to have a larger yield effect that benefits growers and consumers rather than seed producers.

The decision whether to introduce a local or an imported generic GMV, in any context, depends on the difference between the gain and the cost of development. Countries with more advanced crop-breeding capabilities and sufficient public sector resources for developing GMVs are more likely to modify existing varieties, which will lead to preservation of biodiversity. On the other hand, in situations where public sector resources are limited and where investment in GMVs is not profitable for the private sector, the public sector may develop or import a small number of generic GMVs and, in spite of limited adoption, it may lead to a loss of biodiversity.

3. A MODEL

An agricultural sector is producing a single crop in M locations, with varying climatic and agroecological conditions. Let j be a location indicator. Then j takes values from 1 to M. Before GMVs are introduced, farmers at each location use the best traditional variety given their conditions, and we assume that each location has its own distinct traditional variety. For simplicity, we assume that the gross benefits of production are measured in monetary terms. Output price denoted by P is assumed to be constant.

In the analysis, we consider several scenarios with respect to market and properties of genetic materials. Let s be a scenario indicator; $s = 0$ is the initial situation where only traditional varieties are used; $s = m$ is the case in which locally optimized varieties are modified (GMV_j), and $s = g$ is the situation in which only one generic modified variety (GMV) is adopted. Let X_j^s denote output produced at location j under scenario s, which is assumed to be a function of land and genetic materials. The production function $f_j^s(A)$ denotes the output produced on A acres of land in location j and scenario s. The marginal productivity of land for scenario s is $Mp_j^s(A) = \partial f_j^s / \partial A$.

It is assumed that $Mp_j^s(A)$ and $\partial Mp_j^s / \partial A < 0$ (decreasing marginal productivity of land in a location), which may reflect heterogeneity of land quality within a location.

Prior to the introduction of GMVs, when $s = 0$, the demand for land of variety j at location j is denoted by $PMp_j^0(A)$. The marginal cost of land grown with variety *j* is assumed to be constant and is denoted by Mc_j, and the seed cost prior to biotechnology is assumed to be 0.[3] Before the introduction of biotechnology, the acreage allocated to variety *j*, A_j^0, was at the point when marginal benefits equal marginal cost of acreage, or $\left(PMp_j^0\left(A_j^0\right)=Mc_j^0\right)$.

3.1 Assumptions about the GM technology and seed industry structure

Before analyzing several outcomes from GMVs, let us specify assumptions about their benefits and cost structure. We assume that each traditional variety can be modified, but it may also be replaced by a "generic" GMV. The exact outcome depends on costs, constraints, and decision making about seed supply.

For simplicity, we assume that GMVs improve seed productivity. This corresponds to yield-increasing GMVs (e.g., Bt cotton in India or South Africa; see Qaim and Zilberman, 2003).[4] Let $Mp_j^m(A)$ denote the marginal productivity (*Mp*) of land planted with *GMVj*. We will assume that (a) GMVs have higher marginal productivity than the traditional variety, $Mp_j^m(A)>Mp_j^0(A)$ and (b) the marginal productivity gain declines with *A*,

$$\partial\left(Mp_j^m(A)-Mp_j^0(A)\right)/\partial A<0$$

This assumption that better-quality lands gain more from genetic modification is done for convenience, and most results hold for broader circumstances. It corresponds to situations, say, where a certain percentage of the crop is lost to pests. GMVs reduce pest damage and thus provide higher gains to locations with higher potential output.

If traditional varieties are replaced by a generic GMV, then $Mp_j^g(A)=\partial f_j^g(A)/\partial A$ is the marginal productivity of land in location *A*. We assume that the *Mp* of the unmodified version of the generic variety is less than that of variety *j*. Because modification increases the marginal productivity of the target variety, it is assumed that the marginal benefit of the generic GMV, while less than that of the GMV_j, is greater than the traditional variety. Hence,

---

[3] Farmers either keep a portion of seeds or receive seeds at no cost from the government. Pricing seeds will add complexity but not change the results.

[4] The model can be adjusted to also allow cost reductions and environmental advantages. The gain in details is costly in terms of complexity and does not lead to additional insights.

$$Mp_j^m(A) > Mp_j^g(A) > Mp_j^0(A)$$

and also that $\partial\left(Mp_j^g(A) - Mp_j^0(A)\right)/\partial A < 0$.

The introduction of the GM technology is associated with several cost categories. The first is the fixed cost to introduce the GMVs at the crop level. It consists of research, testing, registration, and regulatory compliance costs to introduce, say, Bt cotton in South Africa or Bt sweet potato in Kenya. We assume that this cost has been incurred and consider two other costs.

(a) F_j = fixed cost to modify variety j. This cost includes both the technical cost to insert genetic materials into a specific variety and the cost of IPR transactions and biosafety regulation. Countries with a more advanced breeding sector are likely to have lower modification costs. The IPR component depends on specific circumstances. If a private company already controls a local variety, it will not face any IPR cost. On the other hand, it may have significant cost if a competing company controls a specific variety.

(b) V_j = per acre costs of GMVs of variety j. We assume that these variable costs are constant. They include the physical variable cost of producing GMV seeds and the IPR and marketing costs that a manufacturer has to pay in order to be able to sell the seeds. Let V_g be the per acre cost of the generic seed variety when it is used. It is assumed to be smaller than V_j. It is imported from a low-cost production center and does not require payment to owners of the rights for local seeds. The difference decreases as the seed sector producing GMV_j becomes more advanced and when the seed producer does not face IPR costs for the use of the local variety.

The outcomes with GMVs depend on the cost to introduce them as well as the structure of the seed industry and the constraints it faces. We consider two patterns. The first pattern applies to countries where public research institutions develop GMVs, which are sold to farmers by a competitive seed sector. In the second pattern, GMVs are introduced and marketed by a monopolist. This scenario is appropriate for the developed world, where major companies such as Monsanto control the GMVs available in the market. In the case of a monopoly, the seed industry may face IPRs and other constraints that impede its capacity to modify traditional local varieties. Recognizing these constraints, we derive outcomes for four stylized scenarios presented below.

3.1.1 Competitive markets for seeds of GM local varieties

Assuming that the fixed cost of introducing the technologies are covered by the public sector, competitive seed sellers will charge farmers V_j per land unit of GM seeds. The GMVs may be either fully or partially adopted. In the case of partial adoption, some land will continue to be grown with traditional variety *j*. Let A_j^1 be total acreage of variety *j* (traditional and GM), and let A_j^{m1} be the GM acreage. The acreage of the traditional variety *j* will be equal to $A_j^1 - A_j^{m1}$. There are three possible outcomes under competition:

- $C_j^m 1$: No adoption when $PMp_j^m(0) - PMp_j^0(0) < V_j$. In this case, $A_j^{m1} = 0, A_j^1 = A_j^0$.
- $C_j^m 2$: Partial adoption when $PMp_j^m(0) - PMp_j^1(0) > V_j >$ $PMp_j^m(A_j^0) - PMp_j^m(A_j^0)$. Here $A_j^{m1} < A_j^0$ where $PMp_j^m(A_j^{m1}) =$ $PMp_j^0(A_j^{m1}) + V_j$ and $A_j^1 = A_j^0$.
- $C_j^m 3$: Full adoption when $PMp_j^m(A_j^0) - PMp_j^0(A_j^0) > V_j$. $A_j^1 = A_j^{m1}$ when $PMp_j^m(A_j^{m1}) = Mc_j + V_j$.

The assumption that the *Mp* gap between the GMV and traditional variety declines with acreage is the key to the above results. If gains from adoption cannot cover the variable cost even for the first acre, adoption will not occur. If marginal gains from adoption are greater than the variable cost, but only for a subset of the acreage, there will be partial adoption; and if the marginal gains from adoption are greater than the variable cost for all acres, there will be full adoption and the acreage under the crop might even increase (recall that we have assumed the total output market is sufficiently large that this change has no impact on output prices).

Fig. 14-1a depicts case $C_j^m 1$ with no adoption when $PMp_j^0(0) + V_j >$ $PMp_j^{m1}(0)$. Fig. 14-1b depicts case $C_j^m 2$ of partial adoption, and Fig. 14-1c depicts case $C_j^m 3$ with full adoption. These figures demonstrate that high variable costs may discourage adoption entirely, while reduction in these costs may result in partial or, eventually, full adoption. In the figures, the same notation should be used as in the text, that is, replace *Mb* by *PMp*.

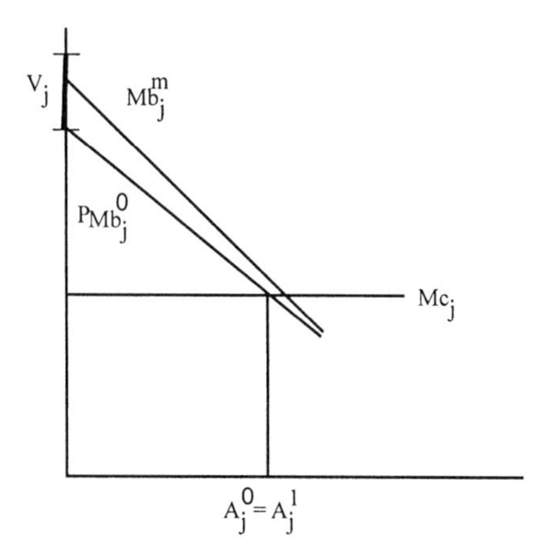

Figure 14-1a. No adoption

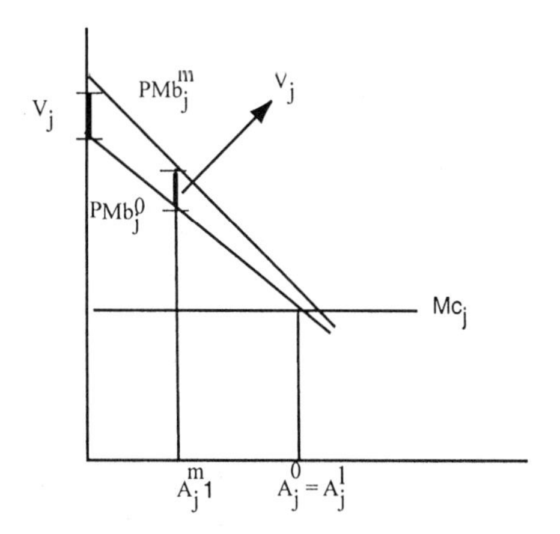

Figure 14-1b. Partial adoption

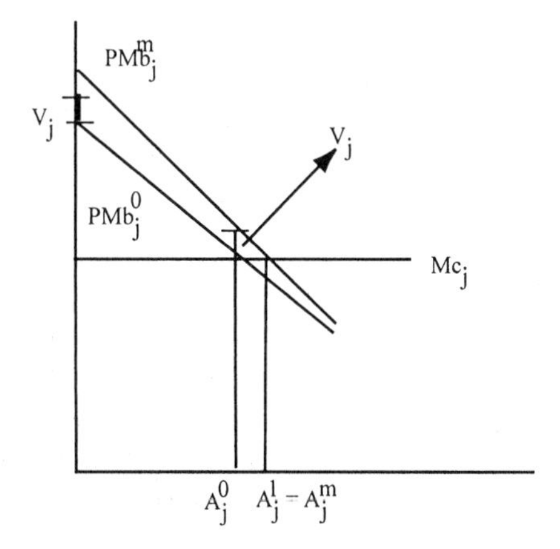

Figure 14-1c. Full adoption

The introduction of GMVs affects output and net benefits of land utilized with variety j. The output when GMV_j is available becomes

$$X_j^{m1} = \begin{cases} f\left(A_j^0\right) & \text{if } C_j^m 1 \\ f_j^m\left(A_j^{m1}\right) + f_j^0\left(A_j^0 - A_j^{m1}\right) & \text{if } C_j^m 2 \\ f_j^m\left(A_j^{m1}\right) & \text{if } C_j^m 3 \end{cases}$$

The net change in social benefit is

$$Nsb_j^{cm} = P\left(X_j^{m1} - X_j^0\right) - V_j A_j^{m1} - F_j \qquad (1)$$

and it represents the difference between the production gain and the variable and fixed cost of the new variety.

Since up until now almost all crop biotechnologies have been developed and commercialized by private companies in monopoly situations, introduction of GMVs by the public sector has rarely occurred in reality. However, the case of Bt cotton in China closely corresponds to this scenario. Although Monsanto and Delta and Pine Land (D&PL) have introduced U.S. Bt cotton varieties in China, the Chinese Academy of Agricultural Sciences has developed and commercialized its own Bt cotton technology, which can be freely used by public and private sector breeders. Due to weak IPR protection, the Monsanto technology also has been incorporated into Chinese cotton varieties by local organizations, without payment to the company.

At present, there are 22 officially registered local Bt varieties and five imported ones available on the market (Pray et al., 2002). Adoption of Bt varieties has occurred in approximately 35% of the Chinese cotton sector and is increasing rapidly. There is no indication that GM technology has a negative effect on cotton biodiversity. On the contrary, the Bt varieties imported from the United States appear to have broadened the local germplasm base.

A similar situation could occur in other countries that do not protect IPRs but have a strong breeding capacity. If foreign GMVs are introduced in these countries, breeders can freely use these varieties to cross-breed the transgenic traits into their own germplasm. The trade-off, however, is that without IPR enforcement, private seed industry development is hampered and technology transfer from abroad is discouraged.

3.1.2 Monopolistic markets for seeds of GM local varieties

Consider the case when GMVs are produced and marketed by a monopolist. The monopolist is assumed to have access to the traditional local varieties and to modify them. Let A_j^{m2} denote the area of GMV_j and let A_j^2 denote total area (traditional and GM) of variety j. In this case the inverse demand function, denoting the maximum price $\left(W_j^m\right)$ farmers are willing to pay per acre for GMV seeds, as a function of acreage, is

$$W_j^m\left(A\right)=D^{-1}(A)=\begin{array}{ll} P\left[Mp_j^m\left(A\right)-Mp_j^0\left(A\right)\right] & \text{if } A< A_j^0 \\ PMp_j^m\left(A\right)-Mc_j & \text{if } A> A_j^0 \end{array} \tag{2}$$

The marginal revenue from the sale of seeds for A acres is

$$MR_j^m\left(A\right)=\begin{array}{ll} P\left[Mp_j^m\left(A\right)-MP_j^0\left(A\right)+A\frac{\partial}{\partial A}\left[Mp_j^m\left(A\right)-Mp_j^0\left(A\right)\right]\right] & \text{if } A< A_j^0 \\ P\left[Mp_j^m\left(A\right)+A\frac{\partial}{\partial A}\right]\left[Mp_j^m\left(A\right)\right]-Mc_j & \text{if } A\geq A_j^0 \end{array} \tag{3}$$

This inverse demand curve indicates that buyers will not be willing to pay more than (a) the difference between the marginal benefits per acre of GMVs and traditional varieties when both are viable ($A < A_j^0$) and (b) the difference between marginal benefits of land with GMV and marginal cost of land when only GMVs are economical.

The possible adoption patterns of GMV_j under monopoly includes

- $M_j^m 1$: <u>No adoption</u> if $PMp_j^m(0)-PMp_j^0(0)<V_j^m$.
- $M_j^m 2$: <u>Partial adoption</u> if $PMp_j^m(0)-PMp_j^0(0)>V j> MR_j^m\left(A_j^0\right) -MR_j^0\left(A_j^0\right)$. In this case, $A_j^{m2}<A_j^0$, at A_j^{m2}; $MR_j^m\left(A_j^{m2}\right) -MR_j^0\left(A_j^{m2}\right)=V_j$, $A_j^2=A_j^0$.
- $M_j^m 3$: <u>Full adoption</u> if $Mc_j>V_j$. At A_j^{m2}, $MR_j^m\left(A_j^{m2}\right)=Mc_j+V_j$.

The monopolist will sell the amount of seeds where its marginal revenue is equal to the variable per unit cost. When the marginal revenues intersect V_j at a quantity smaller than A_j^0 (case $M_j^m 2$), there will be partial adoption and when the marginal revenues intersect V_j at a quantity greater than A_j^0, there will be full adoption. It can be verified that (a) higher gains in marginal productivity $\left(\text{high } PMp_j^m(A) - PMp_j^0(A)\right)$ result in an increase in adoption of the GMV, and (b) adoption rates under monopoly are smaller than under competition. This is so because the monopoly price $PMp_j^m\left(A_j^{m2}\right)$ for GMV_j will be greater than the competitive price, V_j.

The profit of the monopolist, presented in Eq. (4), is smaller than the net social benefits considered by public sector decision makers when determining whether or not to assume the fixed cost of introducing a new variety. Thus, a monopoly outcome will provide a less-than-optimal introduction and adoption of GMV_j. There may be cases when profit does not cover the fixed cost of modification. Then, a monopolist will not introduce GMV_j, even though the net social benefits might be positive.

Fig. 14-2a denotes the monopoly outcome for the case of partial adoption. Curve ABC denotes demand for GMV seeds and has two segments—AB is $PMp_j^m - PMp_j^0$ and BC is $PMp_j^m - Mc_j$. The marginal revenues AE, associated with AB, intersect V_j to establish $A_j^{m2} < A_j^0$. Fig. 14-2b denotes the monopoly outcome for the case of full adoption. With the marginal benefits of using GMV, PMp_j^m's are much higher than those of the traditional variety. Demand for GMV seeds is represented by ABC, and the relevant marginal revenues are EF, which intersects with V_j at $A_j^{m2} > A_j^0$.

The output of the industry in a monopoly situation is

$$X_j^{m2} = \begin{cases} f_j^0\left(A_j^0\right) & \text{if } M_j^m 1 \\ f_j^0\left(A_j^0 - A_j^{m2}\right) + f_j^m\left(A_j^{m2}\right) & \text{if } M_j^m 2 \\ f_j^m\left(A_j^{m2}\right) & \text{if } M_j^m 3 \end{cases} \tag{4}$$

The price of seeds is equal to $P\left[MP_j^m\left(A_j^{m2}\right) - MP_j^0\left(A_j^{m2}\right)\right]$ if $M_j^m 2$ and $PM_j^m\left(A_j^{m2}\right) - MC_j$ if $M_j^m 3$. Thus, taking into account the fixed cost i, net profit of the monopolist is

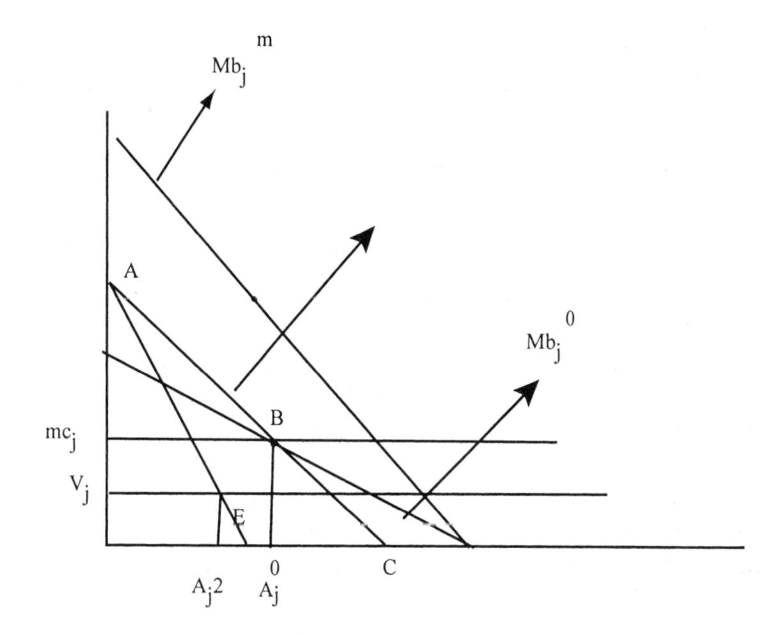

Fig. 14-2a. Partial adoption

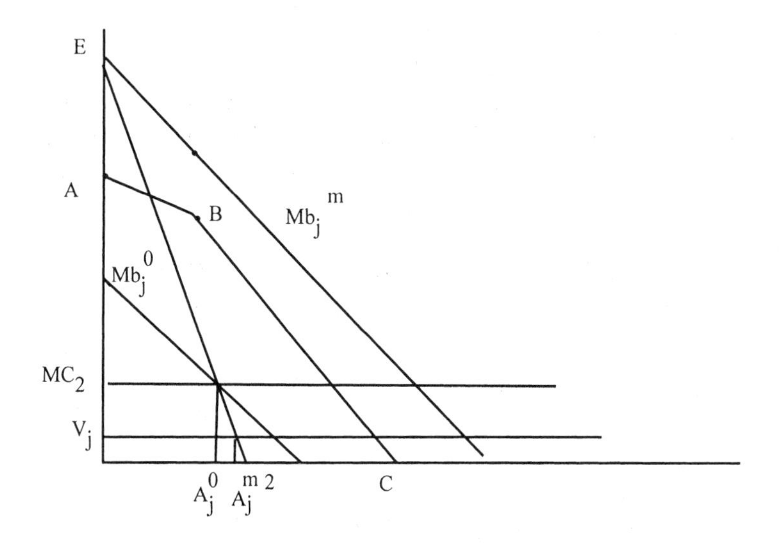

Fig. 14-2b. Full adoption

$$\pi_j^m = \begin{cases} 0 \text{ if } M_j^m 1 \\ \left[P\left[Mp_j^m\left(A_j^{m_2}\right) - Mp_j^0\left(A_j^{m2}\right)\right] - V_j\right]A_j^{m_2} - F_j \text{ if } M_j^m 2 \\ \left[PMp_j^m\left(A_j^{m_2}\right) - Mc_j - V_j\right]A_j^{m_2} - F_j \text{ if } M_j^m 3 \end{cases} \quad (5)$$

Whether the monopolist has access to all local varieties and markets GMVs itself or the technology is licensed to other seed producers does not matter for the scenario outcome. Technology licensing under strong IPRs is rather typical for GMVs in the United States. For Roundup Ready (RR) soybean and RR and Bt corn, biotechnology firms have issued nonexclusive licenses to all breeders and seed companies interested in endowing their own breeding lines with the transgenic traits. Due to high demand and a relatively low fixed cost to modify local varieties, numerous GMVs are available on the U. S. market.[5] The number of RR soybean varieties increased along with increasing technology adoption rates. In 2002, around 200 different seed companies marketed over 1,100 RR soybean varieties, which were adapted to diverse local conditions (Carpenter et al., 2002). On average, this implies an area of less than 20,000 hectares per variety. Likewise, several hundred RR and Bt corn hybrids are available from seed companies of all sizes, with an average area of less than 10,000 hectares per hybrid. Although the patent owners capture a significant share of the rent through monopoly pricing, U. S. farmers and consumers also benefit significantly (Moschini, Lapan, and Sobolevsky, 2000; Falck-Zepeda, Traxler, and Nelson, 2000), as the licensors have not set prices to take advantage of variations in marginal product across regions or varieties.

For RR soybean and Bt corn in Argentina, the scenario is similar. Although IPR protection is weaker than in the United States, the technologies have been licensed to various seed companies that incorporated them into their own breeding lines. Today, there are seven different companies providing 56 RR soybean varieties and four companies providing over 20 different Bt corn hybrids in Argentina (ASA, 2002). Most of this germplasm has been locally bred or adjusted, and the total number of soybean varieties and corn hybrids in Argentina did not change significantly since the introduction of GM technology (INASE, 2000). Especially in the case of RR soybeans, farmers are the main beneficiaries because Argentine

[5] For Bt and RR cotton in the United States, technology release was somewhat different. Instead of issuing licenses to all seed producers, Monsanto entered into an exclusive agreement with D&PL, the dominant seed company in cotton. Although in 2002 there were 33 GM cotton varieties available on the market, the exclusive agreement led to an increase in D&PL's market share at the expense of some varieties sold by smaller seed companies.

legislation allows the on-farm reproduction of seeds (Qaim and Traxler, forthcoming). Thus, average royalty payments are relatively low, so that the situation has some elements of the social optimum scenario.

3.1.3 Competitive markets for seeds of a GM "generic" variety

In some countries, a limited capacity to modify GMVs or cost considerations may lead even the public sector to introduce a generic GMV, imported from abroad, instead of genetically modifying local varieties.

The results for the competitive and monopolistic markets for *GMVj* can be modified accordingly. If the seed industry is competitive, the possible outcome includes

- $C_j^g 1$: <u>No adoption</u> when $PMp_j^g(0) - PMp_j^0 < V_j$.
- $C_j^g 2$: <u>Partial adoption</u> if $PMp_j^g(0) > V_j > PMp_j^g(A_j^0) - PMp_j^0(A_j^0)$. Here $A_j^{g3} < A_j^0$ when at A_j^{g3} $PMp_j^g(A_j^{g3}) = PMp_j^0(A_j^{g3}) - V_j$, $A_j^0 - A_j^{g3}$ in a traditional variety.
- $C_j^g 3$: <u>Full adoption</u> when $PMp_j^g(A_j^0) - PMp_j^0(A_j^0) > V_j$. At A_j^{g3}, $Mp_j^g(A_j^{g3}) = Mc_j + V_j$. Output in this case is

$$X_j^{g3} = \begin{cases} f_j^0(A_j^0) & \text{if } C_j^g 1 \\ f_j^g(A_j^{g3}) + f_j^0(A_j^0 - A_j^{g3}) & \text{if } C_j^g 2 \\ f_j^g(A_j^{g3}) & \text{if } C_j^g 3 \end{cases}$$

and net social benefit is

$$NSb_j^{Cg} = p[X_j^{g3} - X_j^0] - V_j A_j^{g3} - F_g \tag{6}$$

The generic GMV has a lower marginal productivity than GMV_j, but its variable cost is lower. Thus, comparing $C_j^m 1 - C_j^m 3$ with $C_j^g 1 - C_j^g 3$ suggests that more (less) acreage will be utilized with the GMV_j than the generic GMV if the marginal benefit gain $P\left[Mp_j^m\left(A_j^{m1}\right) - Mp_j^g\left(A_j^{m1}\right)\right]$ is greater (smaller) than $Mc_j - \left(V_j - V_g\right)$.

When the variable cost of GMV_j is very high due to an undeveloped crop-breeding sector (and the yield disadvantage of generic variety is not overwhelming), adoption rates of a generic GMV will be higher. In these situations, it may be that, despite lower yield per acre, the actual output of the generic GMV will be greater than those of the GMV_j. In most situations, however, we do not expect the extra cost of GMV_j to be dominant and expect both adoption and output to be higher with a GMV_j. At this stage, there are no empirical examples of a generic GMV being introduced by the public sector. As indicated before, the introduction of most GMVs worldwide has been done by private monopolists, and outcomes with generic GMVs in such situations are discussed below.

3.1.4 Monopolistic markets for seeds of a GM "generic" variety

If a monopolistic firm that controls the GM technology is precluded from access to local varieties, or if the fixed cost of modifying local varieties is too high relative to expected profits, a generic GMV will be introduced to the area grown with variety j. In this case total acreage is A_j^4 and acreage of the generic GMV is A_j^{g4}. The inverse demand $W_j^g(A)$ and marginal revenue functions $MR_j^g(A)$ for this case are defined similarly to the ones in Eqs. (2) and (3), and only the indicator g replaces m. The possible outcome includes

- M_j^g1: <u>No adoption</u> when $PMp_j^g(0) - PMp_j^0(0) > V_j$.
- M_j^g2: <u>Partial adoption</u> occurs when $PMp_j^g(0) - PMp_j^0(0) > V_j > MR_j^g\left(A_j^0\right) - MR_j^0\left(A_j^0\right)$. At $A_j^{g4} < A_j^0$, $PMp_j^g\left(A_j^4\right) - PMp_j^g\left(A_j^4\right) = V_j$ and $A_j^0 - A_j^{g4}$ is the non-GMV area.
- M_j^g3: <u>Full adoption</u> when $MR_j^g\left(A_j^0\right) - MC_j > V_j$. In this case adoption of A_j^{g4} is when $MR_j^g\left(A_j^{g4}\right) - Mc_j - V_j = 0$.

With this notation, the output and the profits of the monopolist are given by

$$X_j^{g4} = \begin{cases} f_j^0\left(A_j^0\right) & \text{if } M_j^g 1 \\ f_j^g\left(A_j^{g4} + f_j^0\right) + \left(A_j^0 - A_j^{g4}\right) & \text{if } M_j^g 2 \\ f_j^g\left(A_j^{g4}\right) & \text{if } M_j^g 3 \end{cases} \quad (7)$$

$$\pi_j^g = \begin{cases} 0 & \text{if } M_j^g 1 \\ \left[Mp_j^g\left(A_j^{g4}\right) - Mp_j^0\left(A_j^{g4}\right) - V_j\right] A_j^{g4} - F_j & \text{if } M_j^g 2 \\ \left[Mp_j^g\left(A_j^{g4}\right) - Mc_j - V_j^g\right] A_j^{g4} - F_j & \text{if } M_j^g 3 \end{cases}$$

Comparison of $M_j^g 1$ to $M_j^g 3$ with $M_j^m 1$ to $M_j^m 3$ suggests that, under a monopoly, adoption of GMV_j is greater (smaller) than adoption of the generic GMV if the gain in marginal revenues of the GMV_j $\left[PMR_j^m\left(A_j^{m2}\right) - MR_j^g\left(A_j^{mg}\right)\right]$ is greater than the extra variable cost. Thus, there may be a situation when adoption of GMV_j will be less than that of the generic variety.

There can be different reasons why rights cannot be freely traded. Lack of IPR protection may prevent the innovator from trading because it is difficult to enforce compliance of licensing agreements. A similar situation might occur when IPRs are strong but high transaction costs hinder a smooth transfer. Due to diverging objective functions, licensing agreements can be difficult to negotiate for the private innovator, especially when germplasm is owned by the public sector.

3.2 The choice of GMV_j or a generic GMV

Thus far, we have analyzed the area to be planted with GMVs in the four scenarios considered. However, a more fundamental question is whether to introduce a generic GMV at a given location or to modify the local varieties. The decision rule for a monopolist is different from that of a public sector entity, which introduces seeds to be distributed by competitive seed companies. We will solve the public sector case first and then consider the monopolist problem.

Let δ_j^m be an indicator variable that assumes the value 1 if option GMV_j is selected and 0 otherwise. The indicator variable δ_j^g assumes the value 1 if option *GMV* is selected and 0 otherwise. The public sector can either (a) modify variety *j*, (b) introduce a generic GMV, or (c) neither. The public sector aims to maximize net social welfare so its decision problem for region *j* will be

$$NSb = \max_{\delta_j^m \delta_j^g,} NSb_j^m \delta_j^m + NSb_j^g \delta_j^g$$

subject to $0 \le \delta_j^m + \delta_j^g \le 1,\ \delta_j^m, \delta_j^g$.

The public sector will select GMV_j if net social benefits with this technology are positive and greater than Nsb_j^g. The generic GMV is chosen when it generates positive social surplus greater than Nsb_j^m.

From conditions (1) and (6), the public sector decision whether to introduce GMV_j or a generic GMV hinges on three factors—revenue differential $P\left[X_j^{m1} - X_j^{g3}\right]$, variable cost differential $\left[V_j A_j^{m1} - V_g A_j^{g3}\right]$, and fixed cost differential, F_j. GMV_j is selected when its extra revenues are greater than the extra variable and fixed cost that its introduction entails. When the production advantage of the local variety is not substantial or the breeding sector is not well developed (so F_j is high), the public sector will prefer to introduce a generic GMV to location *j*.

In cases where the monopolist controls the introduction of GMVs, let δ_j^g and δ_j^m be defined similarly. The optimization problem becomes

$$\max_{\delta_j^m, \delta_j^s} \delta_j^m \pi_j^m + \delta_j^g \pi_j^g$$

subject to thc constraints $0 \le \delta_j^m + \delta_j^g \le 1$.

The monopolist will elect to introduce GMV_j if $\pi_j^m > \pi_j^g$ and $\pi_j^m > 0$. The generic variety will be introduced if $\pi_j^g > \pi_j^m$ and $\pi_j^g > 0$.

A comparison of Eqs. (5) and (7) suggests that in determining whether to genetically modify the local variety or to introduce a generic GMV, the monopolist compares the likely extra revenues of GMV_j with the extra fixed and variable cost its introduction entails. The generic GMV is likely to be introduced (a) in locations with small acreage, where volume of seed sales will not cover the extra cost of GMV_j, (b) in cases when the yield differences between the generic and local varieties are not substantial, and (c) when the variable and fixed costs of modification are substantial. The high fixed cost may reflect cost of access to local varieties or undeveloped local breeding sectors that make it worthwhile to import a generic GMV.

3.3 Case of inelastic demand for agricultural commodities

The price of internationally traded agricultural commodities, such as corn and cotton, is determined according to international demand and supply. Neglecting transportation costs and quality differences, the assumption of price-taking behavior is appropriate for most regions in the world. Yet, there are likely to be situations where commodities, especially food staples, are primarily produced for local consumption within a region. This may occur in regions with high transportation cost to major markets or low degrees of integration to the global economy for other reasons. These regions are not likely to be served by monopolistic seed companies but, rather, by small companies or direct provision of seeds by public extension programs. When GMVs are introduced in these situations, then the output price effect may be significant.

Suppose the demand for output at location j is $X_j = D_j(P_j)$, where P_j is output price in region j. The initial equilibrium with traditional variety j consists of acreage A_j^0, output X_j^0, and price P_j^0. These values are determined solving simultaneously the equilibrium condition in the output market, $P^0 = D^{-1}(X_j^0)$ $(D^{-1}(X)$ in inverse demand$)$; the equilibrium condition in the land market $P_j^0 Mp_j^0(A_j^0) = Mc_j$; and the production function $X_j^0 = f_j^0(A_j^0)$.

Suppose GMVs are sold by small seed companies (or distributed by extension programs) at price V_j. Let P_j^{m1} be output price under competition when GMV_j is introduced. The equilibrium conditions in this case determine output price P_j^{m1}, total output X_j^{m1}, total acreage A_j^1, and acreage with GMV_j, A_j^{m1}.

A procedure that can yield the equilibrium value consists of using the formulas we developed to obtain A_j^{m1}, A_j^1, and X_j^{m1} for cases of fixed output prices for a range of plausible values of P_j^{m1}. P_j^{m1} will clear the output market, so that $P_j^{m1} = D^{-1}(X_j^{m1})$ is the equilibrium output price. The equilibrium price P_j^{m1} also establishes acreage level $\tilde{A}_j^0$ solved from $P_j^{m1}MP_j^0\left(\tilde{A}_j^0\right) = Mc_j$, which is the amount of land that would have been utilized if the initial price was P_j^{m1}. Since the GMVs are assumed to increase yield, it can be shown that their adoption (even partially) will increase supply and, with negatively sloped demand, $P_j^{mi} < P_j^0$, which implies that $\tilde{A}_j^0 < A_j^0$. Thus, in the case of partial adoption, when the total acreage $A_j^1 = \tilde{A}_j^0$, adoption *of* GMV_j *will reduce acreage.* There may be cases of reduced acreage even in cases of full adoption.

The reduction in output price associated with the introduction of GMV_j will lead to increases in consumer surplus, denoted by ΔCS_j^{cm},

$$\Delta CS_j^{cm} = \int_{X_j^0}^{X_j^{m1}} D^{-1}(x)dx \tag{8}$$

The change in net social benefit becomes

$$NCb_j^{CM} = \Delta CS_j^{CM} + P_j^{m1}X_j^{m1} - P_j^0X_j^0 - A_j^{m1}V_j - \left[A_j^1 - A_j^0\right]Mc_j^j - F_j$$

While consumers will gain from the introduction of GMV_j, the impact on farmers is mixed. They produce more, yet receive a lower price. They may plant fewer acres but have to pay a technology fee.

The case with inelastic demand suggests that the introduction of GMVs to a location by the public sector may raise social welfare, reduce farmed acreage (and thus may improve environmental conditions), and improve consumer well-being but not necessarily help farmers.

The analysis of the impact of introducing a generic GMV is similar to that of introducing a modified local variety to location *j*. The generic variety, with its lower yield effect, will have less of an impact on output prices and may lead to a smaller reduction in acreage. Its introduction may benefit farmers more than the introduction of GMV_j and, thus, the introduction of a GMV_j will benefit consumers more. The likelihood of introducing GMV_j increases, the smaller the fixed costs to modify variety *j* and the variable cost to produce GMV_j seeds.

The analysis of the public sector-led scenario is useful to provide some intuition about the private monopoly case, when demand for output is negatively sloped. Private companies that introduce new GMVs globally are aware of the output price effect of the new innovation and its impact on their sales and revenue. We do not analyze the choices formally here but, rather, compare the outcomes when technology is provided by an idealized public sector that maximizes global welfare, versus when it is provided by a monopolist. Under the monopoly, we expect low rates of adoption and, thus, lower aggregate output, and a higher output price. The profit of the monopolist is likely to be less than the aggregate social welfare that also includes consumer and farmer surplus. Thus, the monopolist will not introduce GMVs in some cases where they would have been introduced by the public sector. Furthermore, since generic varieties have a lower yield effect and lower fixed costs, the monopolist is more likely to introduce the generic varieties than modify local ones.

The decision of whether or not to introduce a GMV at a location depends both on expected productivity gains as well as the fixed costs of introduction. Reduction in the costs of introduction, due to improved efficiency of the breeding sector or a decrease in the regulatory costs, is likely to increase introduction of GMVs in general and modified local varieties in particular. In some cases, there may be substantial costs to the public sector to introduce the technologies because of domestic capacity limitations. Then, the private sector may introduce a GMV, even though the private benefits are smaller than the public benefits. In other situations, low private benefit and high costs of introduction by the public sector may prevent the introduction of a GMV to a region. In these situations, if the public benefits of the GMV are sufficiently high, it may be introduced eventually as a result of policies that will enhance the capacity of the public sector and reduce its technology introduction costs.

4. IMPACT ON CROP BIODIVERSITY

The effect of the introduction of GMVs on crop biodiversity depends on the extent that traditional local varieties are replaced by a small number of GMVs. Continued planting of local varieties, even in GM form, can be a mechanism for preserving crop biodiversity. Our analysis has identified a wide array of circumstances in which local varieties may be preserved after the introduction of agricultural biotechnology. In situations where the revenue gains from genetic modification of local varieties relative to a generic variety are substantial, and fixed and variable costs of modification and production are low, local varieties will be modified and biodiversity will be preserved. Even in situations where the introduction of agricultural biotechnology will lead to replacement of areas of local varieties with a generic GMV, adoption of the modified varieties need not be complete. In particular, when a monopolistic private company controls the technology, it will charge a technology fee that may not warrant adoption of the technology on much of the land. Hence, a significant portion of the acreage will continue to be planted with traditional varieties. Full adoption of GMVs rarely occurs, and with partial adoption local varieties may be preserved.

5. DISCUSSION AND CONCLUSIONS

Biotechnology may preserve CGD more than conventional breeding. The reason is that biotechnology allows for separation between the act of developing novel crop traits and the process of breeding plant varieties. As a result, a given biotechnology innovation may be incorporated into a large number of plant varieties. This chapter has shown some of the conditions under which this might happen.

Modern biotechnology is still a fairly recent phenomenon, so that empirical evidence about the actual impact on biodiversity is limited. Table 14-1 shows adoption levels and the number of GMVs available in different countries for selected innovations. So far, widespread adoption occurred only for RR soybeans and Bt corn in the United States and Argentina, and Bt cotton in the United States and China. In all these cases, the technology has been incorporated into a large number of varieties, which supports our general hypothesis that biotechnology can preserve CGD. In other empirical cases, technology diffusion is still at an early stage so that conclusive statements are difficult to make. Biosafety regulations for GM crops can play an important role in this respect. The cost of regulatory compliance has become a major component in the overall budget to develop new biotechnologies. In most countries, only the transformation event is regulated, so that the regulatory cost for each technology occurs only once, regardless of the number of varieties into which it is incorporated later on. However, in countries such as India, each GMV is regulated separately. Such variety-specific approval procedures may foster loss of CGD and can be challenged on this basis.

Table 14-1. Number of available varieties for different GM technologies in selected countries (2001/2002)

Country	Technology	Area under technology (ha.)	Number of local varieties/ hybrids [a]	Number of imported varieties/hybrids
USA	RR soybean	22 million	> 1,100	0
	Bt corn	7 million	> 00	0
	Bt cotton	2 million	19	0
Argentina	RR soybean	10 million	45	11
	Bt corn	0.7 million	15	6
	Bt cotton	22,000	0	2
China	Bt cotton	1.5 million	22	5
India	Bt cotton	40,000	3	0
Mexico	Bt cotton	28,000	0	2
South Africa	Bt cotton	20,000	1	2

[a] Including locally adjusted ones.

Based on the conceptual analysis and empirical observations, we suggest a fourfold classification of situations according to the expected biodiversity outcome.

1. *Strong IPRs, a strong breeding sector, and low transaction costs.* Most situations within the private sector in developed countries, and some advanced developing countries such as Argentina, Brazil, and South Africa, belong in this category. The private technology owner will license the innovation to different seed companies that incorporate it into many or all local varieties, so that CGD is preserved. Adoption will be fairly widespread, and the innovator captures a rent through royalty payments. This outcome is equivalent to case II of the above analysis.
2. *Strong IPRs and a strong breeding sector, but high transaction cost to trade rights.* High transaction costs can occur, particularly when licensing contracts between a private technology owner and a public breeding organization have to be negotiated. Examples are biotech companies trying to reach agreements with international agricultural research centers, or public breeding stations that serve certain regional niche markets. If an agreement cannot be reached, the most likely outcome is that the biotech company will directly introduce GMVs that are not locally adapted. A widespread adoption of these varieties would lead to a loss of CGD, which is the outcome described under case IV of our analysis.
3. *Weak IPRs and a strong breeding sector.* Countries such as China and India belong into this category. Because IPRs are weak, every breeder or seed company can use commercialized GMVs, in order to cross-breed the technology into their own germplasm. Thus, many different GMVs will be available on the market. Due to the competition, the innovator's ability to capture rents are limited, so that farmers and consumers are the main beneficiaries. This outcome almost corresponds to case I, the social optimum.
4. *Weak IPRs and a weak breeding sector.* This situation is typical for most of the least-developed countries in Africa, Asia, and Latin America. In none of these countries have GM crops been commercialized so far. If biotechnology developed by the private sector abroad should reach these countries, the most likely outcome is that foreign GMVs are directly introduced without adaptation. A widespread adoption of these varieties would lead to a loss of CGD, which is the outcome described in case IV of the analysis.

There are different policy implications for each category. In category (1), strong IPRs and smooth transactions ensure that biodiversity is preserved. In category (2), widespread realization of the benefits of biotechnology will likely require international efforts to create an effective mechanism that reduces the transaction costs of trading IPRs. An intellectual property clearinghouse is one model that may be effective, if it is designed in a manner that addresses the needs of the poor as well as environmental concerns, while recognizing that much of the technology will be developed by profit-driven firms in the developed countries (Graff et al., 2003).

In category (3), the outcome is socially optimal in the short run, but the situation might look differently from a dynamic perspective. Lack of IPRs deters international technology transfer and innovation in the private sector. If biotechnology is entirely acquired and provided by the public sector, this might be less problematic. However, this may not be feasible or even advisable in more advanced developing countries. It may require significant amounts of funds and retard the evolution of private seed companies. Our analysis suggests that, in such situations, introduction of IPR protection and enabling sale of rights will be desirable. The alternative will lead to takeovers and concentration in the seed industry, which would be associated with a loss of CGD and underutilization of the economic potential of biotechnology.

For situations in category (4), the implications are somewhat different. Realizing the biotechnology opportunities in least-developed countries will require that there is a minimum absorptive R&D capacity. Since a local private seed industry is hardly existent, funding for the incorporation of biotechnology into local varieties will have to come from noncommercial sources that recognize the total gain in social welfare rather than strictly private financial returns. In practice, this means that the existing system of agricultural experiment stations may increasingly become centers of biotechnology adaptation and application. Also, the international agricultural research centers could play a bigger role in this respect. Lump-sum royalties to private innovators will have to be paid if such technologies are being used. In some cases such royalties might be waived for humanitarian purposes. An international clearinghouse mechanism would also be very beneficial in these situations.

In summary, biotechnology-based innovations in agriculture have the potential to preserve CGD, yet the actual impact will depend on the specific institutional conditions, R&D capacities, IPR policies, and biosafety regulation schemes in the individual countries.

REFERENCES

ASA, 2002, *Cultivares de Maíz y Soja Transgénicos Comercialmente Disponibles en Argentina*, Asociación Semilleros Argentinos, Buenos Aires.

Carpenter, J., Felsot, A., Goode, T., Hammig, M., Onstad, D., and Sankula, S., 2002, *Comparative Environmental Impacts of Biotechnology-Derived and Traditional Soybean, Corn, and Cotton Crops*, Council for Agricultural Science and Technology, Ames, Iowa.

Evenson, R., and Gollin, D., 2003, Assessing the impact of the green revolution, 1960 to 2000, *Science* **300**:758-762.

Falck-Zepeda, J. B., Traxler, G., and Nelson, R. G., 2000, Rent creation and distribution from biotechnology innovations: The case of Bt cotton and herbicide-tolerant soybeans in 1997, *Agribusiness* **16**:21-32.

FAO, 1998, *The State of the World's Plant Genetic Resources for Food and Agriculture*, Food and Agriculture Organization, Rome.

FAO, 1999, *Glossary of Biotechnology and Genetic Engineering*, Food and Agriculture Organization, Rome.

Graff, G., Cullen, S., Bradford, K., Zilberman, D., and Bennet, A., 2003, The public-private structure of intellectual property ownership in agricultural biotechnology, *Nature Biotech.* **21**:989-995.

INASE, 2000, *Yearbook 1999*, Instituto Nacional de Semillas, Secretaría de Agricultura, Ganadería, Pesca y Alimentación, Buenos Aires.

Moschini, G., Lapan, H., and Sobolevsky, A., 2000, Roundup Ready soybeans and welfare effects in the soybean complex, *Agribusiness* **16**:33-55.

Pray, C. E., Huang, J., Hu, R., and Rozelle, S., 2002, Five years of Bt cotton in China–The benefits continue, *The Plant J.* **31**:423-430.

Qaim, M., and Traxler, G., forthcoming, Roundup Ready soybeans in Argentina: Farm-level and aggregate welfare effects, *Agri. Econ.*

Qaim, M., and Zilberman, D., 2003, Yield effects of genetically modified crops in developing countries, *Science* **299**:900-902.

Traxler, G., Falck-Zepeda, J. B., and Sain, G., 1999, Genes, germplasm and developing country access to genetically modified crop varieties, paper presented at the 3rd ICABR conference, Rome.

Chapter 15

ESTABLISHING EFFECTIVE INTELLECTUAL PROPERTY RIGHTS AND REDUCING BARRIERS TO ENTRY IN CANADIAN AGRICULTURAL BIOTECHNOLOGY RESEARCH

Derek Stovin[1] and Peter W. B. Phillips[2]
[1]Professional Research Associate, Department of Agricultural Economics, 51 Campus Drive University of Saskatchewan, Saskatoon, SK S7N 5A8; [2]Department of Agricultural Economics, 51 Campus Drive University of Saskatchewan, Saskatoon, SK S7N 5A8

Abstract: Governments face a second-best situation when they consider the appropriate policy for realizing the optimal level of agricultural research. As public research effort is constrained, private research is required. In absence of any effective way of excluding others from using intellectual innovations, a public good problem will exist. While intellectual property rights help to reduce the public good problem through granting market power to innovators seeking to commercialize their inventions, these rights also create deadweight losses. This paper examines the tradeoffs inherent in IPR policies and uses a Canadian example to show how some of the unwelcome effects can be minimized.

Key words: holdup; intellectual property; licensing; strategic behavior.

1. INTRODUCTION

The innovation, knowledge, and technology derived from recent advances in the basic science of biology have been of such universal importance that approximately 40% of the world's market economy is now based upon biological products and processes (Gadbow and Richards, 1990). One of the fundamental drivers of this transformation is the introduction of new or strengthened individual rights over intellectual property. These rights provide an increased incentive for the private production and adoption of new ideas. However, in apparent contradiction, these rights can also be an impediment to technological change by providing a means of establishing a barrier to entry in the idea production sector. Accordingly, we first develop a model of the social decision with regard to choosing an intellectual property regime. We then discuss key determinants of the size of the associated welfare loss, the degree to which they are present in Canadian agricultural research, policy options to mitigate the welfare loss, and implications for those countries currently choosing an intellectual property regime.

2. BACKGROUND

2.1 Property rights

Property rights are social constructs that confer exclusive use-rights of a specific item upon an individual. Although these use-rights are exclusive, they are not unlimited. An individual who owns property rights to a specific item is constrained in its use by the usual libertarian criterion that one cannot infringe upon another individual's civil rights. Thus, for example, an individual has the right to possess a firearm, but the uses to which it may be put are limited. An individual's use rights are further constrained by society's reservation of the right to removal for the social good. Precedents for this removal of individuals' property rights include expropriation of land for public infrastructure building and seizure of goods allegedly obtained through illicit means. An individual's remaining use rights can, however, be protected in many ways. These include maintaining important related secrets, developing brand identities, and acquiring patents, trademarks, or copyrights. The legal methods of protection are the subject of this paper since they are a state- sanctioned institution that, at times, can conflict with society's aforementioned right of removal.

Contrary to popular belief, intellectual property rights (IPRs) are no different than traditional property rights. IPRs, like traditional property rights, are a social mechanism to formally recognize ownership. They confer upon an individual who purchases a specific item, whether that purchase is through monetary expenditures or through expenditures of mental and physical labor, the right to be compensated for the use of that item. Confusion abounds, though, because the specific items purchased under these different nomenclatures are often fundamentally different. Traditional property rights usually refer to an item that is embodied in some physical object (such as land), whereas IPRs often refer to an item that is fundamentally an idea and is, therefore, often disembodied.

The fundamental difference between embodied and disembodied items is often characterized as being rival or nonrival in use. A physical object, such as a computer, is characterized as a rival good because only one person is able to use it at any given time. It is, therefore, quite easy for the owner of the property right to exclude others from using the item. A disembodied item, such as the knowledge of how to build a computer, is characterized as nonrival because more than one person can use that knowledge at the same time. Since these disembodied items are often "new ideas," once they are produced, excluding others from using them is often difficult because monitoring and enforcing the exclusive use-right is very costly. Establishing IPRs in law is one possible way to reduce these transaction costs. Many individuals will not willingly break the law, thus monitoring and enforcement become unnecessary in many situations. However, choosing to employ the legal system involves assuming its own unique set of transaction costs, and the outcome of the relevant cost/benefit analysis is far from assured.

In effect, establishing effective IPRs transforms an inherently non-excludable, nonrival "new idea" into the same circumstances as an excludable, rival good. This is important because without excludability, it is difficult to extract rent in return for the production of the good. Thus, IPRs are necessary to provide the incentive that motivates the employment of private resources in the production of the inherently nonrival good. In other words, IPRs allow the market mechanism to "pull" the production of new ideas. However, even with IPRs, not all new ideas will be produced privately—particularly those with high sunk costs and high monitoring/enforcement costs. In Sections 3 and 4 a model is developed of the social decision in choosing an IPR regime, and the associated welfare implications are discussed.

2.2 Intellectual property rights and agricultural biotechnology

Canada's IPR regime changed fundamentally in August of 1990 when the Plant Breeders' Rights Act was implemented. By 1997, public funding of Canadian agricultural research had fallen in real terms to $374 million from $419 million in 1990 (-11%) (Canadian federal and provincial public accounts). In contrast, private expenditures in Canadian agricultural research rose in real terms to $29 million in 1997 from $21 million in 1990 (+38%) (Canadian federal and provincial public accounts). Although multicellular life forms are still not patentable material under Canada's Patent Act, plant breeders' rights confers ownership upon the final plant product thereby giving value to the necessary production process patents.

In order to produce a new genetically engineered crop variety, access to two broad categories of inputs is required: Human Capital; and Tools of the Trade. These can be further divided into at least six components that are integral to the genetic engineering (GE) process as illustrated in Fig. 15-1 below. The four components under the "tools of the trade" category are readily patentable under Canadian law.

The two institutions of a society's right of removal and state sanctioned individual IPRs come into conflict where the bestowed monopoly rights of patents are abused by the owner. Assume a legally protected key piece of technology is desired for use, by either public or other private agents, in an area of research that the owner has no intention of pursuing. If the IPR holder refuses access to that technology, then the owner is causing harm analogous to the firearm example in Section 2.1. Society's intent in bestowing a monopoly privilege is to allow the owner to extract rent when other individuals use their intellectual property. It is to encourage innovation that, when disseminated, can spill over to other firms or sectors and contribute to increasing returns to scale at the aggregate level. If the monopoly privilege, rather than being used to acquire fair financial compensation, is

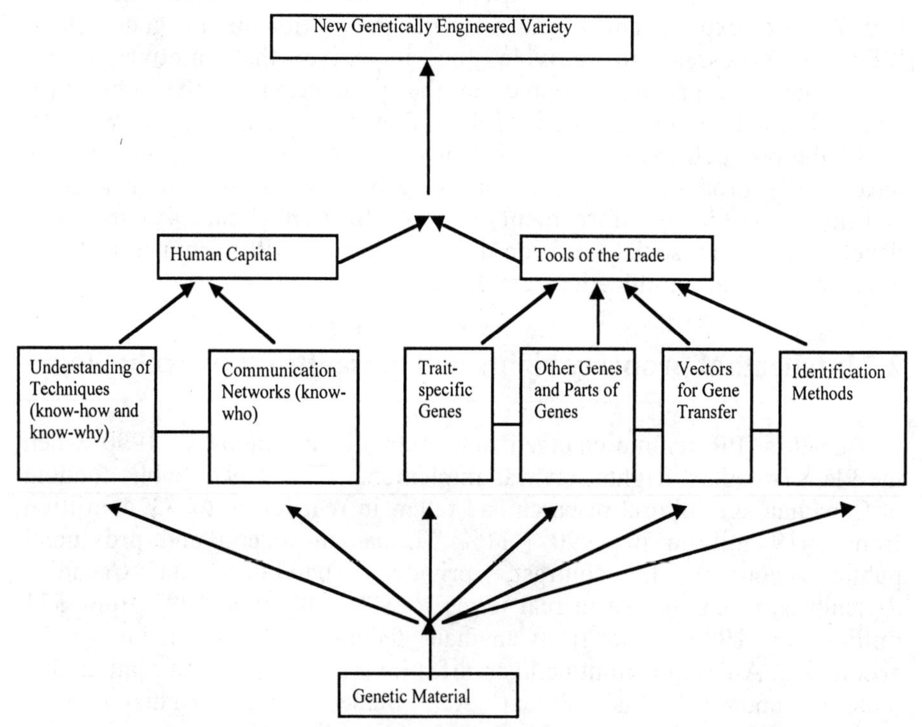

Figure 15-1. Agricultural plant biotechnology research requirements

Source: Authors.

used to erect a barrier to entry, then the IPR is causing economic harm to the other private agent and to society in general. An IPR that is used in this manner has ceased to be a method for society to encourage innovation and has become a private tool for preventing the development of new ideas by the owner's potential competitors. The existence of a strong IPR regime may, then, result in some organizations not having access to key production inputs and, thus, not having the "freedom to operate" in the industry. If this type of strategic behavior is employed, then it is appropriate to follow a public policy that exercises a society's right of removal.

Public policymakers face a conundrum typical of the second-best world. The IPR model, which is developed next, shows that a deadweight loss (DWL) is incurred in moving from a fully funded public research system to a private IPR driven system. An IPR that is employed strategically as described above is a barrier to entry that causes the total DWL to expand. However, given that societies are generally unwilling or unable to fully fund a public research system, an IPR-driven system is the only option. In other words, monopoly rights over intellectual property are necessary but distorting. Thus, complementary institutions need to be developed to discourage behaviors that increase the DWL and encourage behaviors that mitigate the DWL (i.e., minimize the distortion). The second part of the paper, sections 5 and 6, discusses the potential for the above-mentioned strategic behavior and the extent to which it may be present in Canada's agricultural biotechnology sector. Finally, we conclude with suggestions for institutions to improve the efficiency of an IPR regime and draw lessons for the international IPR trading system.

3. THEORETICAL FRAMEWORK

A disembodied (nonrival) item is often difficult to exclude because, once the initial cost that produced the item is borne, it can often be transferred between individuals at virtually no cost.[1] In the case of a new idea, it is not produced until it is communicated for verification, the very act of which ensures nonexcludability. The marginal cost of producing more than one unit of a particular new idea is approximately zero (the cost of communication). The theoretical framework is divided into two scenarios: private production of new ideas under no legally enforceable IPRs and the private production of new ideas under legally enforceable IPRs. These two scenarios are discussed in turn.

---

[1] A good example of this is computer software. It can be copied almost infinitely on media (compact or CDs) that is very inexpensive with little expenditure of time.

3.1 Scenario one

The case of perfectly competitive private production of a new idea in the absence of legally enforceable IPRs is illustrated in Fig. 15-2 below.

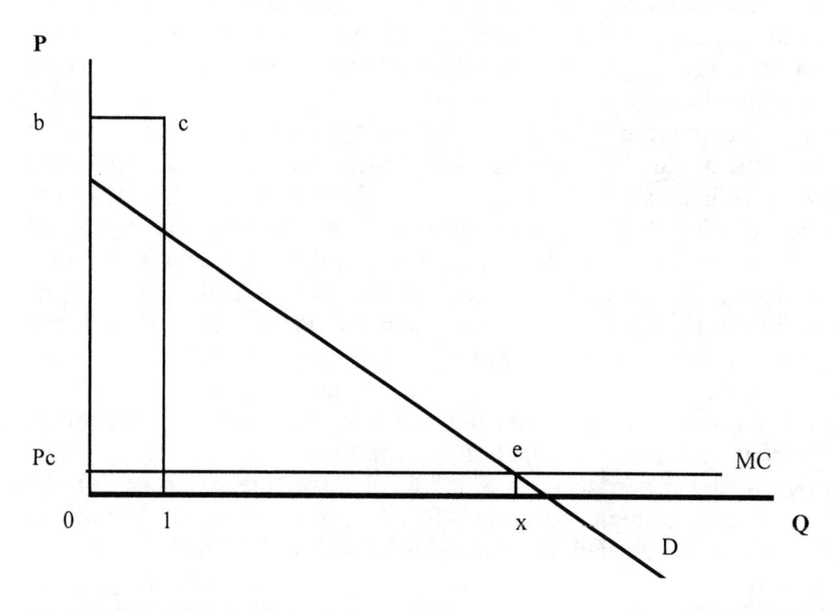

Figure 15-2. Failure of private production of a new idea under perfect competition and no legally enforceable IPRs

In Fig. 15-2, area *0bc1* represents the sunk cost that is necessary to incur in order for the potential new idea to be produced. The marginal cost curve (i.e. the supply curve) after the discovery is flat and virtually zero (*PceMC*), reflecting the low and constant costs of disseminating knowledge. Given a downward-sloping demand curve (*D*) that intersects with the supply curve (*PceMC*) at point *e*, the market price of extra units of the new idea would be *Pc*, which is approximately equal to zero. Since line *PceMC* virtually lies on top of the x-axis (because *MC* equals approximately zero), the area enclosed by *0Pcex* represents the total revenue (approximately zero) derived from the new idea. Therefore, because area *0bc1* (expected total private cost) is larger than area *Pcex* (expected total private benefit), the new idea will not be produced through the actions of private agents.

3.2 Scenario two

When effective IPRs are present, some disembodied items can earn sufficient economic rent to make the associated expenditure of private resources worthwhile and, therefore, production will occur. However, even with effective IPRs, some worthwhile disembodied items will not be privately produced. Fig. 15-3 illustrates this situation.

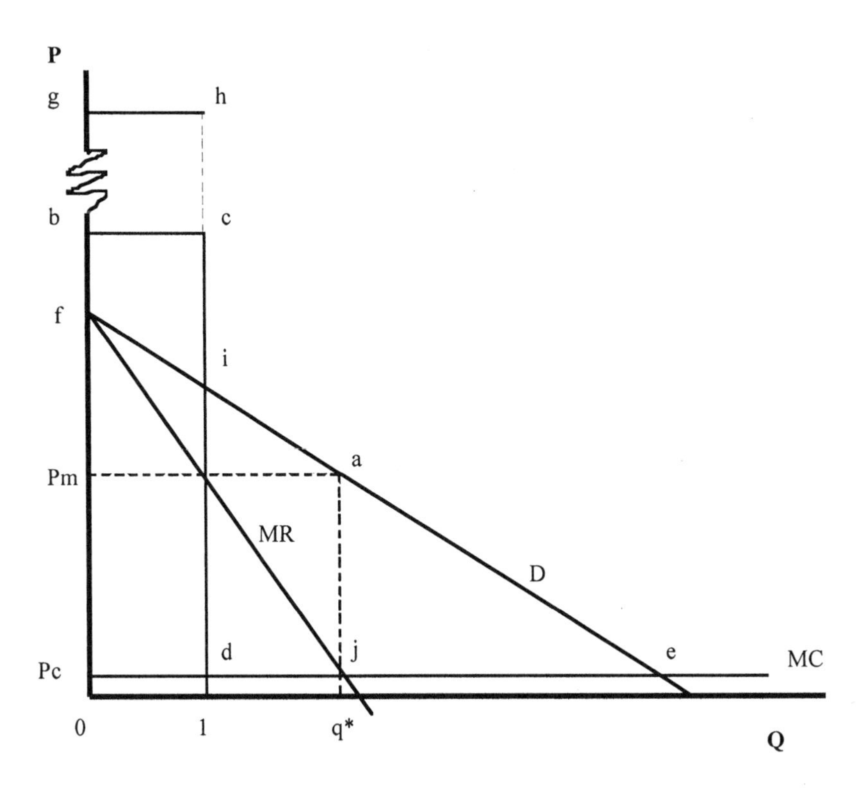

Figure 15-3. Private production of a new idea under legally enforceable IPRs

Because the legally enforceable IPRs endow a firm with monopoly rights over the new idea produced, the monopolist will price off of the MR curve with the result that, when it optimizes profits where $MR = MC$, q^* will be produced and sold at Pm. Since area $0Pmaq^*$ is greater than area $0bc1$, the firm can earn a return on the initial sunk cost investment, and the new idea will be privately produced. Thus, under an IPR regime that provides a reasonable assurance of monopoly rents, individual firms are willing to undertake the initial investment required for the production of some new ideas.

Even under strongly enforceable IPR regimes, however, some disembodied items will not be privately produced. This will occur when the sunk cost investment is so large that it cannot be recovered even with monopoly rents. This is depicted in Fig. 15-3 where the sunk cost is *0gh1* and its area is larger than area *0Pmaq**. Disembodied items will not be privately produced, even if the monopolist is able to perfectly price discriminate, if area *0gh1* is larger than area *0faq**. This lack of private production does not mean, however, that society would not gain from the production of the new idea. The idea will make society better off as long as area *0gh1* is less than or equal to area *Pcfe*.

4. WELFARE IMPLICATIONS OF CHOOSING A LEGALLY ENFORCEABLE INTELLECTUAL PROPERTY RIGHTS REGIME

Consider again Fig. 15-3. The new idea can be produced through one of two methods[2]: (1) fully publicly funded research and no IPRs or (2) privately funded research and strong IPRs. Because no IPRs are attached to the new idea in method (1), assume extra units are available at the *MC* of production (price equal to *Pc*, which is nearly zero). The social welfare derived from production of the new idea is, then, equal to the consumer surplus generated (area of triangle *fePc)* minus the cost of the investment (area *0bc1*). With private property rights (second method), assume the private funding agent is able to perfectly price discriminate. There is, then, a producer surplus generated by the granting of IPRs equal to area *Pcfaj* minus area *Pcbcd*. For the subset of all potential new ideas that would be produced under either method, choosing private property rights over public investment causes both a transfer of benefits from consumers to producers (area *diaj - fbci*) and a net welfare loss to society (equal to the triangle *aej*).

Assume further that, for the rest of the potential new ideas, government funding is always available when, in Fig. 15-3, area *0gh1* is less than area *Pcfe* (i.e., the sunk cost is less than the total consumer surplus generated) and it is never available when area *0gh1* is greater than area *Pcfe* (i.e., the sunk cost is greater than the consumer surplus generated).[3] There is, then, always a net welfare loss from choosing private property rights. Choosing a legal IPRs regime is, therefore, only

[2] There are, of course, other combinations but the extremes are used for clarity.

[3] These assumptions are equivalent to assuming that no government failure exists in the allocation of public research funds.

economically justifiable if public research dollars are limited and the net welfare loss is deemed to be worth incurring in order to capture the consumer surplus associated with shifting the limited public resources to research activities that private agents are not willing to perform.

An important implication of this theory is that if a society follows a public policy of a strongly enforceable IPR regime, then it ought to not directly perform research that private agents desire to undertake. Public resources should be directed towards the activities depicted in Fig. 15-3 where area *0gh1* is larger than area *Pcfaj* but less than or equal to area *Pcfe*. In other words, public organizations should not "crowd out" private activity and should only perform research that private agents are unwilling to undertake. This is also implied by Diamond (1999), who found that judicious public expenditures on basic research can actually "crowd in" private research expenditures.

The size of the DWL will, then, increase in response to the following two situations: public expenditures in areas of private activity and private strategic behavior in the employment of the publicly bestowed patent privileges. The following two sections explore the second situation in the context of Canadian agricultural biotechnology.

5. THE FREEDOM-TO-OPERATE PROBLEM

As noted in the introduction, biotechnology applications have been used for centuries. The term biotechnology was first used to describe traditional activities that involved various techniques for using living things to make products or provide services (Grace, 1997). These activities had agricultural, medical, environmental, and other applications. It is, therefore, an umbrella term that includes aspects of engineering, basic sciences, humanities, commerce, and other disciplines. Agricultural biotechnology has this same multidisciplinary aspect but its applications are of particular interest to consumers since their lives are impacted on a daily basis through the food supply. Agricultural biotechnology has also been applied for centuries through "conventional" crop and animal breeding, which over time has drastically altered the traits displayed by the crops and animals that farmers choose to produce. The production of new crop varieties is, however, increasingly relying on "nonconventional," or recombinant, techniques to alter traits. It is this GE of plants critical to the food supply that is currently causing consumers concern and not biotechnology *per se*. A related concern among many consumers is that IPRs may cause power to accrue to groups in society who have preferential access to the technology. The following subsections draw upon Fig. 15-1 for exploring in greater detail the six technical requirements of GE in order that the potential for strategically employing these items can be determined.

5.1 Human capital

5.1.1 Understanding of biotechnology techniques

An understanding of recombinant techniques is, obviously, a necessary condition for genetically transforming plants. This is a very important component of the process since a successful biotechnology application requires the researcher to have a good grasp of an entire system. This knowledge is very difficult to acquire because an individual must have an extensive background in many different academic disciplines. Nonetheless, many students acquire the relevant skills and abilities in laboratories at the undergraduate and graduate levels. Thus, they are readily available for expanded Canadian agricultural biotechnology research.

5.1.2 Communication network

Participation in relationship networks with a wide range of scientists is necessary for having timely access to new information and application techniques across the spectrum of required disciplines. The GE of crops is such a small profession that “everyone knows everyone” (at least those doing work on the same species). If an individual holds a plant breeder position, then they are able to access the relevant networks. It is not very difficult for a scientist, through these networks, to acquire the knowledge, skills, and materials that are needed for a specific application. This has been called the "clone by phone" (McHughen, 2000) research method.

5.2 Tools of the trade

5.2.1 Trait-specific genes

Trait-specific genes control specific plant characteristics. Examples of some such characteristics are cold and drought tolerance, insect resistance, and herbicide tolerance. Novel genes are readily patentable and ownership of the property rights is highly concentrated. For example, from 1986-1997, approximately 270 patents related to novel *Bacillus thuringiensis* (Bt) genes were granted in the Organisation for Economic Co-operation and Development countries (Krattiger, 1997). The five major company groups (Bayer, Dow, DuPont, Monsanto, and Syngenta) hold about 60% of these patents (Lindner, 1999). Although the remainder of the Bt-related gene patents are widely held, this does not necessarily translate into market contestability because only a very small portion of the Bt-related gene patents are useful for any specific application. Thus, for commercialization purposes, access to specific genes that are necessary for specific applications are likely to be legally controlled by one or more of the above company groups. However, since genomics research is progressing at a rapid pace, the possibility of "inventing around" such a holdup before the patent protection expires is very good. Accessibility of high quality germplasm is also necessary in order to have material in which to insert the novel genes. In the past, germplasm was readily available through public storage and dissemination facilities but, with the advent of the patentability of whole plants in some jurisdictions, the best-quality germplasm may increasingly be held privately.

5.2.2 Other genes and parts of genes

Parts of genes or other nontrait-specific genes are also necessary for genetically engineering plants. Promotors are used to control expression of the trait-specific gene in plants. Gene silencing or regulating technologies are used to suppress or modify gene expression in plants. Virtually all known promotors are protected by patents, so commercialization of a new transgenic variety requires securing legal access to the IPRs of the relevant promoter sequence. CaMV 35S, the patent on which is held by Monsanto, is one of the most widely used promotors. It has been freely used for research purposes and has been licensed to several companies for use in the development of commercial transgenic crops. Some public institutions have, however, had difficulty in obtaining permission to use it in their transgenic crops for commercial release (Lindner, 1999).

5.2.3 Vectors for gene transfer

Currently, there are two widely used transgenic methods: the *Agrobacterium tumefaciens* approach and the biolistic approach. Both of these approaches to transferring genes are patent-protected intellectual property. Other methods have been developed but do not yet have a high enough success rate for commercial use. There is much concern that the lack of alternative transformation technologies is a major holdup problem in the development of new transgenic crops. The likelihood of "inventing around" this problem appears to be low and "... it may be necessary to wait the remaining years until the technology comes 'off patent'" (Lindner, 1999).

5.2.4 Identification of transformation methods

A method for identifying the plant cells that have been successfully transformed is necessary for transgenic research. Genes, called selectable markers, are used to perform this function. While there may be many options available to scientists in this category, commercialization can be held up due to the increased costs associated with not having legal access to the most efficient techniques.

6. STRATEGIC BEHAVIOR IN CANADA'S AGRICULTURAL BIOTECHNOLOGY SECTOR

6.1 The potential for holdups in Canadian agricultural biotechnology research

There is no convincing evidence that research activities have been seriously held up in Canada as a result of IPRs. Public research organizations exist and are active in plant breeding through both conventional and nonconventional techniques. Even if private agents cannot get a license to use key biotechnologies, those who desire to engage in nonconventional plant breeding research can do so by partnering with one of many public organizations and, particularly, by supporting graduate student research. The Canadian Patent Act has a research exemption that encompasses a broad range of activities and has consistently been liberally interpreted. There are two sections that are relevant: Section 19 (Use of Patents by Government) and Section 55 (Liability for Patent Infringement).

> *USE OF PATENTS BY GOVERNMENT*
> *Government may apply to use patented invention*
> *19. (1) Subject to section 19.1, the Commissioner may, on application by the Government of Canada or the government of a province, authorize the use of a patented invention by that government.*
>
> *Conditions for authorizing use*
> *19.1 (1) The Commissioner may not authorize the use of a patented invention under section 19 unless the applicant establishes that*
>
> > *(a) it has made efforts to obtain from the patentee on reasonable commercial terms and conditions the authority to use the patented invention; and*
> > *(b) its efforts have not been successful within a reasonable period.*

Exception
(2) Subsection (1) does not apply in cases of national emergency or extreme urgency or where the use for which the authorization is sought is a public non-commercial use.

Liability for patent infringement
55. (1) A person who infringes a patent is liable to the patentee and to all persons claiming under the patentee for all damage sustained by the patentee or by any such person, after the grant of the patent, by reason of the infringement.

Exception
55.2 (1) It is not an infringement of a patent for any person to make, construct, use or sell the patented invention solely for uses reasonably related to the development and submission of information required under any law of Canada, a province or a country other than Canada that regulates the manufacture, construction, use or sale of any product.

Idem
(2) It is not an infringement of a patent for any person who makes, constructs, uses or sells a patented invention in accordance with subsection (1) to make, construct or use the invention, during the applicable period provided for by the regulations, for the manufacture and storage of articles intended for sale after the date on which the term of the patent expires.

For greater certainty
(6) For greater certainty, subsection (1) does not affect any exception to the exclusive property or privilege granted by a patent that exists at law in respect of acts done privately and on a non-commercial scale or for a non-commercial purpose or in respect of any use, manufacture, construction or sale of the patented invention solely for the purpose of experiments that relate to the subject-matter of the patent.

Section 19.1, subsection 2, is an exception to the exclusivity rights granted to a patent holder that allows governments to use a patent for public, noncommercial purposes. Research, teaching, and other scholarly activities clearly fall within the scope of this exception. Section 55.2, subsection 6, clarifies that nothing should be interpreted in such a way as to remove the research exception and extends the right to experiment to private agents.

6.2 Freedom to operate in the Canadian agricultural biotechnology industry

Although agricultural biotechnology research is not held up by strategic employment of IPRs, this is not necessarily the case for the commercialization of research. This holdup is for two reasons. First, IPRs in agricultural biotechnology research are tightly held. The private owners in some cases would appear to have chosen to strategically employ their most important IPRs in a manner that creates a barrier to entry, which is in contravention of the intent of an IPR regime. Second, the transaction costs for potential private entrants of assembly of the relevant IPRs can be very high and uncertain.

Most large biotechnology companies are unwilling to perform research on small crops because, for them, the market is too small to justify the expense. Further, they are unwilling to license their IPRs, ostensibly because of the potential public relations costs related to the possibility of licensees' misuse of their IPRs.[4] Other private agents or public organizations are free to use patent protected intellectual property to develop genetically altered small crops but, in this manner, are prevented from commercializing their research results. Under the Canadian Patent Act, section 65 (2), abuse is deemed to have occurred:

> *(c) if the demand for the patented article in Canada is not being met to an adequate extent and on reasonable terms;*
> *(d) if, by reason of the refusal of the patentee to grant a license or licenses on reasonable terms, the trade or industry of Canada or the trade of any person or class of persons trading in Canada, or the establishment of any new trade or industry in Canada, is prejudiced, and it is in the public interest that a license or licenses should be granted;*
> *(e) if any trade or industry in Canada, or any person or class of persons engaged therein, is unfairly prejudiced by the conditions attached by the patentee, whether before or after the passing of this Act, to the purchase, hire, license or use of the patented article or to the using or working of the patented process; or*
> *(f) if it is shown that the existence of the patent, being a patent for an invention relating to a process involving the use of materials not protected by the patent or for an invention relating to a substance produced by such a process, has been utilized by the*

[4] Confidential communication.

patentee so as unfairly to prejudice in Canada the manufacture, use or sale of any materials.

An IPR holder's refusal to license their patent on reasonable terms to other agents who possess the requisite relevant knowledge and abilities is an abuse of the monopoly privilege that society has bestowed upon them.

In recognition of exactly this freedom to operate problem, section 66 of the Canadian Patent Act grants powers to the Commissioner in order to rectify the situation in the best interests of the public.

Powers of Commissioner in cases of abuse
66. (1) On being satisfied that a case of abuse of the exclusive rights under a patent has been established, the Commissioner may exercise any of the following powers as he may deem expedient in the circumstances:
(a) he may order the grant to the applicant of a license on such terms as the Commissioner may think expedient, including a term precluding the licensee from importing into Canada any goods the importation of which, if made by persons other than the patentee or persons claiming under him, would be an infringement of the patent, and in that case the patentee and all licensees for the time being shall be deemed to have mutually covenanted against that importation;
(b) [Repealed, 1993, c. 44, s. 197]
(c) if the Commissioner is satisfied that the exclusive rights have been abused in the circumstances specified in paragraph 65(2)(f), he may order the grant of licenses to the applicant and to such of his customers, and containing such terms, as the Commissioner may think expedient;
(d) if the Commissioner is satisfied that the objects of this section and section 65 cannot be attained by the exercise of any of the foregoing powers, the Commissioner shall order the patent to be revoked, either forthwith or after such reasonable interval as may be specified in the order, unless in the meantime such conditions as may be specified in the order with a view to attaining the objects of this section and section 65 are fulfilled, and the Commissioner may, on reasonable cause shown in any case, by subsequent order extend the interval so specified, but the Commissioner shall not make an order for revocation which is at variance with any treaty, convention, arrangement, or engagement with any other country to which Canada is a party; or
(e) if the Commissioner is of opinion that the objects of this section and section 65 will be best attained by not making an order under the provisions of this section, he may make an order

refusing the application and dispose of any question as to costs thereon as he thinks just.

These powers are broad, with the Commissioner's options ranging from doing nothing, to dictating licensing terms, to revocation of the property right. Further, section 70 establishes that if the Commissioner dictates the licensing terms, then they are to be considered as though the two parties negotiated privately in good faith. However, this provision has not been entirely successful. Vaver (1997) argues that "proceedings have been prolonged and expensive; appeals are *de rigueur*; patentees, when alerted, often correct the abuse and retaliate against offending applicants. Of the fifty-three applicants who persisted between 1935 and 1970, only eleven got relief ... [and] today hardly anybody bothers trying." Thus, although the writ of Canadian patent law provides for balance between public and private interests, it has arguably been ineffective due to implementation failure of the bureaucracy.

Even if a private agent attempts to perform research for commercial purposes and, therefore, decides that owners of the relevant IPRs will not refuse to sell licenses on reasonable terms or that they can be acquired under the act, a holdup may still exist for a number of reasons related to the assembly of the tools of the trade. First, one may not know a priori which technologies will be necessary for the particular application. This could result in an unforeseen midstream requirement for an IPR that is not readily accessible. However, a public policy to counteract this potential problem is not appropriate as it is an organizational human capital management issue. Second, the legal status of any specific IPR can be in a state of flux since patent offices have been operating in a manner that grants the widest interpretation of claims and allows the courts to decide on the extent of the patent holder's rights. Both of these may result in a holdup because of the potential for post-hoc opportunism on the part of the unforeseen or unknown patent holder. A final reason for a potential investment holdup is that the transaction costs associated with the assembly of IPRs, regulatory compliance of products, and the monitoring and enforcement of IPR licenses can be very large. These high expected transaction costs, since they are a component that is factored into the initial expected sunk cost, can result in the situation depicted in Fig. 15-3 where line *gh* is sufficiently high to preclude private production. These transaction costs, while highly variable in the short run, should fall as the system evolves over time.

6.3 Specific cost of production example

The production of a new genetically engineered crop variety must, as illustrated in Fig.15-1, begin with high quality genetic material. Because the highest quality new cultivars are increasingly being privately held, the successful production of a new genetically engineered cultivar for commercialization requires the ability to generate high quality germplasm. The Crop Development Centre (CDC) at the University of Saskatchewan is widely respected for its "traditional" crop breeding activities. Flax has been chosen as an example since the CDC has recently produced both traditional and genetically engineered varieties.

The production time for a traditional new cultivar at the CDC is between 10 and 15 years. In 1997-98 the flax-breeding program used the services of three secretaries, an internal administrator, external university administrators, a senior research scientist, one permanent technician, six other technicians, two technical assistants, three field managers, two professional research associates, four master of science students, three doctorate of philosophy students, and 11 summer students. Over the years, these resources have varied considerably. The annual expenditures are broken down as shown in Table 15-1.

Table 15-1. Annual cost of team to produce a new flax cultivar by traditional breeding

Category of expenses	Expenses for 1997/98, Canadian dollars
Salaries and benefits	150,742
Supplies and services	12,443
Travel	2,121
Equipment/maintenance/rentals	5,922
Insurance/tax/license	1,000
Building space rental (estimate)	320,625
Land rental (estimate)	7,500
Total	500,353

Source: CDC Annual Report and personal communications.

The total cost of producing the required high quality germplasm, which takes 10-15 years, is, then, between $5,003,530 and $7,505,295. This is a conservative estimate for a number of reasons. First, equipment expenditures are underreported since significant investments using resources from other sources are made in purchasing expensive physical capital outright from time to time. Second, the CDC breeds many other crops and, thus, is able to exploit economies of scale to some degree. For example, the shop assistant, farm assistant, and administrative salaries are shared between breeding programs.[5] Finally, as a public organization, the CDC may have communication networks open to it freely that would not be accessible privately or only at a significant cost. These communication networks also likely result in a shorter production period. It ought to be noted, however, that the cost per new variety would fall

[5] The CDC has four major breeding programs: alternative and specialty crops, barley and oats, winter wheat, and spring wheat. Between 1995 and 1997 the CDC produced eight new varieties of peas, eight of barley, four of beans, four of wheat, two of flax, two of lentils, and two of oats.

rapidly in relation to the length the breeding program is expected to run. Nonetheless, for the purposes of genetic modification, a commercial entity must be willing and able to bear the cost of developing the initial germplasm as well as any resulting production risk.

In addition to the cost of germplasm development, the cost of developing a new cultivar through agricultural biotechnology was estimated in 1998 to be \$1.5 - \$15 million (Canadian) (McHughen, 1998) Most of this cost is incurred for navigating through different international regulatory regimes. McHughen (1998) offered a histogram that illustrates the challenges encountered in securing access for GE products in different countries. The regulatory testing and paperwork to register a conventionally bred variety named CDC Normandy added up to about 40 pages. The portfolio of reports required to gain approvals for the GE variety CDC Triffid added up to more than 2 feet. These transaction costs are likely to be much higher now in light of recent international events that have resulted in some jurisdictions placing a de facto moratorium on approvals for novel agricultural food products.

Further to the cost of germplasm development, the GE process requires securing the "freedom to operate" if it is for commercial purposes. Thus, licenses for relevant IPRs must be acquired. This can be a daunting task. For example, the number of U. S. patents related to Bt was 345 in October of 1999 (Phillips and Stovin, 1999). Furthcr, as was noted above, kcy patents are often tightly held making negotiations difficult. The CDC, as a public research organization, was able to avoid some of the potentially larger sunk costs in licensing fees because of the research exemption. Nevertheless, to commercialize their research, the CDC needed to negotiate freedom to operate. By the time the GM flax was commercially viable, the CDC had received a U. S. patent for the biolistic GM process for flax, which offered the opportunity of cross-licensing. Licensing negotiations would be much more difficult and costly for other small organizations that do not enjoy a research exemption and do not yet have their own IPRs to trade for access.

The experience of the CDC at the University of Saskatchewan highlights the uncertainties involved in commercializing products of agricultural research. It provides support for the idea that, regardless of these uncertainties, agricultural research programs are generally not held up in Canada but that a significant barrier to entry may exist for small, private, niche market agricultural research organizations.

7. CONCLUSIONS

Public policy decision makers face a conundrum when choosing an IPR system. A legal IPR system is necessary for providing the incentives to motivate private resources in the research sector when a society is unwilling or unable to commit to fully fund public research activities. However, a DWL is incurred in a move from a fully funded and efficient public research sector to a private IPR-driven system. This DWL is exacerbated if the remaining public research expenditures compete with (rather than complement) private expenditures or if the monopoly privilege endowed by legal IPRs is employed strategically to create a barrier to entry.

If private investments in research are to be made, legal IPRs must provide a reasonable assurance to private agents of the ability to extract rent from the resulting products. These incentives must, at the same time, be balanced with institutions for the public good that ensure market distortions are not increased by private abuses of the monopoly privilege. The less readily IPRs are licensed, the less legal IPRs provide an effective counter to the problem of limited public resources in research activities. In the Canadian agricultural biotechnology sector, two potential private research investment holdups exist: strategic behavior in licensing of intellectual property and uncertainties in patent ownership. The former is acknowledged by the mandatory licensing provisions in the writ of the Patent Act. However, in order to be effective, a commitment to the appropriate use of these provisions needs to be made. The latter can be dealt with, in part, through increased resources to the patent office. A narrower and more careful granting of monopoly privileges in the first instance will reduce the likelihood of the necessity to remove or redistribute holdings of the intellectual property in the future as well as reduce the a priori investment uncertainties of ownership.

Finally, less-developed countries (LDCs) that, in aspiring to participate more fully in the world trading system, find it necessary to develop legal intellectual property protection should consider Canada's intellectual property model. It has proven effective for motivating private research expenditures yet contains provisions for not only protecting the public interest but also enhancing the public benefit. The mandatory licensing provision, if stated clearly and credibly, provides protection of the public interest. It will not scare off private agents who only desire to make a reasonable rate of return as long as the bureaucratic procedures are timely and fair. The research exemption provisions enhance the public benefit by encouraging locally targeted research, which simultaneously creates the opportunity to capture some human capital spillovers. These provisions are particularly important for regions that lack the financial resources to make the necessary infrastructure and human capital investments. Canada's patent law provides a model that is amenable to both international business interests and LDCs unique needs.

REFERENCES

Canadian Public Accounts, Federal and Provincial Public Accounts (various years).

Crop Development Centre, 1997, Annual Report, University of Saskatchewan, Canada.

Department of Justice of Canada, 2000, Chapter P-4, Patent Act. Retrieved from Consolidated Statutes; http://canada.justice.gc.ca /STABLE/EN/Laws/Chap/P/P-4.html.

Diamond P. (ed.), 1999, *Issues in Privatizing Social Security*, MIT Press, Cambridge, Massachusetts.

Gadbow, R. M., and Richards, T. J. (eds.), 1990, *Intellectual Property Rights—Global Consensus, Global Conflict,* Westview Press, Boulder, Colorado.

Grace, E. S., 1997, *Biotechnology Unzipped*, Trifolium Books Inc., Toronto, Canada.

Krattiger, A. F., 1997, *Insect Resistance in Crops: A Case Study of Bacillus thuringiensis (Bt) and Its Transfer to Developing Countries,* ISAAA Briefs. No. 2, Ithaca, New York.

Lindner, B., 1999, Prospects for public plant breeding in a small country, in: *Proceedings of the 4th International Consortia of Agricultural Biotechnology Research (ICABR): The Shape of the Coming Agricultural Biotechnology Transformation*. University of Rome "Tor Vergata," Rome and Ravello, p. 17.

McHughen, A., 1998, *Proceedings of the 5th International Symposium on BioSafety Results of Field Tests of GM Plants and Micro-Organisms,* Biologische Bundesanstalt für Land-und Forstwirtschaft, Braunschweig, Germany.

McHughen, A., 2000, *A Consumer's Guide to GM Food: From Green Genes to Red Herrings,* Oxford University Press, Bath Press Ltd., Bath, Avon.

Phillips, P., and Stovin, D., 2000, The economics of intellectual property rights in the agricultural biotechnology sector, in: *Agricultural Biotechnology in Developing Countries: Towards Optimizing the Benefits for the Poor*, M. Qaim et al., eds., Kluwer Academic Publishers, New York.

Vaver, D., 1997, *Intellectual Property Law: Copyright, Patents, Trade-Marks,* Irwin Law, Toronto, Ontario.

Chapter 16

ADOPTION OF BIOTECHNOLOGY IN DEVELOPING COUNTRIES

Holly Ameden,[1] Matin Qaim,[2] and David Zilberman[3]
[1]*Ph.D. Candidate, Department of Agricultural and Resource Economics, 207 Giannini Hall, University of California, Berkeley, CA 94720;* [2]*Professor, Department of Agricultural Economics and Social Sciences, University of Hohenheim, 70593 Stuttgart German;* [3]*Professor, Department of Agricultural and Resource Economics, 207 Giannini Hall, University of California, Berkeley, CA 94720*

Abstract: This chapter identifies the factors that lead to the adoption of genetically modified varieties in developing countries and the sources of differences in the impacts and patterns of adoption of biotechnology between developed and developing countries. We present the finding of our model analyzing the profitability of pest-controlling biotechnologies. The model shows that in locations with mild pest issues, adoption of GMVs is likely to result in reduced pesticide use while in areas with high infestation levels, as is the case in many developing countries, adoption of GMVs will have both a pesticide-reducing and a yield-enhancing effect. Thus, successful adoption of biotechnologies in developing countries will depend on the availability of technologies appropriate for local agricultural conditions, and policies that enhance the ability of poor farmers to obtain these technologies such an affordable pricing schemes and credit programs. Following the conceptual model, the chapter provides some of the empirical findings on adoption of biotechnology for both developed and developing countries, discusses adoption and biosafety issues and, in the last section, synthesizes our results and provides further policy conclusions.

Key words: biotechnology; developing countries; technology adoption; technology diffusion.

1. INTRODUCTION

Agricultural biotechnology has been widely used in the developed countries of the North. However, its value and benefit to the developing countries of the South are subject to debate. What will be the impact of these technologies on productivity in developing countries? Will farmers in the South adopt these technologies? This chapter addresses these questions through both theoretical analysis and review of empirical evidence.

The first part of the chapter provides a conceptual framework for analyzing adoption and diffusion patterns of agricultural biotechnology. The next section presents the findings of our model analyzing profitability of pest controlling biotechnologies, which are the technologies that have been introduced thus far. Details of the model are provided in the Appendix to this chapter. There is growing evidence on adoption and impact of agricultural biotechnology both in developed and developing countries. Thus the conceptual section is followed by a discussion of some of the empirical findings on adoption of biotechnology for both developed and developing countries. The last section synthesizes our results and provides policy conclusions.

2. DIFFUSION AND ADOPTION: MODELING OVERVIEW

The study of adoption of agricultural innovations was spurred both by the failure of some very promising innovations, which in the lab seemed highly capable of being diffused among farmers, and the unexpected success of other innovations that did not seem as promising. Sociologists were the first to systematically study the spread of new innovations. They distinguished between two concepts—diffusion and adoption. Diffusion is the extent to which a given population utilizes a technology. One measure of diffusion of, say, tractors is the percentage of the farmers that utilize the tractor or percentage of land that is cultivated with the tractor. Adoption occurs when a particular individual utilizes a given technology or when a technology is utilized on a given field. Thus, one can use diffusion as a measure of aggregate adoption.

Statistical studies have found that diffusion is a dynamic process consisting of three stages: an early period of technology introduction where the diffusion rates are low, a second period of take-off, and a third period of saturation. Thus, diffusion curves are S-shaped functions of time. For almost every technology, there is also a final period of decline where the technology is replaced.

The survey by Sunding and Zilberman (2001) distinguishes between two major types of economic models of diffusion. The first assumed diffusion to be a process of imitation and technologies one modeled to spread in the same manner as infections (Mansfield, 1981). A key feature of the diffusion process is contact between individuals. In an early period, a small number of individuals are introduced to the technology and utilize it and, as more people are exposed, the diffusion process advances more rapidly until most of the population uses the new technology. Mansfield (1981) argues that the speed of imitation depends on factors, such as profitability, farm size, and industry structure. Econometric applications of these models were spawned by Griliches (1957); these studies have proven to be extremely useful in estimating diffusion patterns under many circumstances. They can be modified and applied extensively in marketing and economics. Although these models are useful statistically and provide evidence that economic considerations of profitability as well as farm size and other variables affect adoption rate, they lack an explicit microeconomic understanding of the working of the diffusion process.

An alternative approach, developed by Davis (1979), is the threshold model. This approach consists of three elements: (1) a microeconomic behavior, (2) a source of heterogeneity, and (3) a dynamic process affecting the microeconomics and driving adoption. Potential adopters consist of economic decision makers who are heterogeneous. The sources of heterogeneity may be such factors as the size of a farm, human capital and knowledge, time, risk preference, etc. These decision makers are assumed to pursue profit or to have objective functions that integrate profit as they choose between distinct technologies, for example, traditional and modern seed varieties. Since adoption decisions require investment, the decision-making criterion may aggregate economic benefits over several periods, for example, net present value of investment in new technology.

A key aspect of the threshold model is that constraints faced by producers are crucial in understanding and modeling adoption choices. Producers may be limited in their ability to finance new innovations. In addition, there may be comprehension and learning constraints such as individuals' difficulties comprehend with complex new systems. Because of heterogeneity, at each moment a subset of the population will choose the new technology, while another subset will stay with the traditional one.

Moreover, several dynamic forces drive the adoption process. One is learning by doing—the cost of the technology may decline as manufacturers improve their efficiency in producing it. Another is learning by using—the farmers adopting the technology become more adept in using it and, thus, the technology becomes more profitable over time, relative to the traditional technology. Changes in supply may affect the prices of output and, in some cases, enhance adoption, while in others, slow it. Most importantly, farmers adapt their perception and assessment of the technology as more information is accumulated, and if their perception of the technology improves relative to where it was initially, they will adopt it. The threshold model can generate S-shaped diffusion curves and the parameters of these curves depend on the distribution of the source of heterogeneity as well as the parameters of the dynamic processes driving the system. With the introduction of discrete choice estimation methods, the application and use of threshold models has proliferated.

3. THE ECONOMICS OF ADOPTING PEST-CONTROLLING BIOTECHNOLOGY

This section presents the primary elements and findings of a conceptual model that analyzes the profitability of pest-controlling agricultural biotechnologies. This model, which follows the spirit of the threshold model, evaluates how profitability varies among producers with different resources and constraints. First, the model analyzes how pesticide use, output levels, and, ultimately, adoptions decisions vary according to economic conditions (prices of pesticides, output, and the technology) and environmental conditions (severity of infestation). The model then analyzes the impacts of credit constraints, risk considerations, human capital, size of operation, and location effects. The details of this model of farm level choice of pest-control strategy are presented in the Appendix.

In the model, we consider pest management and seed technology choice at the field level and later consider issues of scale. Following the damage control approach (Lichtenberg and Zilberman, 1986), we distinguish between actual output and potential output—the difference is the result of pest damage that depends on pest infestation levels, use of chemical pesticides, and whether a pest-controlling genetically modified variety (GMV) is adopted. With traditional varieties, the only way to reduce pest damage is by using chemical pesticides that require fixed costs for application time and equipment and variable costs for materials. Two GMV varieties are considered—a modified local variety and an imported generic variety. Introduction of a generic variety may lead to a yield loss. The adoption of both GMVs also increases fixed cost per unit of land, as the farmer has to pay a license fee for GMVs.

Farmers make two choices: ***a technology choice*, whether to adopt a GMV or the traditional variety, and a *pesticide-use level choice* with each variety.** The analysis of the pesticide-use choice, detailed in the Appendix, shows:

- *Pesticide-use levels are higher when output prices and pest damage are higher, application costs are low, and pesticides are cheap.*
- *The introduction of a GMV will always reduce pesticide use and sometimes eliminate it.* The pesticide-saving effect is more substantial when the relative price of pesticides and the fixed costs of applying pesticides are relatively low and when the GMV has high pest-control

efficacy. These conditions are more likely to occur in developed countries and where chemicals are highly subsidized.

- *Adoption of GMVs is likely to result in significant increases in yields in locations with significant pest damage over the traditional variety.* The high-yield effect is more likely in locations with high pesticide prices, high pesticide application costs, high infestation levels, and low efficacy of chemical pesticides. This may be the case in Africa and some parts of South Asia. The overall yield effect for a generic GMV depends on the gains from reduced pest damage versus the losses associated with the variety switch.

The analysis of the technology adoption choice in the Appendix shows that *GMVs will be adopted in situations where the gains are high because of high pest damage, high cost of pesticides, or pesticide applications, and sufficiently low price for the use of the GMVs.* Adoption of generic GMVs may be less than that of the locally modified GMVs because of the yield effect, if fees are not adjusted. Regional heterogeneity in pest infestation suggests that the adoption levels will increase under discriminatory pricing of GMVs. Regions with higher infestation levels, higher costs of pesticides, and higher output prices are likely to adopt GMVs sooner.

3.1 Credit constraints

Many peasants and farmers lack the resources to fully pay for purchased inputs, such as pesticides and GMVs. They may need to borrow to finance these choices. Farmers' ability to obtain funds, and the price paid for these funds, affect adoption choices. As shown in the Appendix, when the GMV is costly, lack of credit may disallow some farmers from adopting this technology, or may induce them to adopt the generic GMV in cases where it is much cheaper. On the other hand, credit considerations may prevent or limit applications of chemicals and may induce farmers' interest in GMVs. Essentially adoption will suffer if the extra cost of seeds exceeds the extra cost of pesticides use. Adoption of GMVs can be enhanced by subsidization or provision of credit for new seeds, or by removal and restrictions of credit subsidies for pesticide use whenever they exist.

There is significant heterogeneity of credit availability among regions and farmers. Poorer farmers, and those located in regions farther away from financial centers, are more likely to face strict credit constraints. This may slow adoption by these groups of producers in some cases, and enhance it when pesticides are very costly. Comparing the impacts from adoption of GMVs across regions differing in their credit constraints, our analysis suggests that *the GMV will have more of a pesticide-saving effect on regions with more lax credit constraints and more yield-increasing effects where the credit constraint is more severe.*

3.2 Risk considerations

Farming activities are subject to risk, both in terms of production and market conditions, affecting adoption (Just and Zilberman). Pest infestation is one of the major sources of risks that farmers face. GMVs are forms of insurance against this pest risk. They protect the farmers both from increases in randomness of pests and the associated extra costs of pesticides, as well as the extra damage associated with major infestations. As shown in the Appendix, for a given average level of pest infestation, an increase in the randomness of pest infestations increases pesticide use when farmers are risk averse. Pesticide use increases as risk aversion, variance of damage, output price, and potential output become larger. An increase in the variance of pest infestation will increase the likelihood of adopting GMVs. Since farmers with smaller farms are likely to be more risk averse, risk considerations may lead to more adoption of GMVs by smaller farms.

Potential output may also be a source of risk affecting adoption of a GMV, especially when it is a generic, not a local, variety. Farmers are familiar with the yield distribution of local varieties and have accumulated knowledge on how to address their unique features. The relative lack of knowledge about the management and performance of a generic variety under local conditions increases the subjective yield risk. Just and Zilberman's (1988) analysis suggests that this increase in output risk will reduce pesticide use and the expected gain from adopting the generic GMV. Thus, adoption rates and economic welfare may be improved if prior experimentation and adjustments of production practices are conducted before the introduction of a generic GMV to a region.

Combined risk and credit considerations are likely to have a significant impact on farmers' choices. Failing to repay a loan because of pest damages or low prices may lead to bankruptcy. Farmers are likely to make technology choices that will reduce the likelihood of bankruptcy. For example, high levels of pesticide use will increase farmers' debt and income requirements to avoid bankruptcy, while low pesticide use increases the probability of crop failure. Thus, GMVs are likely to be adopted if they provide the same risk-reducing effect but are relatively lower cost compared to pesticides.

3.3 Human capital

Schultz (1975) distinguished between two categories of human capital: worker ability (the ability to perform tasks more effectively) and allocative ability (the ability to deal with new situations and learn new techniques). Allocative ability is closely related to intelligence and formal education and training. Pest management requires understanding of natural systems and good decision-making capacity, and there is evidence that effectiveness of pesticide use is related to allocative ability (Weibers, 1993). Under plausible assumptions, it can be shown that increase in human capital tends to reduce pesticide use and reduce pest damage. It tends to increase the relative benefits of the traditional, pesticide-intensive variety. *Thus, adoption of GMVs will be relatively more beneficial to individuals with lower human capital, as the pest-control aspect of the GMV substitutes for pest management skills. Moreover, adoption of GMVs will have a relatively higher "yield-increasing effect" for individuals with lower human capital and greater "pesticide-saving effect" for individuals with higher human capital.* Both aspects of human capital may also be positively related to potential output. The net effect of human capital on adoption of GMVs depends on the relative importance of the positive impact on potential output due to adoption of GMVs and the negative impact on pesticide productivity.

3.4 Size of operation

Larger operations tend to have several advantages that affect their technological choices. Their use of pesticides with the traditional variety may be more economical for several reasons. (1) Volume discounts on pesticide purchases benefit larger operations. (2) The fixed cost of improved application equipment enables only larger units to purchase them. Thus, larger units may apply pesticides with a tractor, rather than manually. That will reduce the application cost per unit of land. (3) Some individuals within a larger organization will specialize in pest management, and develop the human capital needed to improve pest-control productivity. The lower cost and higher efficiency of pesticide use of larger farms is likely to make GMVs less appealing to larger farms and to contribute to a higher adoption rate by smaller farms. On the other hand, several factors may make GMVs more appealing to larger operations. Larger farms may have better access to credit needed to purchase GMVs and may receive volume discounts on purchases of GMVs. The net effects of these factors will determine when scale will have a positive or negative effect on adoption of GMVs.

3.5 Location

Climatic and agroecological variations result in differences in pest infestation and potential output across locations. Our analysis suggests that locations with both high potential output and pest infestation levels are more likely to be earlier adopters of GMVs technologies, especially if the process of GMV seeds is not adjusted to reflect differences in value of the GMVs across locations. Rogers (1995) showed that diffusion rates of modern technologies are likely to be higher in villages located closer to regional centers. Distance from the center slows diffusion as it increases transportation costs and that, in turn, result in higher input costs and less frequent (or no) contact with extension agents or sales people who promote modern technologies. Distribution networks for inputs associated with new technologies tend to be established first in regional centers and then farther away. The impact of distance on adoption depends on the nature of the technology. The more complex a technology is to adopt, the lower diffusion across locations. Chemical pesticides require specialized equipment for application and care in handling and storage. GMVs are embodied in seeds, and *Bacillus thuringiensis (*Bt) crops require no changes in production practices. Thus, GMVs may "travel" faster than chemical pesticides and reach some remote locations where chemical pesticides have not been adopted. *These results suggest that adoption of GMVs may have a relatively greater yield-increasing effect at more remote locations and a higher pesticide-saving effect closer to the centers.*

4. DIFFERENCES IN GMV USE AND IMPACTS BETWEEN DEVELOPED AND DEVELOPING NATIONS

The results from our conceptual model enable us to analyze patterns of pesticide use across locations. We can distinguish between countries according to several factors including their level of pest infestation and their pricing of pesticides and pest-control technologies. The levels of pest infestation vary across locations. More humid regions are subject to higher levels of infestation than regions with dry climate. Thus, they have higher potential for yield loses. The humid, more pest-prone regions tend to be closer to the Equator, while the dryer, cooler regions are closer to the Poles. Many of the developing countries in these areas of Africa, South Asia, and South America are subject to higher pest pressure than the developed countries with temperate and even cold climates.

The cost of pest-control technologies also varies across locations. Generally speaking, we expect the fixed cost-to-output-price ratio (fixed cost of pesticides divided by output price) to be smaller in developed rather than developing countries due to output price subsidies and the larger scale of operations in developed countries that reduce application costs per unit. We expect the pesticides-to-output-price ratio (variable cost of pesticides divided by output price) to be higher in most developing countries than in the developed ones.

These differences in basic parameters will lead to different impacts of biotechnology in developing and developed countries. Given these considerations, *we expect that when biotechnology is adopted in developed countries much of its impact will be in terms of pesticide-use reduction.* The relatively low cost of pesticide use will tend to lead to intensive use of pesticides. *The yield effects will be low,* given that initial infestation levels are relatively mild and prior use of pesticides has controlled pest damages. Only when GMVs address a pest problem without prior treatment do we expect a significant yield effect. The strength of the breeding sector in countries like the United States suggests that they will attempt to modify many varieties, and the yield losses due to the transition from local varieties will be small. In contrast, the high pest pressure in many developing countries and the high cost of pesticides suggest that the *adoption of GMVs, while reducing pesticide use, will have a strong yield effect.* These effects are likely to be smaller in China, where pesticides are subsidized, and be very high in Africa where application rates of pesticides are relatively small.

Thus, certain policies are likely to increase the adoption of GMVs in developing countries. For example, discriminatory pricing recognizing the differences in impacts across regions is likely to enhance adoption. Moreover, the availability of local GMVs and affordable pricing of the new seeds are likely to be key factors for adoption in developing countries.

5. EMPIRICAL EVIDENCE: ADOPTION IN DEVELOPED COUNTRIES

The empirical research on genetically modified (GM) technology adoption in the developed world focuses on several areas including identifying the factors that lead to farmer acceptance of these new technologies, the patterns of adoption, and associated impacts. This research supports the findings of theoretical adoption models concerning the influence of the nature of innovations and farmer heterogeneity (e.g., farm size, access to credit, uncertainty, and risk) on farmer acceptance and in turn, on the distribution and diffusion of technologies in both the developed and developing world.

5.1 Farmer acceptance

Many researchers are performing empirical analyses of farmer acceptance of new technologies. As predicted, the nature of innovations—specifically, whether technologies are divisible or indivisible, whether or not sophisticated knowledge is needed for adoption, and whether or not the technologies require complementary inputs—is of primary importance in adoption decisions. Moreover, farm size, as predicted by theory, turns out to be an important determinant of adoption and, finally, farmers' perceptions and risk come into play.

Researchers have found that divisible technologies have high adoption rates because they have low fixed costs—both in terms of actual cost for the technology and minimal initial investment in improved human capital—and are simple to use (i.e., these technologies are laborsaving). Carpenter and Gianessi (1999) found that since its introduction in 1996, an herbicide-resistant soybean and weed control program has been rapidly adopted in the United States. While cost reduction is one reason for adoption, the primary reason that farmers switch to this program is the simplicity and flexibility of use. Fernandez-Cornejo, Daberkow, and McBride (2001) suggest that adoption rates are high because this kind of technology does not require significant adaptation of the production system (i.e., low fixed costs). Similarly, Bullock and Nitsi (2001) found that herbicide-resistant soybean is adopted by farmers in the United States because of lower treatment costs and, more importantly, because of simplicity of adoption and use and greater flexibility in the timing of treatment (i.e., lower labor costs). Farmers put in less time and effort scouting for weeds and determining how to treat them. Haung et al. (2001) also found that, unlike certain Green Revolution varieties, GMVs have a higher rate of adoption and are adopted by smaller farms because they are simple and convenient to use.

Theoretical predictions concerning farm size (scale), farmer perceptions, and uncertainty and risk are supported by recent research. Darr and Chern (2002) analyze adoption of GM soybeans and corn in Ohio (which is dependent on small family farms). They find that adoption of GM soybeans is not as extensive among smaller farms and also farmers who are not familiar with GM technologies. However, adoption is greater among those farmers who believe that the technologies are cost saving and yield increasing. These findings are supported by Alexander, Fernandez-Cornejo, and Goodhue (2002), who evaluate adoption patterns of genetically modified organisms (GMOs) in Iowa between 1999 and 2000. They find that larger operations producing corn or soybeans have a higher tendency to adopt while smaller farms are less likely to adopt.

Uncertainty and risk also play primary roles in farmer acceptance and adoption. Some studies have found significant disadoption. One explanation may be an expected decline in infestation. Growers who are more concerned with insect infestation would be more likely to adopt GMVs. Alternatively, uncertainty about acceptance of GMVs in Europe may reduce adoption. In fact, growers who feed their grain to livestock are more likely to adopt than growers who sell or export their grain. Moreover, while the first generation of GM crops, which are mostly pest-controlling varieties, has had generally rapid diffusion rates in the United States, the second generation of GM crops with improved output characteristics may have lower rates of adoption due to uncertainty about the benefits of the technologies. Jefferson, Traxler, and Wilson (2001) analyze the potential impact of these value-enhanced crops (VECs) based on experiences with field trails. They predict slower growth for VECs because of the uncertain yields (production risk) and product prices for these crops. Furthermore, VECs are likely to require more complex marketing arrangements than first generation technologies.

5.2 Adoption patterns

Both theoretical and empirical research studies have shown that farmer heterogeneity leads to differences in expected incomes and, therefore, to differences in the associated pattern of farmer acceptance and adoption. Certain distribution patterns may occur because a new technology benefits large farms that can adjust to new practices more easily than smaller farms. In addition, owners of low-quality, marginal land may benefit more from land quality-augmenting technology than owners of high-quality land. For example, corn acreage expanded to the sandy soils of Washington and western Nebraska with the introduction of center-pivot irrigation (Lichtenberg, 1989). Similarly, drip irrigation spread California grape and avocado production to areas with sandy soils. Fulton and Keyowski (1999), who analyzed adoption of herbicide-resistant canola in Canada, found that benefits from adoption vary across locations according to land quality and pest problems.

These impacts vary across regions and within regions. For example, adoption of new irrigation technologies shifted tomato production to California, leading to lower prices and losses for growers in Ohio and New Jersey. In addition, in their study of adoption of Bt cotton in the southeastern United States, Marra, Hubbell, and Carlson (2000) found that farmers in the lower South have a higher willingness to pay for the new technology than farmers in the upper South. They suggest that the price of the technology and the expected change in income are important determinants of adoptions.

5.3 Impacts of adoption

Studies on the impacts of adoption of GMVs in developed countries support our conceptual results that pesticide use declines. Hubble, Marra, and Carlson (2001) found that adoption of Bt cotton in the United States results in a reduction of about two pesticide applications per acre, with most of the reduction occurring in the lower South. They note that Bt adopters not only use less insecticide, but they also use proportionately less of the predominant type of cotton pesticide, potentially reducing pesticide resistance to these pesticides. In fact, studies by Frisvold, Sullivan, and Raneses (2003) as well as Marra, Hubble, and Carlson suggest (2001) that adoption of Bt cotton has drastically reduced pesticide applications in cotton (60% and more), though the yield effects were on average small (below 10%). Given that adoption of pest-resistant varieties reduces pesticide use, not only do farmers' input costs decline, but environmental health and farm-worker health may also improve.

The benefits of adopting GMVs vary across farmers, consumers, and seed companies as well as countries. Falck-Zepeda, Traxler, and Nelson (2000) studied the yield-increasing and pesticide-reducing benefits associated with adoption of Bt cotton in the United States. They found that U. S. farmers benefit the most, receiving approximately 59% of the estimated $240 million in benefits per year. Benefits to U. S. consumers are approximately 9%, while seed companies and Monsanto, the seed developer, received 5% and 21%, respectively. Moschini, Lapan, and Sobolevsky (2000) analyzed the impact of adoption of herbicide-resistant soybeans. They found that farmers in the United States gain substantially relative to farmers in other countries, and that this advantage is reduced as exports of the technology increase. In addition, they found that the innovating company receives much of the welfare gain, and consumers benefit globally. In contrast, in their study of the adoption of herbicide-resistant soybeans and Bt and herbicide-tolerant cotton, Price, Lin, and Falck-Zepeda (2001) find that U. S. farmers realized much less that half of the total benefits. Most of the benefits went to the gene supplier, seed companies, U.S. consumers, and thc rest of the world. Price, Lin, and Falck-Zepeda also note that results from these studies vary greatly depending on the farm-level effects and supply and demand elasticities for domestic and world markets.

6. EMPIRICAL EVIDENCE: ADOPTION IN DEVELOPING COUNTRIES

Empirical research concerning biotechnology adoption in developing countries is limited. It is clear, however, that theoretical findings concerning farmer heterogeneity and adoption are supported by this research. For example, researchers have found that divisible technologies that are simple to use and that have limited fixed costs (new GM seed varieties and tissue culture technologies) hold the most promise for adoption by small, poor farmers. In addition, adoption levels vary across countries, depending on many factors including a country's research capacity, input and output markets, intellectual property rights (IPR) regulation and enforcement, farm structure, and biotechnology approval and biosafety programs.

The experiences of different developing countries and regions with biotechnologies are instrumental in identifying constraints to adoption and determining how the potential benefits offered by biotechnology can be realized in all developing countries. We briefly examine adoption in China, Latin America, and Africa where certain GM crops have been approved for use. We then consider the case of India, where GM crops were adopted illegally, prior to biosafety approval. Finally, we discuss the challenge of biosafety regulations.

6.1 China

Of all developing countries, China is the most aggressive country in terms of biotechnology research and adoption of GM crops. It was the first developing country to commercialize a transgenic crop (virus-resistant tobacco). Currently, Bt cotton is the primary commercial GM crop grown in China, although GM varieties of tomato and sweet pepper are also approved for use. The spread of transgenic crops in China was supported by extensive research and development capabilities on the part of government research institutes and foreign companies and also by aggressive efforts on the part of local officials and extension agents to push Bt varieties when they became commercially available (Huang et al., 2001).

Haung et al. (2001) studied the impact of Bt cotton in Northern China and found that small farmers received substantial benefits. Farmers who adopted this technology greatly reduced the use of pesticides and reported pesticide poisonings without reducing the quality or quantity of cotton produced. Haung et al. also found that farmers benefited, instead of government research institutes or the foreign firms that developed these varieties, because of weak IPRs. In a follow-up study that covered a wider area of China, Huang et al. (2002) found that in all areas, adoption of Bt cotton improved yields and reduced pesticide and labor inputs, thereby increasing farmer income. Moreover, use of Bt cotton had positive environmental and health impacts.

6.2 India

In India, Bt cotton has been perceived by industry and government as a means to reduce pesticide use and increase productivity by combating the American bollworm, a major pest in India. Yet, its introduction faced resistance for environmental reasons. Herring (2003) argues that farmers' experience with the technology and its performance led to its official commercial introduction in 2002. The high costs of chemical pesticides, and their declining efficacy, increased the appeal of Bt cotton to farmers, many of whom suffered severe financial loss, and even bankruptcy, because of pest damage.

The results of the early trials (Herring, 2003) show yield effects of 30%-50% and substantial pesticide cost saving. Moreover, the result of trials in 2001 (Qaim and Zilberman, 2003) shows yield effects of 80%, which is consistent with other findings (Herring, 2003) because of a high infestation rate that year. An unauthorized introduction of Bt cotton to local varieties in Gujarat resulted in high performance compared to traditional hybrid corn during bollworm infestation and played a crucial role in commercial introduction of the technology in 2002. Three varieties were introduced commercially that year. Two were very successful, with yield effects of 30% and pesticide cost saving. However, one of the varieties was not appropriate to the local conditions and resulted in excessive wilting. The results of these studies indicate that in India the benefits of Bt cotton vary according to the pest infestation and that switching away from the traditional variety may be a source of significant loss of yield, and thus use of GMVs should proceed with caution.

6.3 Latin America

In Latin America, although many countries are conducting field trials of transgenic crops, only Mexico and Argentina are active in producing GM crops for commercial use. Argentina is the second largest producer of GM crops in the world behind the United States primarily due to its open biosafety regulations. Argentina's research capacity and funding targeted at local needs are limited (Qaim, 2002). In contrast, science and research capacities and funding in Mexico for adapting transgenic crops to local conditions are growing, experience with biosafety procedures is expanding, and seed markets are large enough to attract private sector interest (Traxler et al., 2001). In addition, biotechnologies have been used to address agroecological issues in Mexico. For example, Traxler et al. studied the impact of Bt cotton in Mexico where one-third of the country's cotton area was planted with Bt cotton in 2000. Focusing on a particular region that has had serious pest problems, they estimate $600/ha. of net benefit during years of pest pressure and equal profitability in years when pest populations were low. However, the growing gap between small, poor farmers and large, multinational agricultural corporations, as well as negative public perception of GM crops, continues to be significant constraints in all Latin American countries.

6.4 Africa

In Africa, although research and field-testing of transgenic crops is being actively pursued in countries such as Kenya and Egypt, South Africa is the only country where transgenic crops are grown commercially (Bt maize and Bt cotton). The challenges faced by potential adopters in Africa can be characterized by the findings of Ismael et al. (2001) in their study on the adoption of Bt cotton in a South African province for the 1998 and 1999 seasons. The farmers in this area are primarily small landholders—rural households and farms on land allocated by their tribal chiefs. The landholders face many difficulties. Tenure arrangements are uncertain, high quality land is scarce, and land is unfenced and threatened by livestock damage. In addition, pests, excessive rain, and drought are significant concerns, and labor is constrained when younger men leave the rural area and migrate to towns to seek work.

Ismael et al. (2001) found several factors affecting adoption. The most experienced farmers and those that owned more land were more likely to adopt the technology, probably because these farmers can more easily obtain credit or afford higher seed costs. Farmers who adopted in the initial year also planted Bt cotton in the second year. When asked for their main reasons for adoption, farmers cited expected savings in input costs (chemicals and pesticides), pest problems, increases in yields, and saving labor. Overall, the authors found adoption of Bt cotton by the surveyed farmers had a positive impact—Bt adopters had higher yields and higher gross margins than nonadopters. The increase in yields and reduction in input costs outweighed the higher cost of seed. Given the results of this study, benefits of transgenic crops can be expected to spread to other regions, assuming that most issues faced by farmers included in this analysis are common to farmers in other regions of Africa.

6.5 Adoption and biosafety

Addressing biosafety and trade concerns can severely hold up adoption of GMVs. For example, even though Brazil spends more on agricultural research than any other Latin American country (Janssen, Falconi, and

Komen, 2000), and despite the significant potential for GM crops in Brazil, a controversial judicial decision concerning biosafety held up commercial production of these crops until very recently. However, even before GM crops are granted biosafety approval, farmers seem eager to use them. This was true in India, where a seed company illegally distributed GM cotton to farmers prior to approval by India's biosafety committee (Jayaraman, 2001). Farmers embraced the technology, although they did not know the seed was transgenic, and were more than willing to pay a higher price. Furthermore, farmers paid the seed company in advance for a supply of seeds the following season. The government, in response to the biosafety violation, ordered the illegal cotton to be burned. These incidents illustrate the willingness of farmers to adopt GM crops and ease in doing so, as well as the primary challenge for all developing countries—to balance the potential benefits from GM technologies with appropriate biosafety measures and trade concerns (with Europe and Japan in particular).

7. SYNTHESIS AND POLICY CONSIDERATIONS

The Green Revolution, with its high-yielding varieties of crops, brought great promises of alleviating hunger and poverty by making food easier to grow. Although the wave of new technologies led to significant improvements in some areas, in others, i.e., those areas that suffered most, the benefits were not realized. Green Revolution technologies required that farmers have specialized knowledge about chemical fertilizer use and new irrigation techniques as well as access to inputs; therefore, these new technologies were relatively costly and difficult to adopt in the poorest regions.

In contrast, the "first generation" of agricultural biotechnologies are relatively simple to use, divisible, and are scale-neutral technologies that do not require significant investment up front or drastically alter local farm and cultural practices (depending on seed prices and type). These technologies have been highly successful in developed countries and are likely to be of even greater importance in the developing countries of the South. Ease of adoption means that the gap between educated farmers owning large farms and uneducated farmers with smaller farms is likely to diminish. Furthermore, these technologies can address the obstacles that act as primary constraints to the poorest farmers in developing countries—such as drought, high-saline soils, pests, and disease—and may go far just in terms of increasing production of basic, traditional crops. Therefore, simple, supply-enhancing, pest-resisting technologies have great potential for helping poor farmers in developing countries, even in the most challenged areas.

Considering insect-resistant technologies specifically, such as Bt cassava and Bt corn, farmers in developing countries where pests and disease are far more damaging may benefit even more from these technologies than farmers in developed countries. In fact, in the medium to longer term, the benefits of agricultural biotechnologies may be even greater for the South than what has been experienced by the North.

To reach the full potential that biotechnology adoption holds for developing countries, decision-makers must first adopt policies that enhance the development of appropriate technologies—those that meet the specific needs of farmers and consumers in developing regions. Research should continue to focus on "first generation" innovations that have low fixed costs, that are compatible with the human capital constraints in these countries, and meet the specific input needs of local farmers (e.g., salt-tolerant and pest-, pesticide-, and drought-resistant). In addition, the reduced environmental and health impacts, and resource-conserving characteristics of these technologies, must also continue to be developed.

Now that the development of first-generation technologies is well underway, researchers are now turning their attention to developing "second generation" technologies. These technologies have improved output characteristics that meet the needs of consumers. Although these VECs, such as high-vitamin or high-oil content crops, may initially pose more risk to the producer, they are vital for addressing the serious nutritional needs of developing countries.

Research on appropriate technologies is most effective when public national research centers from the targeted areas are involved. These centers are the best equipped to integrate research with local farmer needs through participatory approaches. Furthermore, research can be facilitated through private-public research collaborations that draw on expertise and resources from both the developed and developing world.

Of course, development of technologies appropriate for adoption by poor farmers is not enough. To fully realize the benefits of these technologies, appropriate and effective institutions that reduce barriers at the farm level, and make these new technologies available to the targeted farmers, must be put in place. Extension specialists play a significant role. Often they are the primary sources of information and tools, especially for poor farmers in remote areas. The extension/farmer relationship is based on trust; therefore, the extension agent needs to be accurately informed about the benefits and costs and should have first-hand experience with the practicalities associated with adoption of new technologies.

The literature suggests that limited access to credit in developing countries may slow or reduce adoption of agricultural biotechnology by smaller farmers. Thus, institutional policies that both reduce the cost of credit and increase its availability are needed. For example, adoption of some chemical solutions to pest-control problems has been limited because of the extra cost involved in equipment and material costs. Installment plans that require payment of biotechnology "fees" on a season-by-season basis, depending on whether a farmer utilized GM seeds, rather than significant up-front fees, will improve adoption substantially for small poor farmers. If the price of seeds is sufficiently low and credit channels are expanded, small farms that did not adopt chemical pesticides will adopt biotechnology seeds.

While creative credit solutions may be effective for commercially viable farms, further solutions are needed for the poorest subsistence farmers. These could come in the form of price discrimination structures aimed specifically at poverty alleviation. Seed companies could be assured access to commercial seed markets in one area in return for offering seed at or below cost in other areas where the poorest farmers are found. Of course, introduction of credit and pricing policies will require monitoring of farmer behavior and markets and effective enforcement capacity. Extension services and the public sector are challenged to cooperate with the private sector to introduce such mechanisms.

REFERENCES

Alexander, C., Fernandez-Cornejo, J., and Goodhue, R., 2002, Determinants of GM use: A survey of Iowa corn soybean farmers' acreage allocation, in: *Market Development for Genetically Modified Foods,* R. Evenson, V. Santaniello, and D. Zilberman, eds., CABI Publishers, London.

Bullock, D., and Nitsi, E., 2001, Roundup ready soybean technology and farm production costs, *Amer. Behavioral Scientist* **44**: 283-1301.

Carpenter, J., and Gianessi, L., 1999, Herbicide tolerant soybeans: Why growers are adopting roundup ready varieties. *AgBioForum.* **2**:65-72. Retrieved July 15, 1999, from the World Wide Web: http://www.agbioforum.missouri.edu.

Darr, D. A., and Chern W. S., 2002, Estimating adoption of GMO soybeans and maize: A case study of Ohio, USA, in: *Market Development for Genetically Modified Foods,* R. Evenson, V. Santaniello, and D. Zilberman, eds., CABI Publishers, London.

Davis, S. W., 1979, Interfirm diffusion of process innovations, *European Econ. Rev.* **12**: 299-317.

Falck-Zepeda, J. B., Traxler, G., and Nelson, R. G., 2000, Surplus distribution from the introduction of a biotechnology innovation, *Amer. J. Agri. Econ.* **82**:360-369.

Fernandez-Cornejo, J., Daberkow, S., and McBride, W. D., 2001, Decomposing the size effect on the adoption of innovations: agrobiotechnology and precision farming. Selected paper presented at the American Agricultural Economics Association Annual meeting, Chicago, Illinois.

Frisvold, G. B., Sullivan, J., and Raneses, A., 2003, Genetic improvements in major U. S. crops: the size and distribution benefits, *Agri. Econ.*, **28**:109-19.

Fulton, M., and Keyowski, L., 1999, The producer benefits of herbicide-resistant Canola. *AgBioForum* **2**:85-93. Retrieved July 15, 1999, from the World Wide Web: http://www.agbioforum.missouri.edu.

Griliches, Z., 1957, Hybrid corn: An exploration in the economics of technological change, *Econometrica* **25**:501-522.

Haung, J., Qiao, F., Pray, C., and Rozelle, S., 2001, Biotechnology as an alternative to chemical pesticide use: Lessons from Bt cotton in China. Paper presented at the 5th ICABR International Conference on Biotechnology, Science and Modern Agriculture: A New Industry at the Dawn of the Century, Ravello, Italy.

Huang, J., Hu, R., Fan, C., Pray, C., and Rozelle, S., 2002, Bt cotton benefits, costs, and impacts in China, *AgBioForum* **4**:153-166.

Herring, J., 2003, Underground seeds: Lessons of India's Bt cotton episode for representations of the poor, property claims and biosafety regimes. Paper presented at the Transgenics and the Poor Conference, Cornell University, New York, New York.

Hubbell, B., Marra, M., and Carlson, G., 2000, Estimating the demand for a new technology: Bt cotton and insecticide policies, *Amer. J. Agri. Econ.* 82: 118-132.

Ismael, Y., Beyers, L., Lin, L., and Thirtle, C., 2001, Smallholder adoption and economic impacts of Bt cotton in the Makhathini Flats, South Africa. Paper presented at the 5th ICABR International Conference on Biotechnology, Science and Modern Agriculture: A New Industry at the Dawn of the Century, Ravello, Italy.

Janssen, W., Falconi, C., and Komen, J., 2000, The role of NARS in providing biotechnology access to the poor: grassroots for an ivory tower, in: *Agricultural Biotechnology in Developing Countries: Towards Optimizing the Benefits for the Poor*, M. Qaim, A. F. Krattiger, and J. von Braun, eds., Kluwer Academic Publishers, Boston, Massachusetts, pp. 357-380.

Jayaraman, K. S., 2001, Illegal Bt cotton in India haunts regulators, *Nature Biotechnology* **19**: 1090.

Jefferson, K., Traxler, G., and Wilson, N., 2001, The economics of value enhanced crops: Status, institutional arrangements and benefit sharing. Paper presented at the 5th ICABR

International Conference on Biotechnology, Science and Modern Agriculture: A New Industry at the Dawn of the Century, Ravello, Italy.

Just, R. E., and Zilberman, D., 1988, The effects of agricultural development policies on income distribution and technological change in agriculture, *J. of Dev. Econ.* **28**:192–216.

Lichtenberg, E., 1989, Land quality, irrigation development, and cropping patterns in the northern high plains, *Amer. J. Agri. Econ.* **71**:187–194.

Lichtenberg, E., and Zilberman, D., 1986, The econometrics of damage control: Why specification matters, *Amer. J. Agri. Econ.* **68**:262-273.

Mansfield, E., 1981, Composition of R and D expenditures: Relationship to size of firm, concentration, and innovative output, *Rev. Econ. and Stat.* **63**:610-615.

Marra, M. C., Hubbell, B. J., and Carlson, G., 2001, Information quality, technology depreciation, and Bt cotton adoption in the southeast, *J. Agri. Res. Econ.* **26**:158-175.

Moschini, G., Lapan, H., and Sobolevsky, A., 2000, Roundup Ready® soybeans and welfare effects in the soybean complex, *Agribusiness—An Inter. J.* **16**:33-55.

Price, G., Lin, W., and Falck-Zepeda, J. B., 2001, The distribution of benefits resulting from biotechnology adoption. AJAE Annual Meeting, Chicago, Illinois.

Qaim, M., 2002, Personal communication (January 21, 2002).

Qaim, M., and Zilberman, D., 2003, Yield effects of genetically modified crops in developing countries, *Science* **299**:900-902.

Rogers, E., 1995, *Diffusion of Innovations*, Free Press, 4th ed., New York.

Schultz, T. W., 1975, The value of the ability to deal with disequilibria, *J. Econ. Literature.* **13**:827-846.

Sunding, D., and Zilberman, D., 2001, The agricultural innovation process: Research and technology adoption in a changing agricultural sector, in: *The Handbook of Agricultural Economics*, G. C. Rausser and B. Gardner, eds., North-Holland Publishing Co., Amsterdam, pp. 1–103.

Traxler, G., Godoy-Avila, S., Falck-Zepeda, J., and Espinoza-Arellano, J. de J., 2001, Transgenic cotton in Mexico: Economic and environmental impacts. Paper presented at the 5th ICABR International Conference on Biotechnology, Ravello, Italy.

Weibers, U.-C., 1993, Economic and environmental effects of pest management information and pesticides: The case of processing tomatoes in California. Unpublished Ph.D. dissertation, Free University of Berlin.

Appendix

FARM LEVEL CHOICE OF PEST CONTROL STRATEGIES

1. BASIC MODEL

We start the analysis at the field level and assume that output is dependent on crop variety. Let output per unit of land (field) planted with variety i be denoted by y_i. The index i assumes the value o for the traditional variety, m for genetic modification of the local variety, and g for a generic genetically modified variety (GMV). Following Lichtenberg and Zilberman, the production function is

$$y_i = y_i^Q\left[1 - D(N_i)\right] \tag{1}$$

where y_i^Q is the potential output with variety i. We assume that $y_o^Q = y_m^Q$ and $y_g^Q = y_m^Q(1-\gamma)$ where γ is the yield loss of switching away from the original variety.

The damage function $D(N_i)$ depends on the pest population after treatment. $N_i = N \cdot h(x_i) \cdot B_i$ where N is the initial pest population, B_i is the fraction of the pest population surviving the effect of the GMV pest control, $h_i(x_i)$ is the fraction of surviving pests after chemical pest control, and x_i is pesticide application. $B_o = 1$ for the traditional variety, but B_g and $B_m < 1$. We assume $B_g = B_m = B < 1$. $1 - B_i$ is the kill rate of the seed variety. It is 0 for traditional technologies but positive for GMVs. If the biotechnology pest control kills 90% of the pests, then B = .1. Since biotechnology and chemical control operate through different mechanisms, their impacts are compounded. Higher application of pesticides is assumed to reduce survivorship, but the impact is declining so that $\partial h / \partial x_i < 0$ and $\partial^2 h / \partial x_i^2 > 0$. Thus, in absolute terms, the marginal productivity of the pesticides is declining. The damage is assumed to increase with the pest population at an increasing rate ($\frac{\partial D}{\partial N_i} > 0, \frac{\partial^2 D}{\partial^2 N_i} \geq 0$).

Output price is denoted by p and pesticide price by w. Application of a pesticide requires fixed cost F^p per unit of land. Let F_i be the fixed cost of other activities with the variety i and $F_m > F_g > F_o$, assuming that the

GMVs have higher fixed costs, and the local GMV has a higher seed cost than the generic GMV.

A farmer has to determine which variety to plant and how much pesticide to apply to each variety. The farmer first finds x_i^* optimal chemical use with variety i, and π_i is the profit level associated with it. Then the farmer compares the profits of the three varieties, choosing the one with the highest positive profits.

The optimal pesticide use for variety i is derived by solving

$$\pi_i = \max_{x_i, \delta_i} p y_i^Q \left[1 - D\left(N B_i h_i\left(x_i\right)\right)\right] - w x_i - F_i - \delta_i F^P \tag{2}$$

where the variable δ_i is equal to 1 when $x_i > 0$ and pesticides are applied and δ_i is equal to 0 when $x_i = 0$. It enables subtraction of fixed cost of application $\left(F^P\right)$ when chemicals are applied.

1.1 Impacts of key parameters on pesticide use, given pesticides are used $\left(\delta_i = 1, x_i^* > 0\right)$

Consider the case where pesticides are used, $x_i^* = x_i^I > 0$. The first-order condition of Eq. (2) is

$$FOC_{x_i} = -p y_i^Q \frac{\partial D}{\partial N_i}(N_i) N \cdot B_i \frac{\partial h}{\partial x_i} = w \tag{3}$$

Condition (3) states that optimal pesticide application is where the value of the marginal productivity (VMP) of pesticides is equal to its price. Differentiation of (3) with respect to x yields the second-order conditions:

$$SOC_{x_i} = -p y_i^Q \left[\frac{\partial^2 D}{\partial N_i^2}(N_i)\left(N \cdot B \frac{\partial h}{\partial x_i}\right)^2 + \frac{\partial D}{\partial N_i}(N_i) N\, B_i \frac{\partial^2 h}{\partial x_i^2}\right] < 0.$$

Let $E_N^{MD} = \frac{\partial^2 D}{\partial N_i^2} N_i / \frac{\partial D}{\partial N_i}$, $E_{x_i^I}^h = -\frac{\partial h}{\partial x_i^I} x_i^I / \frac{h}{x_i^I}$, and $E_{x_i^I}^{Mh} = -\frac{\partial^2 h}{\partial x_i^{I2}} x_i^I / \frac{\partial h}{\partial x_i^I}$. Introducing these definitions to the SOC results in

$$SOC = -\frac{w}{x}\left(E_N^{MD} E^h_{x_i^I} + E^{mh}_{x_i^I} \right) < 0 \tag{4}$$

where E_N^{MD} is the measure of the relative magnitude of change in the pest damage in response to an incremental increase in the pest population, $E^h_{x_i^I}$ is a measure of the effectiveness of pesticides (the relative reduction of the pest population in response to an increase in pesticides use), and $E^{Mh}_{x_i^I}$ is a measure of the relative reduction in pesticide effectiveness as the volume applied increases. The impacts of changes in w, p, y_i^Q, N, and B_i on pesticide use are obtained by total differentiation of (3) to yield:

(a) An increase in the price of the pesticides will reduce its optimal use level. This marginal response is ($\frac{dx_i^I}{dw} = -\frac{x_i^I}{w} E_w^{x_i^I} < 0$), where $E_w^{x_i^I} = -\frac{dx_i^I / dw}{x_i^I / w} = \frac{1}{E_N^{MD} E^h_{x_i^I} + E^{Mh}_{x_i^I}}$ is the price elasticity of pesticides demand.

(b) An increase in the output price will increase pesticide use. This marginal response is $\frac{dx_i^I}{dp} = \frac{x_i^I}{p(E_N^{MD} E^h_{x_i^I} + E^{Mh}_{x_i^I})} > 0$.

(c) An increase in the potential output will increase pesticide use. This marginal response is $\frac{dx_i^I}{dy_i^Q} = \frac{x_i^I}{y_i^Q(E_N^{MD} E^h_{x_i^I} + E^{Mh}_{x_i^I})} > 0$.

(d) An increase in the pest population will increase pesticide use. This marginal response is $\frac{dx_i^I}{dN} = \frac{x_i^I}{N}\left(\frac{1 + E_N^{MD}}{E_N^{MD} E^h_{x_i^I} + E^{Mh}_{x_i^I}} \right) > 0$. The marginal response is larger as pest damage increases and the more effective pesticides are in reducing this damage.

(e) An increase in the kill rate of the GMVs will reduce pesticide use. This follows from $\frac{dx_i^I}{dB_i} = \frac{x_i^I}{B_i}\left(\frac{1 + E_{N_1}^{MD}}{E_{N_1}^{MD} E^h_{x_i^I} + E^{Mh}_{x_i^I}} \right) > 0$. Pesticides are substitutes for the GMV pest-control effect and, as it becomes more effective, pesticide use decreases. Since $B_o = 1 > B$, the replacement of a traditional variety by its GM version leads to reduction of pesticide use, and introduction of a generic GMV reduces pesticides even further, $x_o^I > x_m^I > x_g^I$.

1.2 Impacts of key parameters on output given pesticides are used $\left(\delta_i = 1, x_i^* > 0\right)$.

Changes in price affect output through their impact on pesticides use, while changes in y_i^Q, N, and B_i affect output both directly and through the changes in x_i^I. All these impacts are presented below:

(a) An increase in output price will increase output. This marginal response is $\frac{dy_i}{dp} = \frac{y_i^Q D(N_1) E_{N_1}^D E_{x_1}^h}{p(E_N^{MD} E_{x_i^I}^h + E_{x_i^I}^{Mh})}$. This result is obtained by differentiation of (1) with respect to price yielding $\frac{dy_i}{dp} = -y_i^Q \frac{\partial D}{\partial N_1} NB_i \frac{\partial h}{\partial x_i^I} \frac{dx_i^I}{dp}$. Introducing $E_x^D = \frac{\partial D(N_1)x}{\partial x D(N_1)}$ and other definitions to $\frac{dy_i}{dp}$ yields the shorter formula.

(b) An increase in pesticides price will reduce output. The marginal reduction is $\frac{dy_i}{dw} = -\frac{y_i^Q D(N_1) E_{N_1}^D E_{x_i^I}^h}{w(E_N^{MD} E_{x_i^I}^h + E_{x_i^I}^{Mh})}$.

(c) An increase in potential output will increase output supply. The marginal increase is $\frac{dy_i}{dy_i^Q} = 1 - D(N_i) + \frac{D(N_i) E_{N_i}^D E_{x_i^I}^h}{(E_{N_i}^{MD} E_{x_i^I}^h + E_{x_i^I}^{Mh})}$. Increases in potential output may lead to a direct increase of 1 - D units of output and an additional increase because of the increase in pesticide use.

(d) An increase in the pest population will reduce output when $E_{x_i^I}^{Mh} \geq E_{x_i^I}^h$. This condition is likely to be met under the realistic situations that at the optimal solution, the marginal productivity of the pesticides is very low (in absolute terms) and declining. This condition results from the marginal response of output supply to an increase of the pest population, which is equal to $\frac{dy_i}{dN} = -y_i^Q \frac{D(N_1)E_{N_i}^D}{N} \left(1 - \frac{E_{x_i^I}^h (1 + E_{N_i}^{MD})}{(E_{N_i}^{MD} E_{x_i^I}^h + E_{x_i^I}^{Mh})} \right)$.

(e) <u>When $E_{x_i^I}^{Mh} \geq E_{x_i^I}^h$ increase in the kill rate of the GMV will increase output</u>. This likely condition reflects low and significantly declining marginal productivity of the pesticides (in absolute terms) at the optimal solution, and corresponds to a low price of the pesticides relative to output. The condition is derived from $\frac{dy_i}{dB_i} = -y_i^Q \frac{D(N_1)E_{N_1}^D}{B_i}$ $\left(1 - \frac{E_{x_i^I}^h (1 + E_N^{MD})}{(E_N^{MD} E_{x_i^I}^h + E_{x_i^I}^{Mh})} \right)$. It suggests that when the efficacy of the GMVs in controlling pests is declining and, thus, leads to increased use of pesticides, the gain in output because of the extra pesticides does not compensate loss of output because of reduced pest control efficacy of the GMVs. The analysis suggests that when $E_{x_i^I}^{Mh} \leq E_{x_i^I}^h$ reduction of pest control efficacy of the GMV will lead to increased production.

The introduction of GMVs to replace a traditional technology leads to a significant reduction of B_i. Assuming that under both traditional varieties and GMVs $E_{x_i^I}^{Mh} \geq E_{x_i^I}^h$, <u>the introduction of a GMV leads to an increase in output.</u> When the traditional variety is replaced by a generic GMV, the lower B_i will lead to increased output, but the lower y_i^Q will contribute to a reduced production, and the net effect though is not clear a priori.

1.3 Conditions under which pesticide use is zero $\left(\delta_i = 0, x_i^* = 0\right)$

The analysis thus far assumes positive application of pesticides with both varieties. That will not always be the case; some corner solutions with zero pesticides may be optimal. There are two situations where profit maximization results in zero use of pesticides with a seed variety. These situations occur when:

(a) <u>Even the smallest amount of pesticide use does not generate sufficient benefits to cover the price</u>. Specifically, $w > -p \frac{\partial D}{\partial N_i}(NB_i h(0))$.

(b) The gain from pesticide use does not cover its fixed application cost. In these situations if the internal solution where (3) is met, it may not result in extra revenues that will cover both the fixed and variable cost of the pesticides, or $py_i^Q\left[D(N)-D\left(NB_ih\left(x_i^I\right)\right)\right]-wx_i^I-F^P<0$.

1.4 Optimal level of pesticide application for variety i.

Thus, for variety i the optimal level of application x_i^* is determined according to

$$x_i^* = \begin{cases} x_i^I & \text{if } py_i^Q\left[D(N)-D\left(NB_ih\left(x_i^I\right)\right)\right]-wx_i^I-F^P>0 \\ 0 & \text{if } py_i^Q\left[D(N)-D\left(NB_ih\left(x_i^I\right)\right)\right]-wx_i^I-F^P<0 \end{cases} \tag{5}$$

Let the difference between the optimal profits with positive pesticides and without pesticides for variety i be denoted by

$$\Delta\pi_i^P = py_i^Q\left[D(N)-D\left(NB_ih\left(x_i^I\right)\right)\right]-wx_i^I-F^P>0 \tag{6}$$

The gain from pesticide applications with variety i increases as (a) Output price increases $\partial\Delta\pi_i^P/\partial p>0$, *(b) potential output increases* $\partial\Delta\pi_i^P/\partial y_i^Q>0$, *(c) input price declines* $\partial\Delta\pi_i^P/\partial w<0$, *(d) pest population increases* $\partial\Delta\pi_i^P/\partial N>0$, *(e) the kill rate of the GMOs decreases, i.e., pest survival increases* $\partial\Delta\pi_i^P/\partial B_i<0$, *and (f) pesticide application cost decreases* $\partial\Delta\pi_i^P/\partial F_i^B<0$.

With either traditional or GM varieties, farmers will use pesticides when $\Delta\pi_i^P>0$. The gains from use of pesticides are likely to be smaller with the adoption of GMVs. Thus, the adoption of GMVs may eliminate the use of pesticides rather then reduce it from x_o^I to x_m^I. The likelihood of eliminating the use of pesticides with the adoption of a GMV is higher when the generic GMV rather then local GMV is introduced, since the benefits of pesticides are smaller with the generic variety.

2. THE ADOPTION DECISION AND ITS IMPLICATION

The determination of the optimal pesticides for each variety provides the base for the variety choice. Let the gain from adoption of the local GMV be

$$\Delta\pi_o^m = py_m^Q\left[D\left(Nh\left(x_o^*\right)\right) - D\left(NB_m h\left(x_m^*\right)\right)\right] - w(x_o^* - x_m^*) - (\delta_o - \delta_m)F^p - F^o - F^m \tag{7}$$

and the gain from adopting the generic variety be

$$\Delta\pi_o^g = py_g^Q(1-\gamma)\left[D\left(Nh\left(x_o^*\right)\right) - D\left(NB_g h\left(x_g^*\right)\right)\right] - w(x_o^* - x_g^*) - (\delta_o - \delta_g)F^p - F^o - F^g \tag{8}$$

Let i^* be the indicator of the optimal variety, $i^* = m$ if $\Delta\pi_o^m > 0$ and $\Delta\pi_o^m > \Delta\pi_o^{g\cdot}, i^* = g$ if $\Delta\pi_o^g > 0$ and $\Delta\pi_o^g > \Delta\pi_o^{m\cdot}$, and $i^* = o$ otherwise. The farmer will adopt the local modified variety if it is more profitable then the other two (if $\Delta\pi_o^m > 0$ and $\Delta\pi_o^m > \Delta\pi_o^{g\cdot}$). The conditions for the adoption of the generic GMV are similar. We saw that under reasonable conditions the adoption of the local GMV increases output, so from (7) and (8) *the likelihood of adopting the local GMV increases with output price, potential output, the fixed and variable cost of the pesticides, the initial pest pressure, and the effectiveness of GMVs in eliminating pests.* The generic GMV may reduce yield, so its main advantage over the local GMV may be lower fixed costs. Actually, higher output prices and potential output may reduce the likelihood of adoption of the generic GMV, and they make it less profitable relatively to both traditional and GM local varieties. *The likelihood of adoption of the generic GMV increases as the potential output loss due to transition away from the local variety decreases and pesticide-use costs rise, especially the fixed costs.*

A GMV is adopted if the gains in terms of increased value of output and saving of pesticide cost is greater than the extra costs of the seeds. It is important to distinguish between situations when adoption is mostly associated with increased yield vs. situations when it is associated with pesticides or pesticide cost saving. Let $\varepsilon^y = \dfrac{y_{i^*}^* - y_o^*}{y_o}$ be the yield effect of the adoption of GMV, and let $\varepsilon^x = \dfrac{x_o^* - x_{i^*}^*}{x_o}$ be the pesticide-saving effect of adopting a GMV ($x_o^* = 0$ when ε^x=0), and let $\varepsilon^{PC} = \dfrac{w(x_o^* - x_{i^*}^*) + (\delta_o - \delta_{i^*})F^p}{x_o + \delta_o F^p}$ be the pesticide cost- saving effect of adopting GMVs. Both ε^x and ε^{PC} are nonnegative, since adoption of GMVs does not increase pesticide use or expense, but ε^y may be negative if the generic GMV is adopted. When the generic variety is adopted, its impact on yield is $y^{Q}\left[D(h(x_o^*)) - D(Bh(x_g^*)) - \gamma\right]$. Its adoption has a positive yield effect when the yield loss due to the transition from the local variety is smaller than the yield gain due to reduced pest damage.

The behavior of optimal pesticide use and output under the three varieties of technologies suggest that (1*) when the price of pesticides relative to output (w/p) and the fixed cost of pesticides are low, adoption of the local GMV has a small yield effect (ε^y is close to 0), but adoption of GMV has a significant pesticide-saving effect. (2) When the price of pesticides relative to output is high and the fixed cost of pesticides is substantial, the adoption of the local GMV has a substantial yield effect and some pesticide-saving effect.*

3. CREDIT CONSIDERATIONS

Let us consider the case where the farmer is facing an upper-bound constraint on ability to borrow denoted by R per unit of land. Suppose that the farmer needs to borrow funds to pay for his pesticides purchase and any amount of fixed cost behind F_o. Without loss of generality, let us assume that $F_o = 0$. Let

$$\pi_i^c = \max_{x_i,\delta_i}\left\{\begin{array}{l} py_i^Q\left[1 - D\left(NB_i h_i(x_i)\right)\right] - wx_i - F_i - \delta_i F^p \\ \text{subject to } F_i - F_o + wx_i \le R, x_i \ge 0 \end{array}\right\}$$

denote the profit maximization outcome with variety i under the credit constraints. Let the optimal pesticide use with variety i under the credit constraint be $x_i^{c*} \le x_i^*$. The implications of the credit constraint depend on the relative magnitudes of the extra fixed costs of the GMV, the price of pesticides, and the factors that determine the use of pesticides with different varieties. We present some plausible outcomes. In particular:

(a) The credit constraint will prevent adoption when F_g and $F_m > R$ *and* $\Delta\pi_o^m > 0$.

(b) The credit constraint will lead to adoption of the generic GMV instead of the local GMV when $F_m > R > F_g$ and $\Delta\pi_o^m > \Delta\pi_o^g > 0$.

(c) *The credit constraint will enhance adoption when it restricts purchase of pesticides with the traditional variety.* If the GMV is affordable, $F_m < R$, and w is relatively expensive, the pesticide use under the credit constraint is $x_i^{c*} = R/w < x_i^*$, and the gain from adoption under the credit constraint is greater than without it, $\pi_m^c - \pi_o^c > \Delta\pi_o^m$.

4. RISK CONSIDERATIONS

For simplicity, we will assume that the GMV eliminates pest pressure altogether, so with these technologies, $y_m = y_m^Q, y_g = y_g^Q$. The damage is assumed to be a random variable $D = \alpha^N h(x_o)$ where α^N is a random variable with mean μ^N and variance σ^{N2} multiplied by the damage reduction as a function of pesticide use presented by $h(x_o)$. The farmer determines the pesticide use before the true state of nature is revealed. We also assume that the farmer is risk averse, and his decisions are approximated well by following maximization of a linear combination of the mean and variance of profit.[1]

The decision problem with the traditional technology becomes

$$\pi eq_o = \max_{x_i, \delta_i} py_o^Q\left[1 - \mu^N h(x_o)\right] - .5\phi(py_i^Q h(x_o))^2 \sigma^{N2} - wx_i - F_i - \delta_i F^p \tag{9}$$

where πeq_o is certainty equivalence of the expected utility of the farmer and ϕ is a risk-aversion coefficient. The first order of (9) with respect to pesticide use is

$$-py_g^Q\left\{\mu^N + \phi h(x_o)\sigma^{N2}\right\}\frac{\partial h}{\partial x_g} = w \tag{10}$$

This condition states that optimal pesticide use occurs where the marginal benefits of pesticides in increasing mean profits and reducing variance are equal to its price. Compared to the condition under certainty (3), here there are extra marginal benefits of pesticides—the marginal benefits through risk cost reduction. The extra benefit increases pesticide use relative to the case with full certainty.

[1] The mean variance rule corresponds to situations where it is normally distributed, and the farmer has an expected utility maximizer with a negative utility function.

PART V.
Biodiversity, Biotechnology, and Development: Policy Implications

Chapter 17

TECHNOLOGICAL CHANGE IN AGRICULTURE AND POVERTY REDUCTION: THE POTENTIAL ROLE OF BIOTECHNOLOGY

Alain de Janvry,[1] Gregory Graff,[2] Elisabeth Sadoulet,[1] and David Zilberman[1]
[1]Professor, Department of Agricultural and Resource Economics, 207 Giannini Hall, University of California, Berkeley, CA 94720; [2]Visiting Postdoctoral Researcher, Department of Agricultural and Resource Economics, 207 Giannini Hall, University of California, Berkeley, CA 94720

Abstract: Technological change in agriculture has historically been a powerful force for poverty reduction. We explore in this chapter how biotechnology, as a potentially important new source of technological changes in agriculture, could also be made to fulfill this role. We distinguish between direct effects of technology and poverty that affect adopters and indirect effects that affect others through employment, growth, and consumer price effects. We show that agbio-technology has the potential of providing crops with new traits beneficial to the poor through direct and indirect effects. The poor may not benefit from biotechnology for several reasons including exclusion as a consequence of intellectual property rights, concentration of ownership in the industry, research gaps for traits desired by the poor, and unexplored environmental risks. We conclude that agbiotechnology has potential as a tool for poverty reduction, but that it needs complementary institutional innovations that are lagging relative to current scientific progress. These institutional lags affect the generation, transfer, and adoption of agbiotechnology benefiting the poor. We give an inventory of the institutional innovations needed to reduce these lags and to capture the promise of agbiotechnology for poverty reduction.

Key words: agriculture; biotechnology; poverty.

1. THE CHALLENGE FOR AGRICULTURAL TECHNOLOGY

The challenge for developing country agriculture in the next 25 years is enormous, particularly if it is not only to satisfy the growing effective demand for food, but also to help reduce poverty and malnutrition in an environmentally sustainable fashion. Due to population growth and rising incomes, demand in the developing countries is predicted to increase by 59% for cereals, 60% for roots and tubers, and 120% for meat over this period (Pinstrup-Andersen, Pandya-Lorch, and Rosengrant, 1999). This increased supply cannot come from area expansion since this has already become a minimal source of output growth at a world scale and has turned negative in Asia and Latin America. Neither can it come from any significant expansion in irrigated area due to competition for water with urban demand and rising environmental problems associated with drainage, soil salinity, and chemical runoffs. While it will thus need to come from growth in yields, the growth rate in cereal yields in developing countries has been declining from an annual rate of 2.9% during 1967-1982 to 1.8% during 1982-1994, which is the rate needed to satisfy the predicted 59% increase in demand for cereals over the next 25 years. The growth in yields cannot consequently be let to fall below this rate in developing countries without further increasing the share of food consumption that is imported. With 1.2 billion people in absolute poverty (earning less than $1.00 per day, see World Bank, 2001) and 792 million underfed in the developing countries (FAO, 2000), agriculture should also play a major role in reducing poverty and improving food security, particularly since some three-quarters of these poor and underfed live in rural areas where they derive part, if not all, of their livelihoods from agriculture as producers or as workers in agriculture and related industries. More importantly, the real income of poor consumers depends on the price of food.

For poverty to fall and for the nutritional status of the poor to improve at the current levels of food dependency, the decline in growth rate of cereal yields will have to be stopped, and yield increases compared to current trends will have to occur in part in the fields of poor farmers and will have to generate employment opportunities for the rural poor. Since the growth rate in yields achieved with traditional plant breeding and agronomic practices has been declining, the next phase of yield increases in agriculture will have to rely on the scientific advances offered by biotechnology, precision farming, and production ecology, with most of the gains expected to be derived from the first. Yet, while biotechnology has made progress in the agriculture of some of the more developed countries, it has had little actual impact in most developing countries, and particularly in the farming systems of the rural poor. The objective of this paper, therefore, is to explore under what conditions the current biotechnological revolution in agriculture could be helpful in reducing poverty in developing countries. While there are acknowledged ethical and precautionary objections to the use of some particular techniques of biotechnology, it should be kept in mind that failure to develop and capture this potential could further increase the income gap between developed and developing nations and could be a serious setback in the struggle to reduce poverty. At the same time, environmental and consumer risks that may, for example, derive from adoption of genetically modified organisms (GMOs), will have to be carefully assessed and regulated for biotechnology to yield its potential benefits and not to risk creating setbacks to the already limited welfare of the rural poor in developing countries.

2. THE POTENTIAL OF AGRICULTURAL TECHNOLOGY FOR POVERTY REDUCTION

2.1 Direct and indirect effects of technology on poverty

There are two channels, direct and indirect, through which technological change in agriculture can impact on poverty. First, a technological innovation can help reduce poverty *directly* by raising the welfare of poor farmers who adopt the new technology. Benefits for them can derive from increased production for home consumption, more nutritious foods, higher gross revenues deriving both from higher volumes of sales and higher unit value products, lower production costs, lower yield risks, lower exposure to unhealthy chemicals, and improved natural resource management.

Second, technological change can help reduce poverty *indirectly* through the effects which adoption, by both poor and nonpoor farmers, has on the price of food for net buyers, employment and wages in the agricultural sector, as well as in other sectors of economic activity through production, consumption expenditures, and savings linkages with agriculture, lower costs of agricultural raw materials, lower nominal wages for employers (as a consequence of lower food prices), and foreign exchange contributions of agriculture to overall national economic growth.

Through the price of food, indirect effects can benefit a broad spectrum of the national poor, including landless farm workers, net food-buying smallholders, nonagricultural rural poor, and the urban poor for whom food represents a large share of total expenditures. Indirect effects via employment creation are important for landless farm workers, net labor-selling smallholders, and the rural nonagricultural and urban poor. Hence, the indirect effects of technological change can be very important for poverty reduction not only among urban households, but also in the rural sector among the landless and many of the landed poor who buy food and sell labor.

When are there trade-offs in technology between achieving direct and indirect effects? Within a given agroecological environment, if land is unequally distributed and if there are market failures, institutional gaps, and conditions of access to public goods that vary with farm size, then optimum farming systems will differ across farms. Small farms typically prefer farming systems that offer greater value added per unit of land, are capital-saving, and less risky, while large farms prefer farming systems that are laborsaving, and they can afford to assume more risk if they are compensated by higher expected incomes. In this case, heterogeneity of farming systems prevails and there will exist trade-offs between achieving indirect and direct effects if budget constraints in research requires priority setting. The more unequally land is distributed and the more market institutional and government failures are farm-size specific, the sharper the trade-off will be.

The relative magnitude of the direct and indirect effects of technological change in agriculture on poverty can be quantified through computable general equilibrium models. In these models, the direct effects include the change in agricultural profit for adopting farmers, the changing opportunity cost of home consumption for own production, and the change in self-employment on one's own farm. The indirect income effect comes from changes in nominal income from all sources other than own agricultural production. The indirect price effect comes from the change in prices, excluding the effect through the opportunity cost of home consumption.

Table 17-1 presents results from models representing typical poor economies in Africa, Asia, and Latin America.[1] They show that the relative magnitude of these effects varies widely according to the structure of the economy, the sectoral incidence of poverty, and the sources of income for the poor. In a typical African context where the agricultural sector is large and the bulk of the poor are smallholders, direct effects are dominant: They account for 77% of the income gains for the rural poor and for 58% of the income accrued to all poor. Targeting technological change on poor farmers with their particular crops, farming systems, market failures, institutional gaps, and public goods deficits is thus essential for aggregate poverty reduction. In Asia, by contrast, where most of the poor are rural landless, income gains for the rural poor derive mainly (64%) from indirect effects captured on the labor market. Of the total income gained by the poor, 74% is from indirect effects. Hence, targeting technological change toward employment creation is in this case fundamental for poverty reduction. Finally, in Latin America, where poverty is largely urban and a majority of the land is concentrated in the hands of large farmers, the rural poor derive 73% of their real income gains through indirect effects, mainly captured through falling food prices. The total real income gains captured by the poor

[1] For details on the construction of these models, see Sadoulet and de Janvry (1992).

Table 17-1. Direct and indirect effects of technological change by region

Impact of a 10% increase in TFP in agriculture	Africa	Asia	Latin America
Sources of income gains for the rural poor			
From direct effects (%)	77	37	27
From indirect effects (%)	23	64	73
Sources of income gains for all poor			
From direct effects (%)	58	26	14
From indirect effects (%)	42	74	86

Source: Own calculations.

derive mainly (86%) from indirect effects, and this is also the case for the rural poor (73%). In this case, the main role of technological change is consequently to lower the price of food, and this will have to occur principally in the fields of the large farmers since this is where most of the land is located. Clearly, at higher levels of geographical disaggregation, direct effects may also dominate in specific Asian and Latin American regions, requiring region-specific targeting of research budgets across innovations producing either direct or indirect effects.

We conclude that, if there are trade-offs between creating direct and indirect effects due to constraints on research budgets, care must be taken to allocate budgets optimally between these technological options to maximize poverty reduction. While surprisingly little formal analysis has been made of these trade-offs, optimum allocation needs to be determined for each nation and region for which research programs are organized.

2.2 Technology and rural development

Biotechnology may offer a significant potential for poverty reduction in smallholder agriculture. There are, however, four caveats to be considered. One is that potentially cheaper and faster sources of income gains than agricultural technology may not have been exhausted, particularly through greater access to land, improved property rights, investments in irrigation, higher levels of human capital, and access to nonagricultural sources of employment.

The second is that other technological advances than biotechnology may be more appropriate for enhancing smallholder incomes. This is the case for many products of traditional approaches to research that have never been targeted at smallholders. This includes improved farming systems, agro-ecological farming practices, and traditional breeding for the specific, and often highly particular, contexts where they are located. These approaches will often not be substitutes but complements to biotechnology.

The third is that, for any kind of technology to be adopted by smallholders, many market failures that affect the smallholders need to be eliminated, institutional gaps removed, complementary public goods provided, and policies that do not discriminate against the agricultural sector or poor farmers put into place. This includes in particular access to credit and to risk-coping instruments such as mutual insurance and safety nets, and low transactions costs in factor and product markets. Unless these con-ditions are in place, adoption will not happen.

Finally, for technology adoption to result in maximum poverty reduction, the other dimensions of welfare also need to be accessible. This includes in particular the components of basic needs (health, education) and the more qualitative dimensions of welfare such as empowerment and rights.

Hence, to be effectively used for poverty reduction, technology instruments need to be embedded within a comprehensive rural development and poverty reduction strategy for the region concerned that weighs technology against other instruments for income gains, carefully discriminates among alternative technological paths, makes the technological innovation adoptable by the farmers for whom it was intended, and complements income gains with access to the other dimensions of welfare.

3. AGRICULTURAL TECHNOLOGY AND POVERTY IN A HISTORICAL PERSPECTIVE

The history of technological change in developing country agriculture is one where farmers and farming communities have historically been the main innovators, followed by the public sector, which released the technology of the Green Revolution (GR) as a public good. Recently, however, the private sector has been rapidly penetrating due to changes in intellectual property rights (IPR) legislation[2] allowing private entities to capture returns from research in biology, unleashing a new wave of biotechnological innovations as private goods.

3.1 Green Revolution

As noted in more detail in Chapter 3, the GR started with the release of hybrid maize in the United States in the 1950s (Griliches, 1957). It was

[2] IPR policy is national: Countries have sovereign control of the granting of private property rights over intellectual materials within their own borders. However, recent international treaties have strongly influenced many developing countries' domestic IPR legislation, including the Convention on Biodiversity and the World Trade Organization Trade Related Intellectual Property (TRIPS) agreement.

extended to the developing countries with the introduction of semidwarf varieties of rice and wheat in the mid-1960s. The GR in developing countries can be decomposed into two epochs:

GR I (1965-1975): The main purpose of research was to achieve rapid increases in yields through high-yielding varieties (HYVs), and success was immense, creating large indirect effects for the poor via declining staple food prices and rising employment in agriculture and related activities. Direct poverty reduction effects were, however, small and often negative: HYVs were designed for the best areas (irrigation, high soil fertility) with chemical-intensive technology (Byerlee, 1996). They consequently diffused first among commercial farmers, sometimes with backlash effects on nonadopting poor farmers through falling prices (Scobie and Posada, 1978). This first epoch also often had negative environmental effects through genetic erosion and chemical run-offs.

GR II (1975-today): Research was aimed at the broadening of desirable traits to consolidate yield gains and to extend the benefits of the GR to other crops, areas, and types of farmers. This allowed the increase of pest and drought resistance. The benefits of the GR were thus extended toward rainfed areas (Byerlee and Moya, 1993) and small farms, enhancing direct effects on poverty. These technological innovations were, however, not able to prevent a steady decline in the growth of yields, reducing the pace of gains in poverty reduction through indirect effects compared to GR I.

3.2 New technological revolutions and IPRs

Three major scientific developments are creating a new generation of technological change in agriculture: The information technology revolution that opened the field of precision farming, the better understanding of ecological systems that underlies production ecology, and the genetic revolution that launched biotechnology. While intellectually separate, these three technological advances should be seen as complementary in the domain of applications.

i) Precision farming is one of the major impacts of the information revolution on agriculture (Wolf and Buttel, 1996). It is based on information derived from global positioning satellite systems and electronic monitoring, and processed through a geographical information system. This allows farmers to take into account the heterogeneity of their fields over space and time, and to adapt cultural practices to that heterogeneity through variable rates in planting densities, chemical applications, and irrigation doses, and through just-in-time application of treatments. This increased precision is applied to the use of traditional agricultural technologies: chemical fertilizers, synthetic pesticides, tractor-based mechanization, and genetically uniform HYVs. Fine tuning in the use of these technologies has postponed decreasing returns and reduced pollution where there was overuse of chemicals.

In the industrialized countries, precision farming allows farmers to deal with heterogeneity in spite of scale, recuperating the informational advantages of small-scale farming at a larger scale. Hence, information technology has been used to disaggregate large heterogeneous farms into locationally differentiated management practices. In the developing countries, information technology has been used to aggregate heterogeneous small-scale plots into homogenous (spatially disconnected) mega-environments to which common technological practices can be applied (CIMMYT, 2001). While monitoring in the industrialized countries is done at the farm level, it is done through centralized services in developing countries such as weather stations, satellite monitoring of biomass, and regional intelligence on insect infestations.

ii) Production ecology uses the concept of the agroecosystem as the fundamental unit of analysis (Harwood, 1998). Such systems are characterized by complex biological processes and relationships through which a multitude of species interacts. Production ecology starts from the analysis of these processes, and defines a set of interventions to modify them to achieve desirable outcomes. Interventions thus include the management of carbon flows and biota, increased nutrient cycling from soil to crops, integrated pest management and ecologically based pest management, diversified farming with crop rotations and multiple cropping, the provision of ecosystem services (hydrological cycling, wildlife habitat, preservation of animal and plant diversity, and landscape management), and use of carbon sinks to improve atmospheric chemical balance. The approach has been successfully pursued in agroforestry systems (e.g., by ICRAF, the International Center for Research on Agro-Forestry) and agroecology for smallholders (e.g., by CLADES, the Latin American Consortium on Agroecology and Sustainable Development). Except in the organic agriculture movement, it has not yet gained mainstream recognition but offers considerable promise.

iii) Biotechnology is based on the understanding of how biological organisms function at the molecular level, and manipulation of organisms at the cellular and molecular level, including the DNA molecules that constitute organisms' genetic code, to achieve desirable outcomes (Chapter 12).

4. MAIN FEATURES OF AGBIOTECHNOLOGY FOR THE POOR

4.1 Traits: Potentials and risks

The advent of applications of biotechnology to agriculture offers the possibility of amplifying the achievements of traditional breeding that sustained the GR (Chapter 12). We summarize the reasoning behind this possibility into three categories:

(1) It broadens the spectrum of potential new products and traits through genetic engineering (recombinant DNA techniques, insertion of genetic materials) of plants and animals, including both wide crossings (gene transfers within species from wild relatives of the crop) and transfers of foreign genes (gene transfers across species).

(2) It accelerates the pace of plant breeding through use of selectable gene markers, promoters, and new scanning devices.

(3) It lowers the cost of conducting research and development due to productivity gains in research.

For the sake of smallholders, biotechnology offers the possibility of bringing specific new traits and improvements directly to the best local plant varieties that they already use. Yet, for the poor, biotechnology offers both potential benefits and potential risks. Some of the most important are the following:

- Potential benefits of agbiotechnology for poverty reduction.

i) Yield increases in crops, trees, and animals (including fish) produced in the agroecological and structural conditions of developing countries: tropical and semitropical and arid and semiarid environments, and in peasant farming systems (see Chapters 13 and 16).

ii) Arable area expansion into less-favored lands: varieties tolerant to acidic, saline, and lateritic[3] soils and varieties tolerant to flood and drought.

iii) Multiple-cropping allowed by shortening plant maturation periods.

iv) Cost reduction via resource-saving effects: chemical-saving substitution of fertilizers with nitrogen fixation, low nitrogen tolerance, substitution of chemical pesticides with insect resistance (Hubbell, Carlson, and Marra, 2000; Klotz-Ingram et al., 1999; Pray et al., 2000; Traxler and Falck-Zepeda, 1999); seed-cost saving through the possibility of exact reproduction by

[3] High content in iron and aluminum compounds.

farmers of seeds of high-quality or specific genetics, including hybrids (through the process of apomixis[4]).

v) Risk reduction: lower susceptibility to biotic stress—such as insect resistance (e.g., *Bacillus thuringiensis* (Bt) crops[5]) and virus resistance—and to abiotic stress—such as improved tolerance of saturation (flood), dehydration (drought), extreme heat, or frost. Use diagnostics to detect and identify diseases or infestations, for instance, on seeds purchased or in soils (see Chapter 13).

vi) Improved storability: post-harvest insect resistance, delayed maturation (reduces transport and marketing costs by reducing damage to product, need for refrigeration).

vii) Nutritional improvements of food and feed: quality protein maize, improved micronutrient content ("Golden Rice" with high beta carotene/vitamin A content).

viii) Health benefits for humans and animals: reduced exposure to chemicals (Pray et al., 2000), new vaccines.

ix) Environmental benefits: reduced application of synthetic chemical pesticides and fertilizers, preservation of biodiversity through lower marginal cost of genetic improvements to a wide range of local varieties (see Chapters 3 and 14).

- Potential risks of agbiotechnology for the poor:

i) Staple food crops produced in tropical and semitropical and arid and semiarid environments and by smallholders are bypassed by research, leading to loss of competitiveness.

ii) Terminator genes used to enforce IPRs raise cost of access to latest technologies by preventing reproduction of open-pollinated seeds. Do note, however, that poor farmers can choose to continue to maintain and have access to older unmodified open-pollinated varieties.

iii) Traits pursued in private sector research are for nonpoor consumers (improved industrial processing, delayed ripening) to the neglect of poor consumer needs (more nutritious foods).

iv) Labor displacement by diffusion of labor-substituting (such as herbicidc-tolerant) plant varieties.

v) Production in more-developed countries (MDCs) of substitutes for crops previously produced in less-developed countries (LDCs), particularly labor-intensive and/or smallholders' crops such as sugar and vanilla, creating trade substitution effects.

---

[4] Essentially the growing of seed that is an exact genetic copy of its parent.

[5] Crops that produce a protein in their tissue from an inserted gene derived from the naturally occurring soil microorganism *Bacillus thuringiensis* (Bt). The protein has highly specific toxicity to some insect pests and serves as an "in plant" biopesticide.

vi) Consumer risks: allergies, unknown long-term health effects.

vii) Environmental risks: insect and virus resistance to commonly available and cost-effective means of biological control, gene flows to wild relatives (potentially creating "superweeds"[6]), and destruction of useful insects and species.

4.2 Current progress of agbiotechnology

In contrast to GR research, which was conducted in the public sector and delivered international public goods had occurred intentionally in developing countries (importantly through CGIAR[7] centers' research), most research in biotechnology has been done in developed countries (see also Chapter 3). This research using enabling process technologies privately protected under patents that are now mainly owned by a few large multinational corporations, on commodities that are principally for animal feed and fiber, with traits favorable to large capital-intensive commercial farms and, thus far, without many benefits for final food consumers.

The data in Table 17-2 indicate the global status of this technological revolution as of 2000 (James, 1998 and 2000). They show that expansion of the area planted in transgenic crops has been extraordinarily rapid, rising from zero in 1995 to 44.2 million hectares in 2000 and covering as much as 36% of the area in soybeans, 16% in cotton, 11% in canola, and 7% in maize (James, 2000). While the rate of area expansion declined after 1998, it remains high, still reaching 11% in 2000. There is, however, significant unevenness in diffusion among countries, crops, and traits. As much as 76% of the world area planted in transgenics is located in the developed countries, with the United States alone accounting for 69% of the total. Herbicide-tolerant soybeans and Bt corn (mainly for feed) are the dominant crop-trait combinations, followed by insect-resistant and herbicide-tolerant cotton. In Argentina, the developing country by far most advanced in ag-biotechnology, the main transgenics are herbicide-tolerant soybeans, Bt corn, and Bt cotton. The global status of transgenic crops clearly shows developing countries lagging far behind and the purpose of transgenics directed at nonfood crops and principally laborsaving technological change. Observation of the frequency distribution of GMO field trials across countries indicates that several developing countries have advanced research capacity in DNA

[6] Close relatives to crop varieties that acquire traits (such as herbicide tolerance) cannot be managed by preferred practices (e.g., a specific herbicide, in this case) and become more difficult to control.

[7] The Consultative Group for International Agricultural Research (CGIAR) is an informal association that supports a network of 16 international agricultural research centers, primarily sponsored by the World Bank, FAO, and the United Nations Development Programme (UNDP).

techniques, notably China, Argentina, India, Brazil, Mexico, and Egypt, followed by countries with modest capacity, such as Indonesia, the Philippines, and Kenya (Pray, Courtmanche, and Brennan, 1999).

4.3 Main differences between agbiotechnology and Green Revolution II for poverty reduction

If the potential for poverty reduction offered by agbiotechnology is to be seized and the potential risks of the approach are to be avoided, the specificity of the technology and how it is made available need to be understood in contraposition to the technology of GR II, the last important

Table 17-2. Global status of transgenic crops, 1996-1998

Areas in million hectares	1996	1997	1998	1999	2000
Area in transgenics crops by country					
Total industrialized countries	1.6	9.5	23.4	32.8	33.5
USA	1.5	8.1	20.5	28.7	30.3
Canada	0.1	1.3	2.8	4.0	3.0
Australia	<0.1	0.1	0.1	0.1	0.2
Total developing countries	1.2	3.3		7.1	10.7
Argentina	0.1	1.4	4.3	6.7	10.0
China	1.1	1.8	N.A.	0.3	0.5
World total	1.7	11.0	27.8	39.9	44.2
Area in transgenics crops by commodity					
Soybean	0.5	5.1	14.5	21.6	25.8
Corn	0.3	3.2	8.3	11.1	10.3
Cotton	0.8	1.4	2.5	3.7	5.3
Canola	0.1	1.2	2.4	3.4	2.8
Area in transgenics crops by trait					
Herbicide tolerance	0.6	6.9	19.8	28.1	32.7
Insect resistance	1.1	4.0	7.7	8.9	8.3
Insect resistance and herbicide tolerance		<0.1	0.3	2.9	3.2
Transgenic area as percentage of global area					
Soybeans					36
Cotton					16
Canola					11
Maize					7

Source: James (1998 and 2000).

technological epoch for developing country agriculture. The main differentiating features of agbiotechnology that have implications for poverty reduction are discussed in the following sections.

4.3.1 Technological features of agbiotechnology

i) Research on traits separated from research on varieties. Compared to traditional breeding GR I and II), where research on trait identification was confounded with variety development, biotechnology dissociates research on traits (functional genomics) from product development (insertion or activation of genes corresponding to traits in selected varieties). Results of agbiotechnology research on traits may consequently be used over a wide range of local conditions. Hence, if the technology on relevant traits, derived through functional genomics, exists and can be accessed through markets (e.g., Chapter 18), contracts, or as public goods, and if the process technology to insert or activate these traits in local varieties is widely available, developing countries can produce improved varieties without the need to engage in fundamental research. This has powerful implications for the division of labor in research between developed and developing countries and the type of capacity building needed in the latter, in this case principally to screen and adapt these technologies to their own needs.

ii) Potential environmental externalities and consumer risks. New varieties under GR research were achieved by natural crossings. Biotechnology, and particularly genetic modification (the creation of GMOs), creates new varieties by artificial gene transfers, with yet poorly known risks for the environment and consumers. As a result, experi-mentation on and the diffusion of agbiotechnology innovations need to be accompanied by specific regulatory procedures to safeguard environmental and consumer safety that carefully weigh risks against benefits of innovations. Experimentation on biosafety is a public good in which private firms cannot be expected to invest sufficiently. This is an area where public research is an essential complement to private research since private biotech products cannot be released unless their biosafety implications are known and appropriate regulatory procedures designed. This offers a powerful rationale for private firms to make coalitions to fund public and CGIAR research. As the recent slowdown in the spread of GMO crops has demonstrated, investments in private research that get ahead of advances in complementary public research can create a huge waste of resources.

iii) Biodiversity as the source of research materials. New genes to be inserted in cultivated varieties are expected to be found in the stock of global crop germplasm biodiversity. The option value of preserving biodiversity *in situ* and *ex situ* will thus be enhanced (Koo and Wright, 2000). Incentives to establish property rights over biodiversity and to invest in biodiversity conservation are thus important side effects of progress in agbiotechnology. The large collections of native seeds (landraces) held in trust by the CGIAR are important international public goods (respecting farmers' rights). Their maintenance should be secured by permanent endowments instead of depending on the annual budgets of repository research centers, with the risks that this implies.

4.3.2 Role of IPRs

i) IPRs and access to biotechnology materials for the LDCs. The current policies on the patenting of life forms in the United States allow for the private appropriation of knowledge that makes up the basic "raw materials" of biotech research (Wright, 1998). There is serious concern that such appropriation is creating hurdles for access to the relevant materials for agricultural research in developing countries, public sector institutions, and the CGIAR and for downstream product development. Some of the patents that have been granted are very broad and can be used to block others from accessing related discoveries. Evolution of patent law in the United States is, however, in full progress, as it is modified by case law without being submitted to open national debates. Governments, both in industrialized and developing countries, are pressed by public concerns with biosafety, and by their own interests in preserving the competitiveness of the industry. They are also constrained by World Trade Organization (WTO) requirements to introduce IPR legislation on life forms, potentially leading to changes in current national IPR systems.

ii) Market failures for IPRs and industry concentration. A large number of technological innovations are involved in the development of a final product, and ownership of these innovations is often scattered over many institutions. Rapid concentration of patent ownership in the corporate sector through acquisitions and mergers evidences these technological complementarities in product development and existence of serious market failures in the acquisition of patented materials needed for product development (Graff, Rausser, and Small, 2003). As the Bt example shows (Fig. 18-1), university and public institutions held 50% of the stock of patents in 1987, independent biotech companies and individuals held 77% of the stock in 1994, and the industry's "Big 6" firms (AstraZeneca, Aventis, Dow, DuPont, Monsanto, and Novartis) held 67% of the stock of patents in 1999. As can be seen from Fig. 18-2, 75% of the patents controlled by the six largest firms in the industry in 1999 had been obtained via acquisitions of subsidiary biotech and seed companies. Concentration fueled by market failures for IPRs shows that LDCs and the CGIAR will have considerable difficulties engaging in biotech-enabled research and development until an effective means of accessing the rights to utilize biotech knowledge, such as a IPR licensing clearinghouse (Chapter 18), becomes available.

iii) IPRs and access to GMO seeds. Property rights over seeds can be established by (1) providing hybrid seeds, which because of their natural biological mechanism are undesirable for replanting after the first generation, (2) introducing terminator genes in open-pollinated varieties, thus artificially creating a biological mechanism similar in effect to that of hybrid seeds, and (3) enforcement of the legal prohibition created by IPRs to reproduce seeds of open-pollinated varieties. Failure to provide property rights over seeds via legal means can be expected to (1) limit research on biotechnology to hybrids and terminator-charged varieties, (2) limit insertion of new traits to a narrow range of local varieties, implying suboptimal seeds for the poor and loss of biodiversity through oversimplified farming systems, (3) increase reliance on contract farming by seed producers, with increased concentration of control over the industry, and (4) raise the price of seeds as producers attempt to recoup the cost of research and development in one single sale, increasing liquidity constraints for smallholders exposed to credit market failures. In any form, IPRs give

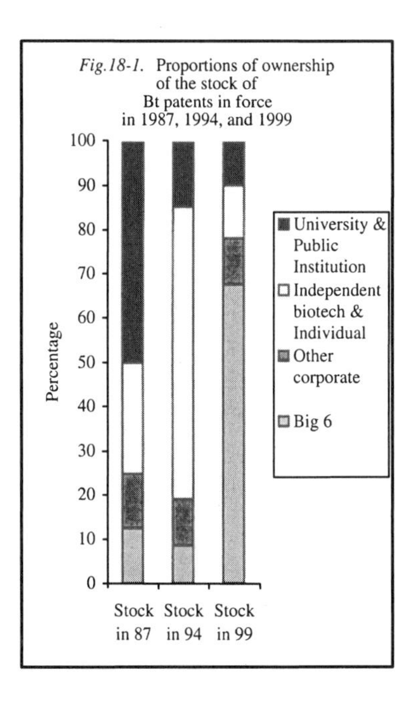

Fig.18-1. Proportions of ownership of the stock of Bt patents in force in 1987, 1994, and 1999

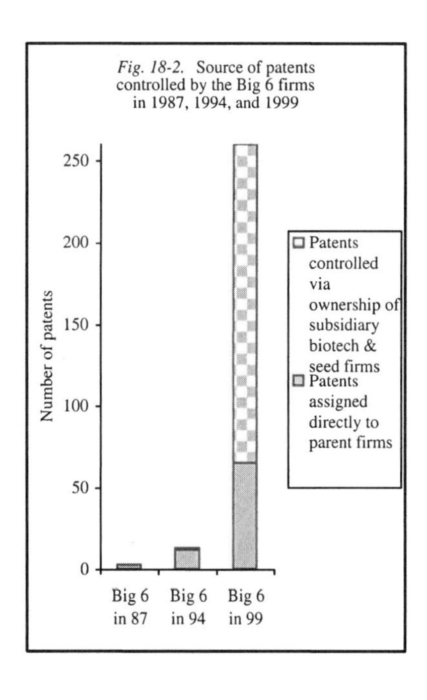

Fig. 18-2. Source of patents controlled by the Big 6 firms in 1987, 1994, and 1999

seed producers some leverage to raise the cost of seeds for farmers who have historically reproduced their own seeds, raising the issue of access to the necessary liquidity to acquire the improved seeds.

iv) Role of IPRs in accessing biodiversity. Since the biodiversity of crop germplasm is another "raw material" for agbiotechnology research, and much of the relevant natural biodiversity of crops is located in developing countries and peasant communities, protecting access to this biodiversity under some form of IPR and selectively granting access can potentially be used as a source of leverage in negotiating access to agbiotech innovations held by MDC interests. Such leverage can apply to scientists in developing countries, including the CGIAR, who need access to patented research materials for their own work, and to farming communities that desire access to seeds improved by biotech research. Again, such exchanges could, in principle, be effectively mediated by a neutral IPR-licensing clearinghouse.

4.3.3 Research and development on GMO technology in LDCs

i) Current research gaps for the poor. Because agbiotechnology innovations are generated principally in the industrialized countries for major crops produced in these countries, for a clientele of large farmers with few market failures, and for relatively high-income consumers, there are important gaps that need to be filled in order to make biotech innovations relevant for poverty reduction in developing countries (Nuffield Foundation, 1999). They include research on staple foods for tropical and semitropical and arid and semiarid environments, labor-intensive technologies, and traits desirable for smallholders that operate under tight land constraints, extensive market failures, institutional gaps, and policy biases. Very importantly, they also include nutritional improvements of significance for poor consumers. Institutional mechanisms need to be devised to fill these research gaps, including defining the roles in biotech research of the developing countries' national agricultural research institutes (NARIs) and the CGIAR.

ii) Structure of research costs and access by LDCs. Biotech changes the structure of research and development costs: It increases the costs of fundamental research, introducing new large fixed costs, but lowers the marginal cost of product development. If fundamental research relevant for LDCs and smallholders is done in the MDCs, followup development in LDCs of agbiotechnology products for poverty reduction can be made cheaper. If for any reason such basic research is not being done (principally because effective demand is lacking due to poverty), the cost of generating agbiotechnology products for poverty reduction may be significantly higher than under traditional breeding, enhancing in particular the role of the CGIAR in bringing the benefits of biotechnology to the poor.

iii) Complementarity between agbiotechnology and traditional breeding. Biotechnological research is complementary to traditional breeding since new traits conveyed by gene transfers need to be inserted into the best possible local varieties in order to deliver to farmers the myriad of traits, which come from using locally optimized crop varieties that cannot be conferred by gene transfers. An effective traditional breeding program thus creates scale effects for biotech research by enabling transfers or development of traits in a wide range of local varieties.

iv) Complementary roles of public and private research. Many biotechnological problem-solving approaches or "paradigms" have originated as inventions made in public sector research seeking basic knowledge about mechanisms underlying general problems and then have been refined for specific application by start-up biotechnology companies. These companies have generally spun off from universities, been financed by venture capital, and turned to large multinational corporations to commercialize their products. Analysis of the granting of patents in agbiotechnology shows sequential shifts in the relative roles of the universities and public sector, the start-up biotechnology firms, and the multinational corporate sector in research and development. Using Bt technology as a case study, Fig. 18-3 shows that university and public institutions generated 60% of the patented research in 1976-1986; start-up biotech firms and individuals, 77% in 1987-1995; and large corporate firms, 55% in 1999. Continued support to public sector research is thus essential for the flow of new innovations to be continually replenished. For this sequential division of labor to be effective, linkages between these institutions is important for research to yield useful products, particularly through offices of technology transfer in universities and public institutions, venture capital for biotech firms, and efficient trading or licensing of property rights among all these institutions.

v) Public-private research partnerships. With some 75% of world investment in agbiotechnology research coming from the private sector, the public sector and the CGIAR are increasingly seeking to develop research partnerships with the private sector (Herdt, 1998). Design of these partnerships is complex since the objectives of partners are at odds: The private sector pursues profits, while the public sector and the CGIAR are, in principle, pursuing the delivery of public goods. Best practice for the negotiation and design of these contracts needs to be established to maximize the synergies they provide in research, but also to protect the present and future interests of public institutions (Rausser, Simon, and Ameden, 2000).

vi) Participation of smallholders to research priority setting on traits. Genetic engineering widely increases the range of potential new traits for resistance to pests, tolerance to stress, improved food quality, and environmental sustainability. Some of these traits are favorable to the poor while others offer risks. As the range of trade-offs rises, who sets priorities for research on traits will be key in determining the impact of biotechnological innovations on poverty. Failures to include participation of the poor in priority setting increase the risk that they will be bypassed by technological progress. New schemes of participatory breeding thus become all the more important in customizing research outputs to the heterogeneous needs of the poor (Ashby and Sperling, 1995).

4.3.4 Institutional context for diffusion

i) Biosafety regulation with weak institutions. Biotechnology takes breeding science into unchartered territories and raises well-founded concerns by consumers, environmentalists, and their respective advocacy groups over possible human health and biosafety effects of under-tested biotechnologies. Hence, the need for regulation of environmental and food safety effects is enhanced. Regulation poses a set of specific problems for implementation in developing countries and among large numbers of poor smallholders. It is also a double-edged sword since costly regulatory procedures operate against smaller firms and farms, inducing concentration in industry and farming. Releasing genetically engineered crops in developing countries that are centers of origin and diversity of these crops (such as maize in Mexico, wheat in the Middle East, and potatoes in Peru) is thought to create higher risks of gene flow in nature and undesirable weediness by close crop relatives. The need for strict biosafety regulations is consequently greater precisely where they are more difficult to implement, calling for innovative approaches in institutional design. Contracting with communities to enforce biosafety regulations is an area of institutional design that needs to be explored.

ii) Trait insertion into local varieties and biodiversity. Biotech allows an increase in the range of varieties of a crop to which new traits can be applied. Hence, the benefits of research on trait improvement that were confined to major varieties under the GR have greater potential to be extended to varieties used in peasant farming systems and in niche farming. If incentives and means can be given for broad improvement of local varieties, this offers the potential of better serving smallholders and preserving biodiversity.

iii) Gene stacking and new farm management. The current state of knowledge in biotech processes only allows the inclusion or "stacking" of a few traits into a single variety by gene transfer. Hence, the question of which functions are to be achieved by gene transfers and which by traditional means (chemical pest management, integrated pest management, precision farming, production ecology, etc.) needs to be assessed for each particular set of circumstances. Use of biotechnology in heterogeneous farming conditions requires the ability to assemble these technological packages for each particular agroecological and socio-institutional environment, opening the need for a new approach to the science and practice of farm management which relies importantly on the ability to deliver customized technology to each particular category of clients.

iv) Preventative vs. remedial technologies. Biotechnological control of pests and weeds is preventative (*ex ante* relative to infestations) as opposed to chemical pesticides and herbicides, which are remedial (*ex post*). Hence, optimal use of biotechnological tools should be planned as part of the total crop production system, calling upon growers to engage in integrated crop management (ICM). ICM aims at the joint management of soil organic matter and structure, pest and disease resistance, and conservation of the beneficial insect and microorganism population. Instruments for ICM include use of crop rotations, pest- and disease-resistant cultivars, weed- and disease-free seeds, and complementary pesticides and chemicals. ICM thus effectively combines agbiotechnology with traditional agronomy, precision farming, and production ecology. While these fields of research have generally been separate, and often philosophically opposed, their integration to serve developing country clienteles is essential.

4.3.5 Use of GMOs by smallholders

i) Biotechnology, human capital, and effort requirement. By offering "smart seeds" (e.g., plants that self-protect with biopesticides or can adapt to stress), agbiotechnology demands less human capital, effort, and specialized equipment from users than chemicals or integrated pest management. Its relative simplicity may be a major cause for the fast rate of adoption observed in developed countries where it has become available. It is a feature clearly favorable to diffusion among developing country smallholders with low human, physical, and institutional capital endowments.

ii) Structure of production costs and adoption by smallholders. By embodying traits in the seed, biotechnology changes the structure of costs for farmers from variable costs (e.g., purchase of chemical insecticides) to seasonal fixed costs (e.g., purchase of seeds with biopesticide traits). With greater value added in seeds that are protected by IPRs, these fixed costs may be sharply higher. While the new technologies can be beneficial in terms of greater expected value, the changing cost structure has several implications for adoption by poor farmers: Partial and sequential adoption of pest control is prevented, the season's (or the crop rotation's) fixed costs are increased, planting-season liquidity requirements are raised, and net risks are enhanced as seed expenditures are committed irrespective of subsequent stochastic events.

iii) Changing exposure to market failures and institutional gaps at the farm level. Because biotech is resource saving by contrast to GR technologies that were resource intensifying, use of GMOs may reduce exposure to market failures and institutional gaps. The risk-reducing effects of biotech crops also mitigate the costs of insurance and credit market failures on smallholders. However, biotechnology creates other sources of exposure to market failures by displacing forward the structure of production costs (as noted above in 4.3.5*ii*) and requiring imposition of biosafety regulations (as noted in 4.3.4*i*).

5. HOW CAN BIOTECHNOLOGY BE USED TO REDUCE POVERTY?

5.1 Overall conclusion: the role of institutional innovations

Agricultural biotechnology has great promise for poverty reduction, both through direct and indirect effects, with considerable flexibility in striking differential balances between these two sets of effects to reduce aggregate poverty according to regional and agroecological contexts. Failing to capture this potential would be both a serious missed opportunity in the struggle against poverty and a risk that the competitiveness of smallholders in developing countries be further weakened relative to that of other producers and other countries. As the large gaps in the use of agbiotechnology across countries and the biases in crop and trait innovations indicate, the current situation is one of massive market and government failures for potential developing country and smallholder users. However, meeting the institutional requirements to overcome these failures is highly demanding. The effort to use biotechnology for poverty reduction will consequently fail or succeed not so much depending on the ability to progress in biological sciences as on the ability to put in place the necessary public and private institutions for the generation, transfer, delivery, regulation, and adoption of biotechnological innovations favorable to poverty reduction. Since weak institutional development is an integral feature of underdevelopment, and a pro-poor bias in developing country institutions has been notably lacking, this poses particular difficulties in achieving success that needs to be proactively addressed. In what follows, we identify the institutional innovations that are needed for this purpose.

5.2 Generation of biotechnological innovations

Institutional requirements to secure the generation of biotechnologically modified crops and animals with traits favorable to poverty reduction include the following:

i) Participation of poor producers in the setting of priorities for applied research and product development, particularly regarding choice of crops, traits, and farming systems. Effective participation requires proactive information campaigns to empower the poor.

ii) Attention to food consumers when indirect effects are also essential to setting research priorities. Lessons should be taken from experience in MDCs where attention to the demand side of the food system, including acceptance by consumers, seemed to come only as an afterthought in the development of the current generation of agbiotechnology products.

iii) Development of the capacity of LDCs' national academic and public sectors to engage in fundamental research complementary to that of the private sector, to test alternative technological options, adapt technology to their own regional needs, and engage in final product development. The type of national capacity to be developed thus depends on the particular optimum balance between these functions that vary country by country. This should be pursued on a regional basis for the smaller and poorer countries.

iv) Traditional breeding efforts should continue. An increased number of high performance varieties will improve the value of traits introduced by biotechnology. Biotechnology both alters the practice of breeding through the use of markers and tissue culture and increases the payoffs from breeding by providing better local varieties for gene insertion.

v) Enhanced public sector and CGIAR research budgets to work on (1) crops and traits not addressed by private sector research that are important for the urban and rural poor, and (2) a more complete understanding of developing countries' ecosystems in relation to gene flows and biosafety. Declining real budgets for the CGIAR and most developing countries' NARIs should thus be an issue of concern if the potential of biotechnology for the poor is to be captured.

vi) Promotion of collaborative arrangements (partnerships, consortia, contract research, gifts) bringing together corporate, nonprofit, public, and international institutions for the development of biotechnology products favorable to poverty reduction. Experimentation to identify best practice for these arrangements is needed (e.g., as pursued by the International Service for the Acquisition of Agri-biotech Applications (ISAAA) at Cornell University; see Krattinger, 2000).

vii) Identify opportunities for technological spillovers from industrialized countries that do not threaten commercial markets for private sector innovations. Under these conditions, technology transfers may be handled as gifts (e.g., Monsanto's virus-resistant potatoes for subsistence farming in Mexico; see Qaim, 1998).

viii) Institutions to link public and CGIAR research to private sector product development through offices of technology transfer attached to universities and public research institutes, venture capital for the financing of agbiotechnology companies, and mechanisms for the fair and effective enforcement of property rights (Cohen-Vogel et al., 1998).

ix) An IPR regime that does not hamper further research and downstream product development, particularly for public institutions, international organizations such as the CGIAR, and nongovernmental organizations that are concerned with the poor. Questioning the features of current patent systems and guiding their future evolution should thus be an integral part of efforts to maximize the role of biotechnology for poverty reduction.

x) Use of defensive patents on public sector and CGIAR-research innovations that have high potential for poverty reduction, such as apomixis, site-directed mutagenesis, and homologous recombination, with the expressed purpose of keeping them in the public domain for selected clienteles. Due to costs and legal complexities, patents are likely to be taken in joint ventures with the private sector. Identification of best practices for the delivery of international public goods under defensive patents is urgently needed.

xi) IPR regimes that recognize the legitimate ownership rights of traditional farming communities over biological resources and give them leverage in gaining access to the private products of biotechnology. Experimentation with innovative contracts to reconcile farmers' ownership rights over biodiversity with efficient bio-prospecting is needed (e.g., Shaman Pharmaceuticals in California and INBio in Costa Rica).

xii) Development of markets or other mechanisms for the trading of patented materials. A neutral and efficient IPR clearinghouse, based on publicly available information, for the rights to utilize patented biotechnology processes, materials, and products would play a crucial role in protecting developing country and smallholder interests.

5.3 Transfer of technologies and the delivery of products

Institutions to link the results of research to the delivery of products adoptable by developing country farmers and particularly smallholders include the following:

i) Public and nonprofit sector roles in (1) the insertion of new traits in poor farmer crops and varieties with insufficient current market size to provide private sector incentives, (2) the assembly of idiosyncratic technological packages for smallholder farming that combines traits controlled by gene insertion with functions delivered by other approaches such as chemical pest management, integrated pest management, and agronomic practices.

ii) Incentives to the private sector to invest in research for developing country needs when there is insufficient effective demand due to poverty.

This can be done through a guaranteed purchase fund set up by donors, analogous to that for research on vaccines for tropical diseases like malaria (Kremer, 2001).

iii) Coordination of private sector initiatives toward market expansion among smallholders, allowing them to overcome the commons problem typical of such investments.

iv) IPR incentives and availability of low cost technology to insert new traits into a wide range of alternative varieties, allowing better adaptation to local conditions, preservation of biodiversity, and competitive farming (as opposed to generalized contracting by patent holders).

v) Development of a regulatory framework for biosafety and consumer protection that corresponds to each country's preferences for risk and expected income gains, which change with stages of development. Attempts to equalize regulations affecting agbiotechnology in the name of harmonization, for instance, to satisfy WTO requirements, should be scrutinized for their impact on the poor.

vi) Decentralization of the monitoring and enforcement of biosafety regulations to the community level, based on community contracting and verification by regulatory agencies.

vii) Emphasis on simple technologies with low biosafety risks (e.g., Rhizobium inoculation in Kenya) for as long as knowledge of environmental risks and enforcement of regulatory frameworks remain weak.

viii) Discriminatory pricing of genetically modified seeds if market segmentation between poor and nonpoor is possible.

ix) Subsidies to private marketing strategies that promote adoption of new technologies favorable to poverty reduction.

x) Promotion of the private sector to deliver integrated services to smallholders combining GMOs and other technological approaches.

5.4 Adoption by smallholders

Institutions to reduce poverty among smallholders by supporting adoption of favorable technologies include:

i) Organization of credit schemes to face higher and earlier liquidity requirements in the purchase of seeds with improved trait content that are protected by IPRs, and potentially subject to noncompetitive pricing.

ii) Insurance and risk-sharing mechanisms to absorb higher risks associated with committed seed expenses and higher cash outlays.

iii) Development of institutional mechanisms (such as labeling) and production contracts for identity preservation of improved small-farm products.

iv) Promotion of producers' organizations such as service cooperatives in support of contract farming with smallholders for the acquisition of information on GMOs, access to modern inputs, production of improved small-farm products, and biosafety management.

v) Negotiated exemptions for poor smallholders to allow the reproduction of seeds covered by IPR for home use.

REFERENCES

Ashby, J., and Sperling, L., 1995, Institutionalizing participatory, client-driven research and technology development in agriculture, *Dev. and Change* **26**:753-770.

Byerlee, D., 1996, Modern varieties, productivity, and sustainability, *World Dev.* **24**:697-718.

Byerlee, D., and Moya, P., 1993, *Impact of International Wheat Breeding Research in the Developing World, 1966-90*, CIMMYT, Mexico City.

CIMMYT, 2001, *People and Partnerships: Medium Term Plan of the International Maize and Wheat Improvement Center, 2001-2003+*, CIMMYT, Mexico.

Cohen-Vogel, D. R., Osgood, D. E., Parker, D. D., and Zilberman, D., 1998, The California Irrigation Management Information System (CIMIS): Intended and unanticipated impacts of public investment, *Choices* (Third Quarter):20- 21.

FAO, 2000, *The State of Food Insecurity in the World: 2000*, Food and Agriculture Organization of the United Nations, Rome.

Graff, G., Rausser G., and Small, A., 2003, Agricultural biotechnology's complementary intellectual assets, *Rev. of Econ. and Statistics* **85**:349-363.

Griliches, Z., 1957, Hybrid corn: An exploration in the economics of technical change, *Econometrica* **25**:501-522.

Harwood, R., 1998, Sustainability in agricultural systems in transition: At what cost, Department of Crop and Soil Sciences, Michigan State University.

Herdt, R., 1998, *Enclosing the Global Plant Genetic Commons*, The Rockefeller Foundation, New York.

Hubbell, B., Carlson, G., and Marra, M., 2000, Estimating the demand for a new technology: Bt cotton and insecticide policies, *Amer. J. Agri. Econ.* **82**:118-132.

James, C., 1998, *Global Review of Commercialized Transgenic Crops: 1998*, ISAAA Brief No. 8, International Service for the Acquisition of Agri-Biotechnology Applications, Ithaca, New York.

James, C., 2000, *Preview: Global Review of Commercialized Transgenic Crops: 2000*, ISAAA Brief No. 21, International Service for the Acquisition of Agri-Biotechnology Applications, Ithaca, New York.

Klotz-Ingram, C., Jans, S., Fernandez-Cornejo, J., and McBride, W., 1999, Farm-level production effects related to the adoption of genetically modified cotton for pest management, *AgBioForum* **2**:73-84.

Koo, B., and Wright, B., 2000, The optimal timing of evaluation of genebank accessions and the effects of biotechnology, *Amer. J. Agri. Econ.* **82**:797-811.

Krattinger, A., 2000, *An Overview of ISAAA from 1992 to 2000*, The International Service for the Acquisition of Agri-Biotech Applications, Ithaca, New York.

Kremer, M., 2001, Spurring technical change in tropical agriculture, Department of Economics, Harvard University, Cambridge.

Nuffield Foundation (The), 1999, Genetically modified crops: The ethical and social issues, http://www.nuffield.org/bioethics/publication/modifiedcrops.

Pinstrup-Andersen, P., Pandya-Lorch, R., and Rosegrant, M., 1999, *World Food Prospects: Critical Issues for the Early Twenty-First Century.* Food Policy Report, International Food Policy Research Institute, Washington, D. C.

Pray, C., Courtmanche, A., and Brennan, M., 1999, The importance of policies and regulations in the international spread of plant biotechnology research, Department of Agricultural Economics and Marketing, Rutgers University.

Pray, C. E., Ma, D., Huang, J., and Qiao, F., 2000, Impact of Bt cotton in China, paper presented at 4th International Conference on the Economics of Agricultural Biotechnology, Ravello, Italy (August 24-28, 2000).

Qaim, M., 1998, *Transgenic Virus Resistant Potatoes in Mexico: Potential Socioeconomic Implications of North-South Biotechnology Transfer*, The International Service for the Acquisition of Agri-Biotech Applications, Ithaca, New York.

Rausser, G., Simon, L., and Ameden, H., 2000, Public-private alliances in biotechnology: Can they narrow the knowledge gap between rich and poor? *Food Policy* **25**:499-513.

Sadoulet, E., and de Janvry, A., 1992, Agricultural trade liberalization and low income countries: A general equilibrium-multimarket approach, *Amer. J. Agri. Econ.* **74**:268-280.

Scobie, G., and Posada, R., 1978, The impact of technical change on income distribution: The case of rice in Colombia, *Amer. J. Agri. Econ.* **60**:85-91.

Traxler, G., and Falck-Zepeda, J., 1999, The distribution of benefits from the introduction of transgenic cotton varieties, *AgBioForum* **2**:73-84.

Wolf, S., and Buttel, F., 1996, The political economy of precision farming, *Amer. J. Agri. Econ.* **78**:1269-1274.

World Bank, 2001, *Attacking Poverty: World Development Report 2000/2001*, The World Bank, Washington, D. C.

Wright, B., 1998, Public germplasm development at a crossroads: Biotechnology and intellectual property, *CA Agri.* **52**:8-13.

Chapter 18

TOWARDS AN INTELLECTUAL PROPERTY CLEARINGHOUSE FOR AGRICULTURAL BIOTECHNOLOGY*

Gregory Graff[1] and David Zilberman[2]
[1] *Visiting Postdoctoral Researcher, Agricultural and Resource Economics, 207 Giannini Hall, University of California, Berkeley, CA, 94720;* [2] *Professor, Agricultural and Resource Economics, 207 Giannini Hall, University of California, Berkeley, CA, 94720*

Abstract: Much of the critique of patent systems for hindering research has focused on the scope or definition of what is patentable. We suggest, rather, that by focusing on the exchange of existing patent rights, significant improvements in freedom-to-operate can be achieved regardless of the state of patent reform. Historically, in other industries, when IP congestion has threatened productivity, both government and industry groups have intervened, forming collective rights organizations such as patent pools and royalty clearinghouses that have provided freedom to operate with substantial savings for whole industries. Furthermore, today's advances in information technology have created new tools, "IP informatics" and "online IP exchanges," which provide interesting new organizational possibilities for collective intellectual property rights organizations. The goal of an "intellectual property clearinghouse" for agricultural biotechnologies would be to reduce transaction costs and other market failures that hinder the exchange of IP, creating pathways through the patent thicket and giving freedom to operate with proprietary biotechnologies.

Key words: agricultural biotechnology; intellectual property; market failures; market institutions; patent pooling; technology transfer.

1. BOTCHING A DELICATE BALANCE

A fundamental economic tension exists between the public and private economic forces that drive agricultural biotechnology research. On the one hand, in the big picture of human welfare, our collective knowledge about agricultural science and genetics is a vital *common* resource for all of humanity (Herdt, 1999). On the other hand, the research that will advance

---

* This chapter has been revised and updated from an earlier version, published in the online journal, *Intellectual Property Strategy Today*, No. 3-2001 (http://www.biodevelopments.org/ip/index.htm).

our knowledge and our ability to wisely manage the earth's genetic resources depends upon *private* incentives of agricultural markets, which encourage companies to invest at levels unlikely ever able to be matched by public spending.

The granting of patents over the use of biological organisms, materials, and processes—in other words, intellectual property rights (IPRs) over the components of life—provides a very important practical compromise between the fundamental public and private economic forces that drive agricultural biotechnology research. The effectiveness of patents to perform this compromise, however, turns on two key factors:

1. The definition of what is patentable, to clearly demarcate between what should be claimable as private knowledge and what should be placed in the public domain of knowledge and open access genetic resources.
2. The mechanism to exchange patent rights, to efficiently move privately deeded knowledge into the hands of those users who are most able to create value with that knowledge and who, in so doing, can fairly compensate the private inventor of that knowledge or the steward of that genetic resource.

When the common (interdependent or complementary) aspects of agricultural knowledge and crop genomes are divided into multiple competing, overlapping, or mutually blocking private property claims, the value of the public economic benefits that would otherwise arise from these resources is diminished. Furthermore, if patent rights cannot be traded, the inventor-owners of these piecemealed resources will not be able to negotiate or purchase access to other patents needed to make use of their own inventions, in which case the power of the private incentives to innovate will be sapped. The cumulative result of such a crisis in research and innovation productivity has been quite aptly dubbed *"the tragedy of the anti-commons"* (Heller and Eisenberg, 1998).

Such concerns are nowhere more relevant than in agriculture (Enriquez and Goldberg, 2000) for, as research in crop genetics, breeding, agronomy, pest control, agroecology, and related systems becomes more and more intertwined and complex, new agricultural research inevitably depends more and more on access to the proprietary knowledge and biological materials previously claimed by others. Indeed, in many cases, agricultural researchers' *"freedom to innovate"* depends on scores of patents. And while *"research only"* allowances may be granted for basic research in universities and public laboratories, the *"freedom to operate"* commercially for new agricultural products is usually immediately choked by a thicket of blocking patents (Shapiro, 2000).

The current status quo of this anti-commons climate benefits no one. Researchers in both public institutions and in private corporations—in both developed and developing countries—are finding their freedom to innovate and freedom to operate overly constrained. Legal costs and transaction costs for attempts to navigate through the patent thickets are mounting. Firms in agricultural biotechnology appear to have consolidated during the 1990s precisely to streamline access to patented technologies (Graff, Rausser, and Small, 2003). Uncertainty over blocking patents and freedom to operate has added additional burdens to the already challenging process of conducting international agricultural research and transferring agricultural technologies to developing countries (Wright, 2001). Both public sector institutions and private sector firms are spending valuable resources to solve intellectual property (IP) problems that could otherwise be used to guarantee the environmental and health safety of their innovations. The wave of consumer and environmentalist opposition to genetically modified foods, particularly in Europe, is spurred on at least in part by the perceived lack of access, transparency, and outside review that characterize the proprietary technologies that make these products possible. Economists studying this situation are concerned that economic growth, environmental health, and food security—all of which could benefit from advances in the biology of agriculture—are stalled and that the potential social, nutritional, and environmental benefits to the human race and the biosphere we live in are being squandered.

2. UNILATERAL RESPONSES TO THE INTELLECTUAL ANTI-COMMONS

On its own, a company has limited options to pursue its own freedom to operate within a congested intellectual property (IP) landscape. As well, universities and public sector research institutions are finding it necessary to devise IP strategies to cope with a shifting interface between public interests and private economic forces (Byerlee and Fischer, 2001; Kryder, Kowalski, and Krattiger, 2000; Press and Washburn, 2000). Overall, the following IP management tactics constitute the potential unilateral strategies available to individual organizations, both public and private, which allow them freedom to innovate and operate:

- Invent around another's proprietary technology
- In-license another's proprietary technology
- Cross-license one's own proprietary technology for another's
- Strike a strategic collaboration or conditional access agreement
- Strive for organizational integration with other IP holders.

3. GOVERNMENT AND INDUSTRY-LED COLLECTIVE APPROACHES TO SOLVING THE ANTI-COMMONS

According to Robert Merges of the Boalt School of Law at U. C. Berkeley (Merges, 1996), theories on the economic nature of common-pool resources suggest that the roots of this problem cannot be effectively addressed through unilateral strategies; instead, some form of collective solution will be needed. Historically, public-policy collective measures taken to solve the problems of IP congestion include the following:

- Government exercise of intellectual "eminent domain," purchasing key enabling technology patents and placing them in the public domain.
- Government mandate of "compulsory licensing" of patents for a fixed fee.
- Government forced merger of firms holding mutually blocking IP.

Interestingly, however, private institutions or industry-led consortia have on occasion negotiated and organized effective actions themselves, without government mediation:

- Collective copyright enforcement of music compositions and recordings (e.g., ASCAP, BMI).
- Small contract-based patent pools.
- Industry-wide patent pools (e.g., Manufacturers Aircraft Association (MAA) formed in 1917, automobile industry patent pools in the 1920s and 1930s).
- Standard-setting patent pools (e.g., DVD technology).

Merges argues that such "collective rights organizations" are more economically efficient than the government-invoked solutions, especially compulsory licensing. Evidence shows that collective solutions have provided substantial savings for entire industries and for society at large. Despite the difficulties that must be surmounted in forming such a collective institution, time and again all players in an industry have seen it worthwhile to participate and conform to the rules and stipulations of the collective institution. However, horizontal collaboration through patent pools can provide a pretext for unhealthy degrees of collaboration and monopolization among the leaders in those industries and, given various abuses over the years, antitrust authorities view simple private patent pools with some suspicion (U. S. Department of Justice, 1995).

Despite these concerns, a strong case remains today for the formation of a multilateral collective rights organization to provide access to mutually complementary proprietary agricultural technologies and genetic resources. All currently unsatisfied parties—in both the public and private sectors as well as in both the biotechnologically advanced industrial economies and in the biodiversity rich developing countries—stand to benefit from some sort of *"intellectual property clearinghouse."* Furthermore, there are several new options to consider in terms of the potential arrangements for such an institution, particularly as major trends in IP information, management, and marketing are emerging with the advent of database and Internet technologies: tools such as *IP informatics* and *online IP exchanges*. These tools provide new options for collective IP rights organizations to work more like markets and less like cartels.

4. INTELLECTUAL PROPERTY INFORMATICS FOR AGRICULTURE

A first practical step toward solving the problem of the anti-commons is the broad provision of *"IP informatics"* to make information about a set of interdependent technologies and the IP that protects them broadly and freely available to all concerned parties. The common availability of information would help to overcome two serious barriers to fair trade in patented technologies: *"imperfect information"* and *"information asymmetry,"* situations where one or both parties in a transaction lack some of the information on which their decisions to buy or to sell rest. A complete and open flow of information helps individual researchers and organizations to identify actual and potential conflicts among patents already granted. When considering the potential savings and gains that may be achieved by providing such information to all organizations involved in agricultural research, IP informatics is a relatively inexpensive and straightforward investment.

The term "IP informatics" was coined at the *Center for the Application of Molecular Biology in International Agriculture (CAMBIA)*, a nonprofit research institute located in Canberra, Australia, which offers the *CAMBIA Intellectual Property Resource,* an information service that is particularly suited to public sector researchers in international agricultural institutions and developing countries. *CAMBIA* provides, at minimal or no cost to the user, a readily searchable database of U. S., European, and international (PCT) patents covering agriculture and the life sciences, augmented with advisory and educational services (see Table 18-1 for the web address of this and other IP information services).

While most national patent offices, such as the European Patent Office and the U. S. Patent and Trademark Office, provide web-based searches of their respective patent databases, these usually consist just of raw data or the texts of the patents themselves. More extensive supplementary patent information and analyses are sold by a variety of IP information services. The largest of these are the INPADOC databases of the European Patent

Office, which cover patents in 65 countries, providing information on the current legal status of each patent and tracing the "family" of patents issued in different countries for the same invention. The foremost private IP information service is *Derwent*, of *Thompson Scientific*, which maintains the *Derwent World Patents Index*, containing up-to-date patents from 40 different countries, summarized in English and classified according to *Derwent's* own comprehensive technology index.

Table 18-1. IP information sources as of November, 2003

Name	Web address
CAMBIA Intellectual Property Resource	http://www.cambialP.org/
CHI Research	http://www.chiresearch.com/
Thomson Delphion Research	http://www.delphion.com/research/
Thomson Derwent	http://www.derwent.com/
European Patent Office, European Patent Register	http://register.epoline.org/espacenet/ep/en/srch-reg.htm
European Patent Office, INPADOC databases	http://www.european-patent-office.org/inpadoc/index.htm
MicroPatent	http://www.micropatent.com
N.I.H. National Center for Biotechnology Information, GenBank database	http://www.mogee.com/
U.S. Patent and Trademark Office	http://www.ncbi.nlm.nih.gov/
U.S.D.A. Plant Variety Protection Office	http://www.uspto.gov/main/patents.htm
World Intellectual Property Organization, Intellectual Property Digital Library* and PCT Full Text Database	http://www.ams.usda.gov/science/PVPO/PVPindex.htm
CAMBIA Intellectual Property Resource	http://ipdl.wipo.int/

* Includes links to many national patent office databases.

An ideal IP informatics tool includes supporting data and analysis to add additional value to the use of basic patent data. This should include:

- A database search methodology specifically structured and indexed to be user friendly and easily navigated by biologists and other non-IP professionals.
- Analytical tools to determine and display the IP landscape around particular patents, to characterize the differences and similarities among patented technologies, and indicate the positions of different organizations in the related technologies.
- Analytical tools to chart or interpret patents' legal claims to outline best approximations of the legal scope of patents.
- Indicators of patent value.

Such analytical capacities are more costly to provide but are already developed and marketed by private IP data providers (such as *Thompson* or *MicroPatent*) and IP consultancies (such as *Chi Research*).

Beyond patents, other kinds of IP and technology data provide information resources important to agricultural researchers:

- Plant varieties protected by Plant Variety Protection Certificates (PVPCs) in the United States and by similar nonpatent *"sui generis"* plant variety protection systems in other countries in accordance with UPOV.
- Seed bank or germplasm collections data (from the USDA, the CGIAR, etc.).
- Gene sequences and protein sequences claimed in patents (listed in *Derwent's* GENESEQ database).
- Publicly available genomic data on major crops and pests (some already listed in the N.I.H.'s GenBank database).

Additionally, an informatics solution could help augment the flow of public and traditional agricultural knowledge and technologies, providing a centralized, user-friendly, and consistently indexed registry for non-proprietary, *"shareware"*-like agricultural techniques, especially sustainable agroecological, biocontrol, and integrated pest management methods that are not patented but published in articles, reports, or other research outlets. The timely publication and ready availability of technical disclosures assures that the technologies cannot later be patented. A forum for sharing such information could engender something like an "open source" legal environment for many agricultural technologies.

In general, an IP informatics tool answers the initial question, *"Who has innovated or patented what?"* It allows technology users to identify and select a needed technology and then to decide upon appropriate IP management tactics, such as whether to invent around or to negotiate with a patent owner. However, such IP information and the expertise needed to use it effectively are not readily available to all agricultural research organizations. Larger corporations have already invested significant amounts subscribing to and installing some of these in-house IP information analysis and management systems and hiring IP legal counsel.

In the final analysis, any IP informatics service functions to augment individual organizations' internal capacities to manage IP. It thereby informs organizations' unilateral strategies and occasionally promotes bilateral transactions. While the universal availability of IP informatics would be a necessary foundation for more market-oriented patent exchange mechanisms or multi-party collective rights organizations, IP informatics alone cannot solve the tragedy of the anti-commons.

5. ONLINE INTELLECTUAL PROPERTY EXCHANGES

In their recent book on markets for technology, Arora, Fosfuri, and Gambardella (2004) explore the key benefits of markets for technology and the primary reasons that such markets fail to form. In light of their arguments, promising development aimed at solving the market failures caused by information failures and high transaction costs may be found in the institution of "*online intellectual property exchanges*."

Beginning in 1999, a number of entrepreneurial startup ventures emerged on the Internet with explicit business plans for creating virtual trading floors for intellectual assets. These online exchanges for IP were inspired by the basic Internet business-to-business (B2B) model, and their promotional efforts have touted the promises of free-market efficiency. The typical online IP exchange consists essentially of an embellished IP informatics service, or even more simply a list of technologies, augmented by basic services to allow technology owners and technology "buyers" to initiate negotiations for a license. Some of the premier exchanges have designed creative and comprehensive transaction-mediating and transaction-managing services, often integrated with more conventional operations of seasoned licensing professionals. Table 18-2 provides a recent list of active web-based IP exchanges, but consolidation is expected to continue, with ultimately just a handful remaining.

Indeed, several serious concerns arise in considering the potential for patent exchanges to optimally redistribute technologies to those who can make the most valuable use of them for society at large. Exchanges are, in general terms, best suited for highly repetitive, routinized transactions of clearly defined, standardized, and readily priced assets, goods, or contracts, albeit including contracts for services (Kaplan and Sawhney, 2000). Patents and licenses, however, do not very often exhibit such qualities. The specified in patents are highly heterogeneous, are often difficult to clearly or completely define, and may be impossible to evaluate sufficiently until well after considerable experimentation and refinement have taken place (i.e., well *after* the licensing transaction). Furthermore, innately held differences between sellers and buyers in their respective valuations of a technology may be wide enough to make it difficult to arrive at a clearing price for a license. These factors create uncertainties that darken the prospects for spot transactions of patents or licenses on an exchange.

Two relatively rare types of patents, however, do have qualities that should make them more conducive to online promotion. The first are those few patents that cover highly important general-purpose research methods, for which a winning marketing strategy would be to grant as many routine nonexclusive licenses as possible throughout the entire industry (which was the licensing strategy for the famous Cohen-Boyer patents of UC-San Francisco and Stanford). Holders of such general-purpose patents would benefit greatly from the low transaction costs of online promotion and distribution. Second, more numerous patents protecting highly specific and well-defined incremental improvements to familiar downstream products or processes could also be distributed online. These kinds of inventions are often most valuable when exclusively sold or licensed to the one specific potential user who values that innovation the most. Holders of these patents would benefit from the ease of finding and notifying a potential buyer and from the low transaction costs for executing a routine transaction. Finally, however, the bulk of patents that fall somewhere in between these two examples, either in terms of importance or in terms of generality of application, will likely be difficult assets to transact in the online exchange environment.

Table 18-2. Online IP exchanges, as of September, 2003

Name	Web address
2XFR (by the Patent Café IP Network)	http://www.2xfr.com/
Brainhead	http://www.brainhead.com/
Buy Patents	http://www.buypatents.com/
Global Techno Scan	http://www.globaltechnoscan.com/
Intellectual Property License Exchange	http://www.iplx.com/
International Invention Register	http://www.inventionregister.com/
IP Marketplace	http://www.ipmarketplace.com/
Knowledge Express	http://www.knowledgeexpress.com/
New Idea Trade	http://www.newideatrade.com/
Patent & License Exchange (pl-x)	http://www.pl-x.com/
Patent Auction	http://www.patentauction.com/
Pharma Licensing	http://www.pharmalicensing.com/
Tech Tuesday (Technology Source Group)	http://www.techtuesday.com/
TechEx	http://www.techex.com/
Uventures	http://www.uventures.com/
Virtual Component Exchange (VCX)	http://www.thevcx.org/
Yet2	http://www.yet2.com/

Online exchanges face other important difficulties. They currently are squeezed in an economic vice-clamp. On the one side, the business model depends upon attracting numerous buyers and sellers to make licensing transactions and then charging a small flat fee or a percentage commission on each transaction. Yet, to achieve a sufficient volume of transactions, a site must maintain what may be called sufficient *"IP liquidity."* IP liquidity is maintained not simply by listing a large overall number of available patents, but, more importantly, by listing a sufficient *"density"* of available patents within any given industry or field of technology, thereby providing potential customers with a sufficient selection to warrant their entering the site and searching for needed technologies. On the other side, the ability of online exchanges to maintain such liquidity is squeezed by intense competition. Startup costs for establishing a new website to host an exchange are quite low, and a large number of online patent exchanges now exists (see Table 18-2 on the previous page), each scrapping for a relatively small proportion of the total market for patent licenses and each possessing only a very low density of patents in any given industry. The overall market is fractured, and most of the individual online licensing markets are currently too small to operate reasonably as exchanges.

In a specific field such as agriculture, no single online exchange provides access to all of the relevant IP currently available. In particular, searching for listings of "agricultural" or "agricultural biotechnology" patents turns up spotty or empty results even on the most developed online exchanges. Indeed, surveying the many online exchanges is itself a significant search cost for a laboratory researcher or technology manager seeking access to a technology. Those in search of a specific kind of technology have to go site to site, registering numerous times for web site memberships, remembering passwords, and in some cases paying significant fees for membership or pay-per-view for patent listings in which they are not yet sure they are interested. Two things would help to alleviate this problem, at least for a given industry such as agriculture: (1) a drastic consolidation of the online patent exchanges into a unified marketplace or (2) a universal cross-listing of current offerings across all of the online patent exchanges.

Consolidation or universal listings would, however, do little to circumvent the *"matchmaker's dilemma,"* yet another problem to which the online patent exchange business model is susceptible. Once a potential buyer (or licensee) has discovered an interesting patent that has been listed by a seller (or licensor) on an exchange, the buyer-seller pair may find it more economically advantageous and secure to go offline and deal directly with one another, thus dispensing with the hapless matchmaker and avoiding a commissions payment. Much like a dating service, the patent exchange may excel in providing first-time introductions, but it does not want to meddle further in a new technological relationship. The matchmaker dilemma threatens to constantly sap away the volume of technology-bearing, fee-paying traffic needed to maintain the IP liquidity of the market and the revenue base of the online exchange.

The value to society of more efficient technology markets—that is, more efficient mechanisms for getting good ideas deployed in their most valued applications—could be enormous. Where markets for technology are viable and competitive, the economic rule of efficiency calls for private enterprise to handle the creation of such markets. Public exchange services should be considered to support the formation of market mechanisms for IP and technologies only in those areas where wider social benefits can be anticipated and where it is unlikely that a private enterprise could support itself.

In the end, however, regardless of whether an IP exchange is privately or publicly backed, only those patents with characteristics that are amenable to the exchange mechanism will be made accessible. Other patents simply will not be distributed via this channel. Given the strategic (or monopoly power) value of many proprietary technologies, patent holders will likely decide not to offer them in an open market. In even these cases, however, it is possible that some type of mutually enforced agreement will offer holders of strategic patents a way to realize the value of their own patents while at the same time giving broader access to the protected technologies.

6. A COLLECTIVE RIGHTS ORGANIZATION FOR AGRICULTURE

We propose that an IP "*clearinghouse*" might be an effective way to reduce market inefficiencies that hinder the exchange of privately deeded knowledge, allowing researchers to obtain the freedom-to-operate status necessary to commercialize agricultural research. Such a clearinghouse should be based on the basic principles of a "collective rights organization" (Section 3) and utilize all available IP informatics (Section 4) and IP market exchanges tools (Section 5). In such a collective arrangement, multiple technology providers and users would be supported by a professional network and linked to one another through common contractual commitments. This would allow users to quickly identify relevant technologies and, through standardized licensing procedures, fulfill transactions of rights and technologies.

To be effective, a clearinghouse mechanism must provide the following three basic services:

1. The capacity to identify all relevant IP claims over a given technology and, of those claims, to indicate which are and which are not available to be negotiated and, if they are, how they can be accessed.
2. The establishment of a pricing scheme and terms of contract and a royalty disbursement accounting system.
3. An arbitration mechanism for monitoring and enforcing contracts.

To serve as a collective rights organization for agriculture, such a clearinghouse should be specific to agriculture and the particular IP needs of researchers involved in agricultural research. While generalized IP informatics data sources and online IP exchanges (discussed above) do provide many valuable services, by maintaining broad coverage of many technologies, they sacrifice the necessary depth and comprehensiveness in a single field, such as agriculture. Moreover, the organization will need to be founded upon the trust and confidence of all its members, and its actions must maintain its members' confidence. A service created by and for agricultural researcher organizations certainly stands a much better chance of maintaining the trust and confidence of those in the field.

Indeed, an agricultural IP clearinghouse should be independent, neutral, and a catalyst for healthy competition in agricultural markets. If it were to be perceived as a technology user's club or a technology seller's marketing tool, its effectiveness would be diminished. The trust of prospective parties in the clearinghouse who were not in the favored core clientele would be eroded, and they would rightfully be reluctant to enter into transactions due to suspicions about unequal bargaining power. In addition, a collective organization that is not neutrally promoting competition in the industry would likely conflict with current regulations or case law pertaining to IP licensing and antitrust (U. S. Department of Justice, 1995). Antitrust is a particular concern given the precedent of some commercial Internet exchanges for industrial supply commodities suspected of price manipulation and other antitrust violations *(The Economist*, 2000). The financial and governance structures of a collective rights organization must be both appropriately distributed among members and transparent to avoid any conflict-of-interest or collusion problems.

An agricultural IP clearinghouse would need to monitor patent validity, check and verify ownership status, and generally serve as a watchdog against problematic patents that are poorly written, overly broad, or otherwise disruptive to the productive flow of information and property rights in the industry.

In order to offer the collective rights efficiencies of a patent pool without the downfalls of pooling, an agricultural IP clearinghouse could *"bundle"* key combinations of interdependent or mutually complementary technologies together into patent *"micropools,"* each consisting of a set of interdependent or mutually complementary patents offered by the clearinghouse under a single contract. Numerous separate micropools or bundles could be constructed and offered, providing access to different platform technology *"research toolboxes,"* particular *"agronomic systems,"* or specific *"plant systems."* Furthermore, by actively pursuing flexible patent licensing strategies, it might be possible to customize bundled licensing products that could greatly increase the use of inventors' technologies (and thereby licensing revenues) as well as make multi-patent technology systems much more readily available and affordable.

Finally, an agbiotech IP clearinghouse would need to maintain and provide data about the current regulatory approval and biosafety status of new technologies in multiple countries. As the field of agricultural biology rapidly develops, it is crucial to keep track of which components of a technology system have been approved for which uses in which countries. Biosafety regulation is an important restriction on technological freedom-to-operate. It has a very strong influence on the value of a given patent or technology system and is crucial information for determining fair pricing and the terms of exchange for a technology.

7. WHO WOULD USE AN IP CLEARINGHOUSE FOR AGRICULTURAL RESEARCH?

Who are the most likely initial participants in an IP collective rights organization or IP exchange? Everyone involved in agricultural research is to some degree both a supplier and a user of new technologies. First, however, let us examine who is actively patenting biological applications for agricultural use. Here are the names of the top 30 assignees of agricultural biology patents in the United States at the end of 1998, with public sector institutions italicized (Table 18-3).

Table 18-3. Patenting organizations with U.S. utility patents in crop biotechnology

Parent entity (includes subsidiaries)	Number of U.S. patents
Monsanto	612
DuPont	579
Syngenta	295
Bayer	185
Dow	147
Stine Seed Farm	90
University of California	*74*
Savia / Bionova	70
Novo Nordisk / Novozymes	64
USDA	*51*
Cornell University	*41*
BASF	38
Iowa State University	*34*
Michigan State University	*31*
Advanta / Garst Seeds	30
Unilever	26

Table 18-3. (continued).

Danisco	25
Sumitomo	25
DSM	24
North Carolina State University	*23*
Washington State University	*21*
University of Florida	*20*
Salk Institute	*20*
Massachusetts General Hospital (MGH)	*20*
Wisconsin Alumni Research Foundation (WARF)	*20*
Max Planck Institute	*19*
Institut Pasteur	*19*
Weyerhaeuser Company	19
Canadian Ministry of Agriculture and Agri-Food	*19*
Rutgers University	*19*

Data source: Graff et al. (2003).

8. AN IP CLEARINGHOUSE AND TECHNOLOGY TRANSFER TO DEVELOPING COUNTRIES

Equally important questions to ask are the following: *"Who is not in the game?"*; *"Who is in danger of being locked out of the dynamic advance of agricultural technologies?"*; or *"Who is not investing in research because of uncertainties surrounding the validity and enforcement of IP?"* These include:

- Farmers and growers.
- Agricultural co-ops and grower's associations.
- Many of the land-grant and public universities in the United States and abroad.
- International agricultural research centers of the CGIAR.
- National agricultural research services (NARS) of developing countries.
- Medium- and small-scale seed enterprises and nurseries in developed countries and national seed companies of developing countries.
- Agricultural development NGOs.

One of the most important things to consider in exploring options for an IP clearinghouse is that the newly available IP informatics and market-based tools might not merely allow for but could actually encourage the participation of those currently left out of the R&D process. Not only would today's outsiders find themselves able to in-license currently unavailable technologies at reasonable costs and on reasonable terms, but they would be encouraged to develop and out-license their own inventions for fair returns on reasonable terms. Incentives would be aligned to encourage the development of agricultural research capacity. Similarly, other potential technology providers (farmers, coops, university professors, independent inventors, small firms) who currently have the capacity but lack the incentives to undertake certain lines of research for themselves, would come to see the advantage of completing and patenting undeveloped ideas that they could offer to others in an active and healthy technology marketplace.

A number of voices have advocated collective IP solutions for public-sector and international agricultural research, seeking to improve conditions of freedom to operate for academic and not-for-profit international agricultural research institutions through some sort of licensing mediation or IP pooling mechanism (Bennett, 2000; Prakash, 2000). Similar calls have been heard in the related fields of medical biotechnology and genomics (Shulman, 2000, Clark et al., 2000.)

A first credible step toward creating an IP clearinghouse might be to erect a mechanism for the bundling and provision of inexpensive *"humanitarian use licenses"* for the release of new developments from agricultural research dedicated to solving problems of food security, malnutrition, and poverty. *GoldenRice*, the rice line engineered to deliver pro-vitamin A, is the first in what could be a long list of potentially useful technologies developed by public sector researchers, but which need permission from multiple private and public sector patent holders in order to be released and sold in most of the countries where it is needed (Kryder, Kowalski, and Krattiger, 2000). A separate set of multi-party IP agreements could be hammered out each time a new variety comes along, an arrangement that may slowly choke off public sector involvement in such work, or an established clearinghouse could build expertise in negotiating such IP agreements and build upon previous agreements.

The utility of a clearinghouse beyond its role in the coordination of IP philanthropy would quickly become clear. The academic and corporate donors of the humanitarian use contracts might soon approach the clearinghouse with requests to help negotiate complex arrangements for their own needs, for example, to provide freedom to operate for a previously neglected crop that only university-based plant breeders were working on, or for an environmentally beneficial trait whose low expected profit level previously could not justify the costly bilateral licensing negotiations necessary to launch it as a commercial product.

9. CONCLUSION

As a collective rights organization utilizing the available tools of the IP informatics service and the online IP exchange, a proactive, industry-specific IP clearinghouse could level the playing field and free up agricultural research by creating paths through the growing thickets of competing IP claims. A clearinghouse might also help to reverse consolidation in the industry, since it would no longer be necessary to control in-house a complete portfolio of interdependent complementary technologies to maximize value from any single component technology. It could free companies from the innovation-constricting technological platforms to which their in-house patent portfolios currently limit them. It could help to move appropriate technologies out into regional and applied agricultural research systems around the world, providing incentives and means for current outside players to strengthen their agricultural research capacities. Finally, an IP clearinghouse could help agricultural research achieve and maintain a healthy, dynamic balance between the public and private forces that are now haphazardly shaping its future. A clearinghouse will help us to rationally direct these energies more efficiently, more safely, and to the benefit of all.

REFERENCES

Arora, A., Fosfuri, A., and Gambardella, A., 2004, *Markets for Technology: The Economics of Innovation and Corporate Strategy*, MIT Press, Cambridge.

Bennett, A. B., 2000, Intellectual property in agricultural biotechnology: Fueling the fire or smothering the flame. Presented at the conference *Biotechnology and the Public Interest: Prospects of Biotechnology in the Developing and Developed World,*University of California (April 28, 2000), Berkeley.

Byerlee, D., and Fischer, K., 2001, Accessing modern science: Policy and institutional options for agricultural biotechnology in developing countries, *IP Strategy Today* (**1**):1-27.

Clark, J., Piccolo, J., Stanton, B., and Tyson, K., 2000, Patent pools: A solution to the problem of access in biotechnology patents? U.S.P.T.O., white paper, United States Patent and Trademark Office, Washington, D. C.

The Economist, 2000, A market for monopoly? *The Economist* (June 17, 2000):59-60.

Enriquez, J., and Goldberg, R. A., 2000, Transforming life, transforming business: The life-science revolution, *Harvard Bus. Rev.* **78**(2):96-104.

Graff, G. D., Cullen, S. E., Bradford, K. J., Zilberman, D., and Bennett, A. B., 2003, The public-private structure of intellectual property ownership in agricultural biotechnology, *Nature Biotech.* **21**(9):989-995.

Graff, G. D., Rausser, G. R., and Small, A. A., 2003, Agricultural biotechnology's complementary intellectual assets, *Rev. Econ. Stat.* **85**(2):349-363.

Heller, M. A., and Eisenberg, R. S., 1998, Can patents deter innovation? The anticommons in biomedical research, *Science,* **280**:698-701.

Herdt, R. W., 1999, Enclosing the global plant genetic commons, presented at the Institute for International Studies, Stanford University (January 14, 1999), Stanford, California.

Kaplan, S., and Sawhney, M., 2000, E-hubs: The new B2B marketplaces: Toward a taxonomy of business models, *Harvard Bus. Rev.* **78**(3):97-103.

Kryder, R. D., Kowalksi, S. P., and Krattiger, A. F., 2000, The intellectual and technical property components of pro-vitamin A rice (Golden Rice): A preliminary framework-to-operate review, ISAAA Briefs No. 20, International Service for the Acquisition of Agribiotech Applications (ISAAA), Ithaca.

Merges, R. P., 1996, Contracting into liability rules: Intellectual property rights and collective rights organizations, *California Law Rev.* **84**:1293-1393.

Prakash, C. S., 2000, Intellectual capital: Hungry for biotech, *MIT Tech. Rev.* **103**(4):32.

Press, E., and Washburn, J., 2000, The kept university, *The Atlantic Monthly*, **285**(3):39-54.

Shapiro, C., 2000, Navigating the patent thicket: Cross-licenses, patent pools, and standard-setting. Presented at the conference, *Innovation Policy and the Economy*, National Bureau of Economic Research (April 11, 2000), Washington, D. C.

Shulman, S., 2000, Toward sharing the genome, *MIT Tech. Rev.* **103**(5):60-67.

U. S. Department of Justice and the Federal Trade Commission, 1995, Antitrust guidelines for the licensing of intellectual property (April 6, 1995); http://www.usdoj.gov/atr/public/guidelines/ipguide.htm.

Wright, B. D., 2001, Challenges for public agricultural research and extension in a world of proprietary science and technology, in: *Knowledge Generation and Transfer: Implications for Agriculture in the 21st Century*, S. Wolf and D. Zilberman, eds., Kluwer Academic Publishers, Boston, pp. 63-78.

Chapter 19

POLICIES TO PROMOTE THE CONSERVATION AND SUSTAINABLE USE OF AGRICULTURAL BIODIVERSITY

Leslie Lipper[1] and David Zilberman[2]
[1]*Economist, Agricultural and Development Economic Analysis Division, Food and Agriculture Organization of the U.N., Viale delle Terme di Caracalla 0010,0 Rome, Italy;* [2]*Professor, Agricultural and Resource Economics, 207 Giannini Hall, University of California, Berkeley, CA 94720*

Abstract: The paper finds that agricultural biodiversity conservation generates several types of benefits, which are realized by different groups in society and over time. The nature and distribution of benefits is an important basis for prioritizing, designing, and financing conservation programs. Maintaining a high level of agricultural biodiversity has been found to have high use values to farm populations in highly heterogeneous and marginal production areas, and many of these areas will also likely be significant providers of option and existence values from *in situ* conservation. An important means of achieving efficient and equitable agricultural biodiversity conservation is identification of areas where there are high potential productivity gains to be made from increasing and enhancing the diversity available to farmers, as well as those which are likely to provide the highest option values of conservation and targeting these for priority under conservation funding. We have also discussed the effectiveness of various types of payment mechanisms for conservation, depending on the supplier and consumer of the good, as well as its nature. A key theme throughout our discussion has been the importance of recognizing human knowledge as a key component of agricultural biodiversity and, thus, the necessity of incorporating means for knowledge preservation as much as the physical conservation of agricultural biodiversity.

Key words: agricultural biodiversity; *ex situ* conservation; *in situ* conservation; plant genetic resources.

1. INTRODUCTION

An increase in awareness of the importance of the environment and the threats it is facing, as well as an appreciation for the value of ecological, economic, and social services it provides, has led to rising concerns about biodiversity conservation. Biodiversity is an environmental good, as well as an indicator of the presence of other environmental goods, and thus its conservation has assumed great importance in the effort to improve environmental management and ecosystem health. Since much of the most valuable and threatened biodiversity resources are located in developing countries, policies to promote conservation and sustainable use frequently have to be studied within the context of economic development. This chapter discusses some of the major issues related to biodiversity conservation and sustainable use. We will especially focus on agricultural biodiversity, and specifically crop genetic diversity, picking up and amplifying themes raised in other chapters in this volume. In section 2 we discuss various categories of biodiversity conservation and their implications for conservation priorities. Section 3 looks at the different objectives that conservation programs may have and the types and recipients of conservation values. In section 4 we describe various types of program and policy mechanisms through which conservation may be obtained, with section 5 providing a discussion of the most likely and effective payment mechanisms associated the varying means of conservation. In section 6 we focus on issues of efficient targeting and management of conservation funds. We conclude the analysis in section seven.

2. THE VALUE OF BIODIVERSITY CONSERVATION AND PRIORITIES FOR CONSERVATION

The 1994 Convention on Biological Diversity states that biological diversity means variability among living organisms and includes diversity within species, between species, and of ecosystems. In this chapter we are especially interested in agricultural biodiversity, a vital subgroup of general biodiversity. Agricultural biodiversity:

> "encompasses the variety and variability of animals, plants, and micro-organisms on earth that are important to food and agriculture which result from the interaction between the environment, genetic resources and the management systems and practices used by people." (Aarnick et al., 1999).

In contrast to wild biodiversity, agricultural biodiversity contains a large human capital component, where genetic diversity depends on a combination of human and natural selection pressures. Therefore, we adopt a broad interpretation of agricultural biodiversity conservation and sustainable use to include species and ecosystems, as well as the management practices which sustain them. Protection of the human capital required to identify and utilize genetic resources is as important as protecting the resources themselves in designing agricultural biodiversity conservation strategies, as opposed to the conservation of wild biodiversity, where human knowledge is a much less important component of conservation.

The associated benefits of the natural and human resources comprising agricultural biodiversity is a natural means of prioritizing conservation programs. These benefits may be divided into use and nonuse categories (Randall, 2001). The assessment of usefulness is from a human perspective, which has often been criticized, particularly in the context of wild biodiversity. With agricultural biodiversity, since the resource itself is the result of human selection applied in conjunction with natural selection with the intention of providing something useful to humans, assessing the value of the resource from the human perspective is quite appropriate.

Agricultural biodiversity conservation yields several types of use benefits, manifested as both public and private goods. Several studies have shown that higher levels of agricultural diversity provide important services to farmers in the form of insurance against production risks, the ability to spread labor requirements over a production season, adaptation to heterogeneous production conditions, and the possibility of producing for differing final consumption outlets, including market or self-consumption (see Chapter 5, also Smale, 2001). Higher levels of biodiversity may generate reduced pest incidence, improved soil nutritional levels, crop pollination, and hydrological functions (Perrings, 2001). All of these characteristics fall into the category of private goods—the farmer's maintenance of biodiversity impacts their own production and consumption outcomes. However, agricultural biodiversity also provides important services to local and global populations through the maintenance of the gene pool, which is the basis for the development of new crop varieties. This capacity allows farmers and plant breeders to develop varieties to adapt to changing production and consumption conditions over time. Some of these use benefits are known and to some extent quantifiable, but much of the use benefits from agricultural biodiversity are in the form of option values, which have not yet been realized. Option values are associated with the possible future uses of biodiversity resources that may be captured with future knowledge and conditions.

The preservation of biodiversity also generates nonuse benefits. Some individuals have a strong bequest motive in their willingness to pay for preserving biodiversity, but this motive may also imply a preference for a future use in which case it is essentially the same as an option value of biodiversity. In other cases individuals may hold an existence value for biodiversity that is derived from the knowledge that valued species and ecosystems exist.

The use and nonuse values of conservation are expressed in various forms of agricultural biodiversity, which also have implications for targeting criteria under conservation programs. A simple categorization of these forms follows below:

2.1 Species and varieties

Species (including crop varieties, animal breeds) provide both use and nonuse biodiversity values (Randall, 2001). Species can be grouped into those that are known and utilized, those that are known but not utilized, and those that are unknown.

2.1.1. Species that are utilized

By definition, these have a use value, although frequently this value is nonmonetary. This category includes species of plants and animals that are utilized in the production of food, fiber, oils, etc. It includes non-harvested species essential for agricultural production such as soil micro-biota, pollinators, etc., as well as harvested species such as crops and livestock. Wild relatives of domesticated varieties may also fall into this category, as they are frequently an important source of value to rural populations. These species and varieties provide the basis for biological production systems—e.g., the basis of food and agricultural production. They also constitute a storehouse of genes, which enables the development of technologies that allow for increased yields, overcoming disease, adjustment to adverse conditions, etc.

The continued collection of species and varieties and the documentation of their properties have become even more valuable with the development of biotechnology, which allows for transgenic species and variety development. Secondly, biotechnology can identify desirable properties of organisms that may lead to innovations that will benefit a wide array of species.

2.1.2 Known but not utilized species

Many of the species that are documented or cataloged are not a source of economic benefit, and most are not likely to be commercially utilized. Besides their important intrinsic value, some species have the potential to be sources of significant economic benefit in the future, and others may have genetic structures that will be beneficial. Thus, species in this category may have significant option values. However, since conservation is a costly activity, and the number of known species is substantial, not all species will be preserved, especially when the cost of preservation significantly outweighs the benefit. Weitzman (1998) proposed a framework for assessing these trade-offs, in which priorities for species conservation are derived from a formula that includes the distinctness of the species, the utility of the species in terms of value to humans, the degree to which the species' potential for survival is enhanced by conservation activities, and the costs associated with the conservation.

2.1.3 Unknown species

Most species are not known, and they may hold many surprises in years to come. The uncertainty regarding unknown species is such that their current market value cannot be estimated. However, from a social perspective, it is worthwhile to invest resources both in their preservation and in discovering and documenting their properties. One of the biggest challenges is how to target conservation activities to bio-resources with the highest potential benefits given the degree of uncertainty about their value. Diamond (1997) argues that only a minute fraction of all the species in the world have been domesticated and, while domesticated species are crucial to our civilization, their close relatives may have genetic content that provides protection against disease and which can improve the performance of agricultural crops.

2.2 Ecosystems

A functional understanding of biological systems or genetic properties cannot be obtained without understanding how species evolve and interact within ecosystems. Regev, Shalit, and Gutierrez (1983) have shown that the population dynamics and evolution of each individual species is dependent on the well-being of other species that are either consumed or preyed upon by that species. Science is relatively young, and until now much of the effort in the biological sciences has been directed towards obtaining an understanding of microlevel processes. As we document the genetic structure of many species and gain a better idea of how organisms perform individually, understanding the interactions among species will become the main challenge of science and a key for achieving new technological developments. Therefore, the preservation of ecosystems is a targeting concept that should be distinguished from preserving individual species.

2.2.1 Knowledge and practices

Knowing that species exist and even documenting their genetic structure is not very valuable unless their function and benefits are known. Indigenous people, farmers, scientists, and others throughout the world have accumulated knowledge and systems to manage species and crop systems in a beneficial manner, and some of this knowledge is disappearing with modernization. Preservation of this knowledge is sometimes even more urgent than the preservation of species. Adoption of modern technologies and practices may lead to the loss of knowledge of traditional technologies and practices. Features of these practices are very valuable and may provide clues to the future capacity to manage resources sustainably.

3. THE OBJECTIVES OF CONSERVATION AND SUSTAINABLE-USE ACTIVITIES

Programs to promote the conservation and sustainable use of agricultural biodiversity may be intended to meet one or more environmental or social objectives, focused on preserving one or more of its associated values and components. The objective of conservation programs determines both the design of the activities and in the establishment of mechanisms for the financing of such efforts. We summarize some of the main types of program foci below.

3.1 The promotion of sustainable production systems and support of local populations

Programs designed with this objective focus on preserving and enhancing the private benefits associated with agricultural biodiversity. The impetus here is on maintaining biodiversity, and the knowledge associated with it, for the purposes of enhancing farmers' capacity to respond to varying and complex production and market conditions, as well as preserving ecosystem functions which directly impact farm productivity. Such programs are based on the notion that the conservation of agricultural biodiversity is the most effective means to enhance the sustainable production capacity of farm populations, particularly among low-income producers operating under marginal production conditions and facing frequent failures in both input and output markets. It is argued that the degree of heterogeneity and risk present in such environments requires high levels of genetic diversity in crops and animals for successful and sustainable production systems (FAO, 1998). An added advantage of such programs is that they also may generate significant option values from the on-farm conservation of agricultural biodiversity, by preserving a dynamic system of interaction between natural and human selection factors (Smale et al., 1998). Thus, the benefits of such programs would be realized not only by the farm communities involved in the implementation, but also the global community and future generations (Jarvis, Sthapit, and Sears, 2000).

The design of programs falling under this criterion may focus at the species level, such as participatory plant breeding or seed system enhancement programs directed at major subsistence crops, or at the ecosystem level, such as programs designed to enhance crop variety availability in highland or drought-prone environments. They also frequently involve a component of local knowledge preservation. Farmers have developed production systems, including rotations and pest management strategies, that have enabled them to utilize biological resources effectively, and preserving the knowledge of these systems is part of agricultural biodiversity conservation. The documentation of landrace varieties, their characteristics, and use is one means by which local knowledge is conserved (Jarvis, Sthapit, and Sears, 2000).

3.2. Maintaining the option value of biodiversity conservation

This is a primary objective of many conservation efforts. Under this objective, the focus is usually on preserving genetic diversity. As Weitzman (1998) argues, species may be perceived as carriers of genes; thus, preservation activities should emphasize maintaining the broadest base of genetic combinations possible. One way to evaluate a species or variety under this criterion is to evaluate their genetic uniqueness and relative distance from others. From this perspective, species that have close substitutes may be less valuable than those that are genetically unique. However, from another perspective, the value of genetic material is derived from the products that they generate. In the case of agricultural biodiversity, preserving closely substitutable varieties may be valuable because some of the genes that distinguish varieties may have a unique value in controlling diseases or improving food quality.

Knowledge is also an important aspect of conservation programs focused on the option value of biodiversity. Human knowledge is necessary for the identification of useful phenotypic and genetic characteristics of species, as well as for their development into new and useful varieties and breeds. The knowledge required is both science-based knowledge on species characterization and breeding, as well as the knowledge of local communities on the identification and use of species under varying types of environmental interactions. Modern crop varieties developed for monocultural agricultural systems rely on a subset of genetic material that is especially valuable under current technological conditions and under good production conditions. However, future improvements in cultivation practices may reduce the cost of adopting more diverse systems of production, and changes in climatic conditions and/or preferences may require modification of crops and cropping systems. However, future improvements in cultivation practices may reduce the cost of multicropping, and changes in climatic conditions and/or preferences may require modification of crops and cropping systems. The capacity to modify production systems will depend on the availability of the genetic material, as well as knowledge regarding interaction among crops, nondomesticated species, and ecosystems.

Benefits derived through the preservation of knowledge for the development of future varieties and breeds will be realized through product improvement cost reduction to both consumers and producers.

3.3 Preserving the existence value of biodiversity

Programs with this objective are more common for wild biodiversity rather than agricultural biodiversity conservation, due to higher ratio of existence to use values with the former as compared with the latter. However, it is likely to be important in agricultural biodiversity conservation as well, particularly in the preservation of unknown species that have the potential to be useful in future applications. Programs, which involve the conservation of ecosystems and evolutionary processes, have the advantage of allowing for the conservation of unknown as well as known species. By protecting a diverse set of ecosystems and their functions, presumably a wide range of diversity of unknown species will also be maintained.

4. MECHANISMS FOR BIODIVERSITY CONSERVATION

The multiple objectives of resource conservation, and the uncertainty associated with their performance and outcome, has led to the development of a wide range of practices for promoting the conservation and sustainable use of agricultural biodiversity. They differ according to the degree of human intervention in the natural system, ranging from the highly managed *ex situ* gene and seed banks to undisturbed wilderness areas. In this section we describe some major forms of conservation, while in the following section mechanisms for their financing are discussed. Any one system of conservation and sustainable use may adopt combinations of methods with other elements that will allow learning or produce other benefits associated with the conservation efforts.

4.1 Seed banks and gene banks

As discussed in Chapter 8 in this book, gene banks are a relatively inexpensive means of conserving genetic resources with the potential to be an effective means of conserving option values associated with genetic resources of known species. Costs of conservation vary by crop and consist of a large fixed cost component, indicating a need for greater coordination and in some cases consolidation for more effective management (Pardey et al., 1998). An important coordinating mechanism for *ex situ* sites is the International Network of *Ex Situ* Collections managed under the auspices of FAO. This network involved

12 centers of the Consultative Group on International Agricultural Research (CGIAR), which placed most of their collections (some 500,000 accessions) into the International Network. The participants agreed to hold the designated germplasm "in trust for the benefit of the international community," and "not to claim ownership, or seek intellectual property rights, over the designated germplasm and related information."

Ex situ collections range in the degree to which they are accessible to local populations, from small community seed banks, highly dependent on frequent flows of seeds in and out of the community, to government and international collections tending to be more remote but with a much wider scope of coverage. Experience from the field has indicated that seed banks need to be more closely aligned with farming communities, as well as integrated into ongoing research activities carried out by research institutions. A framework for cooperative relationships between public and private gene banks and breeders collections should be established, but the details of such a framework are subject to further research.

4.2 Botanical gardens and experimental stations

Botanical gardens provide for the protection of genetic materials, allowing scientists to monitor progress and control inventory while at the same time enabling some evolutionary processes to occur. At present, there are approximately 1,500 botanical gardens worldwide and the vast majority maintain *ex situ* collection (FAO, 1998). We use the term "experiment station" for research units that have plots and collections of plants (or animals), which they preserve and experiment with. In some cases, experiment stations are affiliated with botanical gardens, while in other cases they may collaborate with gene banks by displaying and experimenting with different types of species and varieties. Experiment stations can play a major role in analyzing the functions of genetic materials and in renewing and expanding the use of resources.

4.3 *In situ* conservation projects

Chapters 6, 7, and 8 in this volume have all discussed the central role farmers play in preserving crop genetic diversity through their selection and planting of crop varieties. In addition, farmers are often important agents of other forms of agricultural and wild biodiversity conservation (McNeely and Scherr, 2001). In recent years several programs have been established to provide incentives to farmers to maintain diverse production systems. In a few cases this has involved direct payments to farmers for maintaining diverse crop varieties, one example being the Global Environment Facility funded project: *A Dynamic Farmer-Based Approach to the Conservation of African Plant Genetic Resources* implemented in Ethiopia from 1992 to 2000 http://www.gefonline.org/projectDetails. cfm). Frequently such programs seek either to increase the availability and productivity of diversity to farmers, or to increase the returns to diverse production systems through the development of markets where some sort of premium would be paid for diversity. Adding value through the development of markets for the products of local varieties is a means by which the returns to farmers of growing diverse varieties can be increased Programs and policies to increase diversity availability are discussed in point 6 below.

In situ conservation programs may also be focused on preventing or slowing processes that lead to the loss of on-farm diversity, which in some situations is likely to be the most effective means of promoting *in situ* conservation (Chapter 5). However, a dilemma is raised when these same processes lead to economic development. The adoption of modern crop varieties and integration into markets have been identified as potentially important drivers of the loss of crop genetic diversity on farm; yet, this same process also yields tremendous benefits to the farming populations (FAO 1998; Tripp and van der Heide, 1996; Duvick, 1984; Harlan, 1972). One proposed solution to achieving dynamic efficiency and equity for *in situ* programs is to enhance the private values of genetic diversity to farmers such as developing markets for diversity-related traits, payments to farmers for maintaining diverse systems, or enhancing the productivity of local varieties. Equally important is reducing the costs of access. In the following section, we look more closely at programs that are intended to reduce the costs of diversity by increasing its availability at various points in the seed system.

4.4 Programs and policies that increase the availability of crop genetic diversity

Improving the performance and availability of genetic resources to farmers, particularly the poorest, is a critical means of achieving food security and reducing poverty (FAO, 2003; Tripp and van der Heide, 1996). In this section we focus on policies which affect the availability of crop genetic diversity, defined as all the genetic materials of plant origin of actual or potential value for any particular crop. Since these resources are embodied in seeds and planting materials, factors affecting seed distribution become relevant as well. At the farm level, the availability of such resources is driven by the type of material developed and released from formal sector plant-breeding systems, the distribution patterns of seeds embodying diverse genetic resources, as well as interactions in the informal system of seed exchange and use among farmers.

4.4.1 Increasing the genetic base of crop breeding

Most of the modern varieties and genetic populations with which plant breeders in the formal sector work consist of elite germplasm, which has been carefully built up over periods of perhaps 10-50 years (Cooper, Spillane, and Hodgkin, 2001). These lines can be destroyed by crosses with unimproved germplasm, providing a disincentive to breeders to introduce new materials, particularly when improvements can be made within the existing populations. In some cases the result is dependent on an increasingly narrow germplasm base for crop improvement (Cooper, Spillane, and Hodgkin, 2001). One possibility for broadening the genetic base of formal sector breeding is public investment into "pre-breeding" or genetic enhancement activities, involving the introduction of new characteristics from crop wild relatives or development of specific selection where desirable inbreds can be obtained (Cooper, Spillane, and Hodgkin, 2001). Such lines could then be made available to plant breeders, resulting in an increase in the genetic base of modern variety development. However, at present insufficient resources are allocated to the diversity-enhancement research, and the overall trend in investments has been one of decline (Traxler and Pingali, 1998).

4.4.2 Better incorporation of local materials and knowledge into formal breeding systems

Modern plant breeding methods have been highly beneficial to farmers operating in favorable environments, or those who can profitably modify their environments to suit new varieties (Evenson and Gollin, 2003; Ceccarelli et al., 2001). However, the results of such breeding programs do not result in superior performance under the unfavorable environmental conditions that most low-income farmers operate under (Ceccarelli et. al., 2001). They also result in the adoption of uniform plant varieties over large areas, and thus the erosion of genetic diversity in planted crop varieties. Limitations imposed by large genotype x environment (GxE) interactions are considered to be among the main factors contributing to the poor performance of modern varieties in marginal areas (Ceccarelli et al., 2001). Decentralizing variety selection and testing to target environments is an important means of promoting the specific adaptation of crop varieties to varying production conditions. One aspect of such decentralization would involve the international agricultural research centers assigning more crop selection work to national programs. Jana (1999) proposes "biodiversity friendly" breeding for gene-rich areas, including mass reservoirs and bulk populations, and Cleveland and Soleri (2002) have proposed a reorientation of the formal breeding process along the lines of GxE interaction, in order to utilize more CGR and provide outcomes more appropriate to farmers. Greater participation of farmers in breeding programs is required, which also allows for better incorporation of local knowledge into breeding strategies. In addition, national programs then need to decentralize research further by extending into farmers' fields, particularly in unfavorable environments. This latter step also generally requires the participation of farmers in breeding programs, which also allows for better incorporation of local knowledge into breeding strategies. Such decentralized breeding strategies result in greater maintenance of genetic diversity both within and among the varieties produced. A high degree of variation between selection environments and users results in a high degree of variation in the selection of breeding material in different selection sites and, ultimately, a high degree of crop genetic diversity made available to farmers.

4.4.3 Harmonization and reform of variety and seed regulations

Crop variety and seed regulatory frameworks are generally designed to promote the development and delivery of high quality and reliable commercial varieties. Regulations usually cover variety testing and release, as well as seed certification and quality control. A major objective of such regulations is the development and distribution of varieties that are distinct, uniform, and stable, as well as seeds that are viable and healthy. At present, most regulatory frameworks are set up at a national level, with little regional integration. Consequently, the flow of varieties between countries in similar agroecological zones is limited by the need for time-consuming trials and evaluations in each country (Rohrbach, Minde, and Howard, 2003; Tripp and Louwaars, 1997). The integration of regulatory systems based on environmental conditions rather than political boundaries offers the potential for substantial technological spillovers, which are sorely needed by the resource scarce national agricultural research systems of most developing countries (Rohrbach, Minde, and Howard, 2003). Such regionalization of varieties offers the possibility of developing a wider range of varieties suited to specific environments (e.g., better capturing the GE effects), resulting in better performing varieties under farmers' conditions, as well as a broader genetic base among released varieties. Seed certification regulations are also made at a national level to ensure the stability and uniformity of a given variety. The content of such regulations varies by country but, in general, results in limits on the number and nature of varieties that can be multiplied (Louwaars, 2001). Such regulations that are strictly enforced can also block the adoption of alternative breeding and seed production strategies, such as the multiplication of landrace varieties or participatory plant-breeding programs. Greater flexibility in seed certification to allow for nonuniform varieties, or exemption of certain types of seed production from certification requirements may allow seed systems to better meet farmers' needs, by increasing the levels of diversity and, thus, choices made available to them.

4.4.4 Emergency seed provision

In disaster-prone areas of the developing world, the provision of seeds has become an increasingly frequent response to emergency situations. Emergency seed supplies are thus an increasingly important source of germplasm to low-income and vulnerable farm populations. The most common form of emergency seed aid is the direct seed provision program, which involves the importation of certified seeds to the country experiencing the disaster. Under these programs, the seeds provided tend to be limited to a narrow range of crops and varieties (Cooper and Sperling, 2003). More recently, there has been a move towards local procurement of emergency seed supplies, including the use of local seed markets and merchants under voucher programs. There has been little evaluation of the impact of emergency seed distribution programs on local seed systems and crop genetic diversity, although concerns about negative impacts have been raised, particularly for areas that have experienced repeated inflows of emergency seed supplies (Cooper and Sperling, 2003). Programs which build on greater reliance of local materials and seed systems and utilize both formal and informal parts of the seed system are likely to reduce the possibility of eroding local genetic variability, and are also more likely to complement long-run development efforts in local seed systems (Alkeminders et al., 1994).

4.5 Complementary resource control

Biodiversity conservation may be threatened by a lack of complementary or supporting resources, in particular, water and environmental quality. Thus, an essential component in the design of conservation systems is the provision of sufficient complementary resources to attain the desired conservation goal. For example, assuring a continuous supply of good quality water or providing access to water reserves in periods of drought can be a critical form of agricultural biodiversity conservation. Similarly, policy mechanisms (zoning, taxation, direct control) or incentives (purchasing of development rights) may be needed to divert or prevent pollution damages in areas designated for conservation and sustainable use.

4.6 Knowledge banks

Knowledge about the functioning of ecological, agricultural, and biological systems are also major objectives of biodiversity conservation programs, and conservation systems need to be designed with mechanisms to obtain, preserve, and distribute such knowledge. As mentioned above, some knowledge conservation programs are integrated with other forms of conservation, such as with community seed bank programs. In other cases knowledge banks are being set up to allow for wider distribution networks. The Center for Indigenous Knowledge for Agriculture and Rural Development (CIKARD) at Iowa State University is one such example. CIKARD focuses its activities on preserving and using the local knowledge of farmers and other rural people around the globe. The goal is to collect indigenous knowledge and make it available to development professionals and scientists (http://www.ciesin.org/IC/ cikard/ CIKARD actprog.html). Nineteen other centers for the preservation and documentation of indigenous knowledge have been set up at regional and national levels.

5. FINANCING BIODIVERSITY CONSERVATION AND SUSTAINABLE USE

The expansion of biodiversity and, in particular, crop genetic diversity conservation efforts, requires well-designed mechanisms for financial support. One feature of many conservation programs is that they require farmers and land-users in developing countries to forego certain production activities of benefit to them, in order to generate benefits to individuals outside their region. The beneficiaries in many cases are corporations, environmental groups, and citizens of developed nations, who generally come from much higher income groups, and are more likely to be willing to pay for conservation benefits, particularly the option and existence value aspects. Conservation programs may have a negative impact on equity, without proper attention to the distribution of the costs and benefits of programs and the appropriate design of financing mechanisms. In many cases conservation efforts require the establishment of financial schemes where the gainers from conservation pay those who bear the costs.

A second important issue to consider in the design of financing mechanisms is that many of the benefits of biodiversity conservation have the properties of a public good. For example, genetic or biological knowledge can be utilized simultaneously by many and, until recently, there were few barriers to access to some aspects or manifestations of this knowledge. Without some kind of intervention, public goods will be underprovided, as no incentives exist to provide a good where no profits can be captured. In the past, the solution was to mobilize the public sector to generate such knowledge, through publicly funded research and development programs. More recently, technological and institutional changes have resulted in the ability to assign property rights to biological and genetic knowledge in the form of intellectual property rights. This has created more incentives for the private sector to generate such forms of knowledge, as they stand to reap significant benefits. However, concerns have been raised about the impacts of assigning intellectual property rights on the accessibility of knowledge, particularly as an input to the development of new varieties and breeds. Several mechanisms for overcoming these types of barriers are being designed or set up, and are discussed in other chapters of this book (see Chapters 10 and 19 in this volume). Another concern about the privatization of biological and genetic knowledge is the impact on agricultural research and development programs and new variety development aimed at poor populations (see Chapters 3 and 20 in this volume; also Smale et al., 2001). Such groups do not represent lucrative markets, and thus their needs will not be targeted under private research programs. As discussed in Chapter 3, this implies a greater need for the public sector to focus on such issues.

Even when biodiversity conservation results in outcomes that exclusively benefit a specific and identifiable agent, the magnitude and timing of these benefits may be uncertain. In many of these cases, there may be a significant lag between conservation efforts and the realization of benefits. For example, the decision not to cultivate a land parcel may preserve species that only years later will become essential for the development of a desired and valuable medical product. When outcomes of conservation activities are highly uncertain, it may be easier to raise funds for their support, if at least part of the payment is dependent on the actual outcomes. For example, a payment scheme for providing a company access for a reserve for bioprospecting may include both a fixed fee as well as a royalty tied to actual benefits derived.

In Table 19-1 below, we present a categorization of selected benefits from crop genetic diversity conservation, the likely suppliers and consumers of such benefits, and the implications for payment mechanisms. The analysis is quite general, with only four broad categories of benefits included. These include genetic diversity as a base for improving agricultural productivity and sustainable production systems, particularly in marginal areas, genetic diversity as a source of inputs to commercialized breeding systems, genetic diversity as the basis for producing differentiated consumer products, and genetic diversity in the provision of options and existence values. Although the first two categories of benefits both involve genetic diversity as an input to current breeding systems, we have differentiated them because of likely differences in the ability (and willingness) of the consuming population to meet the costs of research and development associated with breeding. In the first category, we focus on farmers as the consumers of the benefits from diversity, while in the second we look at another point in the seed system where the consumers are breeders and commercial seed enterprises that use diverse genetic resources in developing products, although eventually these products would be sold to farmers as well. This analysis indicates a wide range of potential payment mechanisms between supplier and consumers of genetic resources, and some indication of their implications for stimulating conservation.

Table 19-1. Selected benefits from crop genetic diversity, likely suppliers and consumers at various points in the seed system, and implications for possible incentive mechanisms to stimulate conservation

Benefit of diversity	Supplier	Consumer	Payment mechanism	Note
Increase in agricultural productivity and sustain-ability of productions systems; particularly in marginal environments	Farm communities	Farmers, particularly in marginal environments	Seed prices in cash transactions	Diversity is a source of new varieties through both in-formal and formal systems of breeding In both cases breeding generates value-added to the germplasm which farmers may pay through seed prices of exchange value.
	International and national research systems		Facilitation of seed and information flows	Facilitation of seed and information exchanges may include seed banks, participatory plant breeding and community seed registers, regulatory measures which facilitate seed flow
Input to breeding new seed varieties for comer-cialization	Farmers with *in situ* collections	Public/private plant breeders; Agribusiness and pharma-ceutical firms	Access fees In-kind technology transfers	Requires information on probability of benefits to set fees; ultimately costs are paid by final consumers of agricultural or pharmaceutical products. Values may be quite low due to high substituta-bility among genetic resources.

Table 19-1. (continued).

	Seed banks–*ex situ* collections	Public sector breeders	User fee	Requires information on probability of benefits t set fees; no fees set at present. Values may be quite low due to high substitutability among genetic resources.
	Seed banks–*ex situ* collections	Private sector breeders: commercial seed companies; agribusiness and pharmaceutical firms	Royalties	Currently these resources are free; for a viable system need to set up a system to track use of *ex situ* materials in commercial seed production
Availability of differentiated products to final consumers	Farmers	Consumers of agricultural products	Price premium for diverse varieties/crops	Need market development for niche markets and consumer education.
Option value maintaining genetic resources for possible future use and for existence value	Mostly developing countries	International organizations on behalf of global public Developed countries	Support to *ex situ* gene banks and *in situ* conservation programs	Conservation funds to support *ex situ* and *in situ* conservation; technology transfers

6. THE OPERATION OF CONSERVATION FUNDS

As the analysis in the previous section indicates, conservation funds are likely to be an important source of finance, particularly where option and existence values of conservation are the key objective. Even where royalties or access fees are assessed, they may be placed into some type of conservation fund, as is the case with royalties from commercialized products under the International Treaty for Plant Genetic Resources. Such funds may be managed by countries, private businesses, nonprofit agencies, and international agencies. In this section we look at some of the key issues which arise in managing such funds.

6.1 Forms of biodiversity conservation payments

Here we will focus primarily on various forms of *in situ* conservation and distinguish between the outright purchase of resources versus periodical leasing or payments for environmental services. In general, outright purchases are appropriate for stimulating efforts that require a long-term investment or commitment, while leasing is more appropriate for measures that require a continuous incentive to maintain. Of course, economic, social, and political conditions will also be key determinants of the most appropriate form of conservation purchases.

Outright purchases are more effective if the new buyer has the ability to enforce rules and control of the resources that they purchase. Several examples of this can be seen in wild biodiversity conservation, where environmental groups purchase the rights to a primary forest from a government. If these groups lack the means to control intrusions into it, and enforce the desired management regime, the program may be ineffective. If, instead they leased the forest for a certain time period, then the local government would have greater incentive to insist on proper conservation because of the future earnings at stake. In the context of agricultural biodiversity *in situ* conservation, outright purchases are not likely to be a widely used option. Conservation groups could purchase farmlands and cultivate diverse varieties; however, since *in situ* conservation involves conserving the interaction between human and natural pressures on genetic populations, it would be difficult to conserve the human side of the equation in what is essentially an artificial socioeconomic environment.

Leasing or periodical purchases of environmental services is more appropriate when a purchaser is interested in behavioral modifications of a given environment which require frequent (say, annual) activities on the part of the seller, or which are easily reversible. For example, if the objective is to have farmers maintain a diverse set of crop varieties in an evolutionary setting, a one-time fixed payment to farmers is not likely to be sufficient to ensure continuing participation of the farmers. Establishing a system where producers are paid according to their actual activities as they occur over time is likely to lead to better follow-up and a more effective result. In the case of *in situ* conservation of crop genetic diversity however, this type of payment system is difficult to implement, due to difficulties in establishing the value of maintaining any one variety in production. In addition, monitoring costs associated with such programs can be quite high. In many cases it may be more effective to fund complementary activities that support the preservation of crop varieties in the field, such as niche market development, participatory breeding programs, and so on.

6.2 Targeting-based quantitative analysis

A primary challenge of conservation funding is how to target purchases to maximize the impact of a given budget. Some of the principles for analysis and data collection required to answer this question have been addressed in the emerging economic literature on the management of bioresource purchasing funds (Wu, Zilberman, Babcock, 2001), which has been used to analyze the Conservation Reserve Program (CRP) and water quality programs in the United States (Ise and Sunding, 1998). The basic premise of this approach is that an agency has a certain amount of money that it must to use to purchase, rent, or modify the use of environmental resources. These resources can be land or water rights. The question is how to target the resources. To solve this question, one needs to take into account the quantification of environmental benefits and the costs associated with changes in behavior.

6.2.1 Quantification of the environmental benefits associated with modification of behavior at various locations

This information can be represented by indices of environmental quality. In the case of the CRP in the United States, indices of environmental quality improvement included such items as the reduction in soil erosion, increases in quality and diversity of native plants, increases in populations of migrating birds, etc. OECD[3] has also developed indicators of environmental quality based upon the pressure-state-response (PSR) model. Work has already begun at FAO on developing indicators of agricultural biodiversity, as part of the Global Plan of Action for Conserving Plant Genetic Resources for Food and Agriculture. The types of indicators being developed include measures of crop genetic erosion and vulnerability, number and kind of threatened species relevant to food and agricultural production, number and kind of wild relatives of species relevant to food and agriculture under conservation programs, areas under *in situ* conservation, and degree of genetic integrity of *ex situ* accessions. In light of the discussion in this paper, it may be useful to distinguish between areas which yield high biodiversity conservation benefits in the form of option values versus those which yield high values in terms of improvements of current production systems, as the payment mechanisms and costs associated with conservation in the two areas will be quite different. A key question is the extent to which the value of genetic diversity in improving cropping systems coincides spatially with the value of genetic diversity as an option value for future development.

---

[3] Organization for Economic Cooperation and Development.

6.2.2 Quantification of costs associated with inducing the desirable changes in various locations

Moving towards changes in behavior requires payment of some kind. As discussed in Chapters 6 and 7 of this volume, the most significant cost to farmers in providing *in situ* conservation services is foregoing agricultural productivity gains that may be obtainable with modern variety adoption. Areas with the highest conservation costs, therefore, may be expected to be those where adoption is possible, but has not yet occurred. Areas with the lowest costs of conservation are those where no option of modern variety adoption exists. These tend to be marginal production areas and cropping systems for which no modern varieties have been developed, or where modern varieties do not perform as well as local varieties. As noted in our analysis above, these also tend to be the areas with the highest potential benefits from genetic diversity as an input to the development of sustainable production systems. In many cases, we may end up with a situation where the private incentives to preserve diversity coincide with the public values. If the most effective way of increasing the productivity and profitability of farmers operating under such conditions is to enhance the availability and performance of genetic diversity, then farmers and the breeding systems serving them will have strong incentives to conserve diversity.

With the development of information technologies, the costs of establishing databases on the costs and benefits of conservation funds is declining over time. Studies on the costs and distribution of environmental benefits of resource conservation efforts suggest that there is significant heterogeneity in the distribution of benefits. For example, 10% to 15% of the land base considered for preservation of native plants in the United States provided up to 90% of the potential benefits (Wu, Zilberman, and Babcock, 2001). Similarly, the costs of purchasing resources vary greatly and, again, a relatively small percentage of the resources may possess most of the economic value and may absorb most of the costs.

Efficient conservation fund management will target funds to locations that provide the highest rate of conservation per dollar spent. Thus, locations that have the highest ratio of per acre benefits to per acre costs would be selected. Sometimes, for convenience or political/economic reasons, fund managers may target the cheapest resources. This support will maximize acreage that may be enrolled in a land-based conservation program with a given budget, but the effectiveness of this strategy depends on the correlation between environmental benefits and the cost of land. If land provides a high level of biodiversity conservation and there is a high positive correlation between cost of purchasing/leasing/payment per land area and environmental quality, this strategy may be inefficient because some of the included land provides very little additional conservation value. On the other hand, if there is a strong negative correlation between the cost of bringing the land into a conservation program and its associated biodiversity level, then acreage maximization and targeting of the lowest cost lands may also maximize environmental benefits purchased with a given budget.

Another targeted approach aims to conserve locations that provide higher conservation benefits per acre, regardless of cost. This approach may be suboptimal if environmental benefits are negatively correlated with economic costs, but it may result in the most efficient outcome when there is a strong correlation between environmental benefit and cost. For example, if land potentially highly productive under modern varieties provides high *in situ* conservation benefits, such that the relative advantage in conservation provision is higher than in production (e.g., the value of the conservation benefits are greater than the opportunity costs of foregone production), then funding farmers to preserve cropping and variety patterns which generate *in situ* conservation will be optimal.

6.3 Pitfalls in managing conservation funds

Wu, Zilberman, and Babcock (2001) argue that in some cases conservation funds may affect the prices of food and other commodities sufficiently so that resources previously not used for production will start being utilized—or "leakage" occurs. Thus, we may have a paradoxical situation where farmers are paid to reduce utilization or intensity of use on certain lands, creating pressures to bring other lands into intensive production. Under this type of scenario, increased levels of agricultural biodiversity conservation could lead to reduced levels of wild biodiversity conservation. The designer of a conservation fund has to recognize this possibility and also provide incentives against extension of production into wilderness areas. Recent empirical research has indicated the complexity of putting such incentives into place, and the need for measures on both macro- and microlevels (Lee et al., 2001; Angelsen and Kaimowitz, 2001).

The design of agricultural biodiversity conservation activities has to recognize and address potential impacts on food availability. Particularly in remote areas in developing countries, which are largely self-sufficient, any reduction of agricultural production due to conservation activities may have a negative affect on food consumption, at least for some part of the population. Thus, mechanisms may be needed to increase the productivity or value of the land that stays in production and to enable increased conservation without affecting the food security and economic well-being of the local population. However, two strategies which have been adopted for addressing this concern—agricultural intensification and integrated conservation and development projects (ICDPs)—have proven to have major problems in achieving the intended goals of both increasing food security and biodiversity conservation. Experience with these programs has shown the critical necessity of assessing the driving forces of land-use management decisions by local populations and their potential responses to changes in technology, institutions, and policies. (Angelsen and Kaimowitz, 2001; Brandon, 2001).

Wu, Zilberman, and Babcock (2001) also argue that conservation funds may be the dominant resource buyer in the region, and minimizing their cost in acquiring resources could result in monopsonistic pricing strategies. In such cases, resource prices will be lower than if there was competition among buyers' resources for environmental and conservation purposes, and the net affect is that the owner of the resources, who may be small farmers, may be compromised. This indicates the necessity for careful assessment of the potential impacts of purchasing funds, especially in regions where such funds play a dominant role in the local economy.

While market power considerations suggest that it is preferable for resource-purchasing conservation programs to be restricted by size, there may be biological considerations which would require a minimum size of land parcels in order to take advantage of increasing returns to scale in the generation of environmental amenities. As Wu and Boggess (1999) have shown, when the scale of conservation projects is sufficiently small, then an increase in the marginal productivity of conservation is associated with an increase in size. Only when size is beyond a certain threshold will marginal benefits from expansion of the project decline. That suggests a lower bound on scale of conservation projects and indicates that small-scale conservation funds may be most effective by specializing in a small number of sufficiently large projects, rather than spread resources among a large number of small projects.

7. CONCLUSIONS

In this chapter we have argued that agricultural biodiversity conservation generates several types of benefits, which are realized by different groups in society over time, and this is an important basis for prioritizing, designing, and financing conservation programs. We have noted that agricultural biodiversity conservation has potentially high use values to farm populations in highly heterogeneous and marginal production areas in terms of generating increased productivity and sustainable production systems, and these areas will also likely be significant providers of option and existence values from *in situ* conservation. An important means of achieving efficient and equitable agricultural biodiversity conservation is identification of areas where there are high potential productivity gains to be made from increasing and enhancing the diversity available to farmers, as well as those which are likely to provide the highest option values of conservation and targeting these for priority under conservation funding. We have also discussed the effectiveness of various types of payment mechanisms for conservation, depending on the supplier and consumer of the good, as well as its nature. We emphasize the wide range of actors who are and potentially could become involved in conservation through the use of a wide range of mechanisms that go well beyond the traditional concepts of conservation activities. A key theme throughout our discussion has been the importance of recognizing human knowledge as a key component of agricultural biodiversity and, thus, the necessity of incorporating means for knowledge preservation as much as the physical conservation of agricultural biodiversity.

REFERENCES

Aarnink, W., Bunning, S., Collette, L., and Mulvany, P., 1999, *Sustaining Agricultural Biodiversity and Agro-Ecosystem Functions,* FAO, Rome.

Alkeminders, C., Louwaars, N., et al., 1994, Local seed systems and their importance for an improved seed supply in developing countries, *Euphytica* **78**:207-216.

Angelsen, A., and Kaimowitz, D., eds., 2001, *Agricultural Technologies and Tropical Deforestation,* CABI International, Wallingford, UK.

Brandon, K., 2001, Moving beyond integrated conservation and development projects (ICDPs) to achieve biodiversity conservation, in: *Tradeoffs or Synergies? Agricultural Intensification, Economic Development and the Environment,* D. R. Lee and C. B. Barrett, eds., CABI Publishing, Wallingford, UK.

Ceccarelli, S., Grando, S., Amri, A., Asaad, F. A., Benbelkacem, A. , Harrabi, M., Maatougui, M., Mekni, M. S., Mimoun, H., El-Einen, R. A., El-Felah, M., El-Sayed, A. F., Shreidi, A. S., and Yahyaoui, A., 2001, Decentralized and participatory plant breeding for marginal environments in: *Broadening the Genetic Base of Crop Production,* H. D. Cooper, C. Spillange, and T. Hodgkin, eds., CABI Publishing, Wallingford, UK.

Cleveland, D. A., and Soleri, D., 2002, *Farmers, Scientists and Plant Breeding*, CABI Publishing, Wallingford, UK.

Cooper, H. D., Spillane, C., and Hodgkin, T., 2001, Broadening the genetic base of crops: An overview, in: *Broadening the Genetic Base of Crop Production*, H. D. Cooper, C. Spillange, and T. Hodgkin, eds., CABI Publishing, Wallingford, U.K.

Cooper, H. D., and Sperling, L., 2003, Understanding seed systems and strengthening seed security: A background paper, FAO unpublished paper.

Diamond, J., 1997, *Guns, Germs, and Steel: The Fates of Human Societies*, W. W. Norton, New York.

Duvick, D. N., 1984, Genetic diversity in major farm crops on the farm and in reserve. *Economic Botany* **38**(2):161-178.

Evenson, R. E., and Gollin, D., 2003, Assessing the impact of the Green Revolution, 1960 to 2000, *Science* **300**:758.

FAO, 1998, *The State of the World's Plant Genetic Resources for Food and Agriculture*, FAO, Rome.

FAO, 2003, *Anti-Hunger Programme: A Twin-Track Approach to Hunger Reduction: Priorities for National and International Action,* FAO, Rome.

Harlan, J. R., 1972, Genetics of disaster, *J. of Environ. Quality* **1**(3):212-215.

Ise, S., and Sunding, D., 1998, Reallocating water from agriculture to the environment under a voluntary purchase program, *Rev. of Agri. Econ.* **20**(1):214-226.

Jana, S., 1999, Some recent issues on the conservation of crop genetic resources in developing countries. *Genome* **42**(4):562-569.

Jarvis, D., Sthapit B., and Sears, L., eds., 2000, *Conserving Agrocultural Biodiversity in Situ: A Scientific Basis for Sustainable Agriculture.* International Plant Genetic Resources Institute, Rome, Italy.

Lee, D. R., Barrett, C., Hazell, P., and Southgate, D., 2001, Assessing tradeoffs and synergies among agricultural intensification, economic development and environmental goals: Conclusions and implications for policy, in: *Tradeoffs or Synergies? Agricultural Intensification, Economic Development and the Environment*, D. R. Lee and C. B. Barrett, eds., CABI Publishing, Wallingford, U.K.

Louwaars, N. P., 2001, Regulatory aspects of breeding for diversity, in: *Broadening the Genetic Base of Crop Production*, H. D. Cooper, C. Spillange, and T. Hodgkin, eds., CABI Publishing, Wallingford, U.K.

McNeely, J., and Scherr, S., 2001, *How Ecoagriculture Can Help Feed the World and Save Wild Biodiversity,* International Union for Conservation of Nature and Natural Resources, Gland, Switzerland.

Pardey, P. G., Skovmand, B., Taba, S., Van Dusen, M. E., and Wright, B. D., 1998, The cost of conserving maize and wheat genetic resources ex situ, in: *Farmers, Gene Banks and Crop Breeding Economic Analyses of Diversity in Wheat, Maize, and Rice,* M. Smale ed., Kluwer Academic Publishers, Norwell Massachusetts.

Perrings, C., 2001, The economics of biodiversity loss and agricultural development in low income countries, in: *Tradeoffs or Synergies? Agricultural Intensification, Economic Development and the Environment,* D. R. Lee and C. B. Barrett, eds., CABI Publishing, Wallingford, UK.

Randall, A., 2001, The value of biodiversity, in: *Ecosystems and Nature: Economics, Science and Policy,* R. K. Turner, K. Button, and P. Nijkamp, eds., Elgar Reference Collection, Environmental Analysis and Economic Policy, vol. 7, Cheltenham, U.K., and Northampton, Mass., pp. 508-512.

Regev, U., Shalit, H., and Gutierrez, A., 1983, On the optimal allocation of pesticides with increasing resistance: The case of the alfalfa weevil., *J. Environ. Econ. and Manage.* **10**:86-100

Rohrbach, R. R., Minde, I. J., and Howard, J., 2003, Looking beyond national policies: Regional harmonization of seed markets, FAO Unpublished report.

Smale, M., 2001, Economic Incentives for conserving crop genetic diversity on farms: Issues and evidence, in: *Tomorrow's Agriculture: Incentives, Institutions, Infrastructure and Innovations,* G. H. Peters and P. Pingali, eds., Proceedings of the 24th International Conference of Agricultural Economists, Ashgate, Oxford, pp. 287-305.

Smale, M., Reynolds, M. P., Warburton, M., Skovmand, B., Trethowan, R., Singh, R. P., Ortiz-Monasterio, I., Crossa, J., Khairallah, M., and Almanza, M., 2001, *Dimensions of Diversity in CIMMYT Bread Wheat from 1965 to 2000.* CIMMYT, Mexico City, Mexico.

Smale, M., Soleri, D., Cleveland, D. A., Louette, D., Rice, E .B., and Aguirre, A., 1998, Collaborative plant breeding as an incentive for on-farm conservation of genetic resources: Economic issues from studies in Mexico, in: *Farmers, Gene Banks and Crop Breeding Economic Analyses of Diversity in Wheat, Maize, and Rice,* M. Smale, ed., Kluwer Academic Publishers, Norwell, Massachusetts.

Traxler, G., and Pingali, P. L., 1998, Enhancing the diversity of modern germplasm through the international coordination of research roles, in: *Farmers, Gene Banks and Crop Breeding Economic Analyses of Diversity in Wheat, Maize, and Rice,* M. Smale, ed., Kluwer Academic Publishers, Norwell, Massachusetts.

Tripp, R., and Louwaars, N., 1997, The conduct and reform of crop variety regulation, in: *New Seed and Old Laws: Regulatory Reform and the Diversification of National Seed Systems,* R. Tripp, ed., Intermediate Technology Publications, London.

Tripp, R., and van der Heide, W., 1996, *The Erosion of Crop Genetic Diversity: Challenges, Strategies and Uncertainties*, Natural Resource Perspectives Number 7, Overseas Development Institute, London.

Wu, J., and Boggess, W. G., 1999, The optimal allocation of conservation funds, *J. of Environ. and Manage.* **38**(3):302-321.

Wu, J., Zilberman, D., and Babcock, B. A., 2001, Environmental and distributional impacts of conservation targeting strategies, *J. Environ. Econ. and Manage.* **41**: 333-350.

Chapter 20

INTERNATIONAL TREATY ON PLANT GENETIC RESOURCES FOR FOOD AND AGRICULTURE AND OTHER INTERNATIONAL AGREEMENTS ON PLANT GENETIC RESOURCES AND RELATED BIOTECHNOLOGIES[1]

José Esquinas-Alcázar*
[1]Secretary, FAO Inter-Governmental Commission on Genetic Resources for Food and Agriculture

Abstract: The chapter describes three international agreements that have been or are being negotiated by countries through the FAO Commission on Genetic Resources for Food and Agriculture, focusing primarily on the International Treaty on Plant Genetic Resources for Food and Agriculture, which entered into force in June, 2004. The economic, technical, and legal issues which arose over the long negotiating process of this multilateral agreement for the conservation and sustainable use of plant genetic resources are described, as well as their implications for the design of the Treaty. The chapter describes the Treaty's multilateral system of access to, and the sharing of benefits resulting from the use of plant genetic resources, including provisions on how it relates to intellectual property rights. It also discusses the role of Farmers' Rights through which governments can protect relevant local knowledge, and recognizes farmers' rights to equitable benefit-sharing and to participate in relevant national decisions providing access and benefits to farmers from plant genetic resources. The chapter includes a discussion of the International Code of Conduct for Plant Germplasm Collecting and Transfer, and the negotiations on a Code of Conduct on Biotechnology as it relates to genetic resources for food and agriculture.

---

[1] Negotiated By Countries Through the FAO Commission on Genetic Resources for Food and Agriculture.

Key words: access and benefit-sharing; agricultural production; agriculture; biodiversity; biotechnology regulation, centers of diversity; Commission on Genetic Resources for Food and Agriculture; Farmers Rights; *in situ* conservation; International Treaty on Plant Genetic Resources for Food and Agriculture; plant genetic resources for food and agriculture; plant germplasm collecting.

1. INTRODUCTION

Agricultural biological diversity, or more specifically, genetic resources for food and agriculture, is the storehouse that provides humanity with food, clothes and medicines. Its management is essential in the development of sustainable agriculture and food security.

The conservation and sustainable use of genetic resources and the management of related biotechnologies may appear to be technical issues, but they have strong socioeconomic, political, cultural, legal, and ethical implications, in that poor management of these matters could put the future of humanity at risk.

According to present estimates, food production in developing countries will have to increase more than 60% in the next 25 years just to keep pace with population growth. The possibilities for expanding the areas used for terrestrial and aquatic farming are relatively limited, and most of the world's wild fish stocks are already overexploited. Production must, therefore, be intensified, productivity increased. and productive natural systems must be optimally managed, all in a sustainable manner. This will require the combined application of new and old biotechnologies, including innovative approaches to plant and animal breeding and to farming practices. Success in this endeavor will depend on the sustainable utilization of a broader range of species, and of the genetic material within each species, including genes from the wild relatives of domesticated species.

In spite of its vital importance for human survival, agricultural biodiversity is being lost at an alarmingly increased rate. It is estimated that some 10 thousand species have been used for human food and agriculture. Currently no more than 120 cultivated species provide 90% of human food supplied by plants, and 12 plant species and five animal species alone provide more than 70% of all human food. A mere four plant species (potatoes, rice, maize, and wheat) and three animal species (cattle, swine, and chickens) provide more than half. Within the so-called "main food species," a tremendous loss of genetic diversity has occurred in the present century. Hundreds of thousands of farmers' plant varieties and landraces that existed until the beginning of the 20th century in farmers' fields, have been substituted by a small number of modern and highly uniform commercial varieties. In the United States alone, more than 90% of the fruit tree and vegetable species that were grown in farmers' fields at the beginning of the century can no longer be found, and only a few of them are maintained in gene banks. Similarly, alarming figures can be given for the genetic erosion of domestic animal breeds and varieties. The picture is much the same throughout the world. For example, in the 1970s a virus was decimating rice harvests across Asia. Scientists turned to crop diversity collections and screened 7,000 rice lines for resistance to the grassy stunt virus. They found genetic resistance in just one species of wild rice collected in India, and bred this into new rice varieties. This species (*Oryza nivara*) can no longer be found in the wild, and if not for its conservation in a crop gene bank, grassy stunt virus would have likely gone unchecked, to the great harm of poor farmers across Asia. Loss of agricultural biological diversity has drastically reduced the capability of present and future generations to face unpredictable environmental changes and human needs.

The rapid process of globalization and economic integration is creating an increasing interdependence between nations and regions. This can raise important ethical questions. One of the oldest forms of interdependence, starting in the Neolithic period and continuing today, is in relation to agrobiodiversity, involving the spread of crops from their centers of origin to destinations throughout the world.

Essentially no country on the planet is today self-sufficient with respect to the genetic resources for food and agriculture they are using, and the average degree of interdependence among countries with regard to the most important crops is 70%. Paradoxically, the countries which are poorest from the economic point of view, and which are in general located in tropical or subtropical zones, are also richest in terms of genetic diversity needed to ensure human survival. International cooperation is needed to develop a more fair and equitable sharing of the benefits derived from the use of genetic resources, and provide incentives to ensure that countries continue developing, conserving, and making available to humanity their genetic diversity.

There is also interdependence between generations. Agricultural biodiversity is a precious inheritance from previous generations, which we have the moral obligation to pass on intact to coming generations, allowing them to face unforeseen needs and problems. Until now, how-ever, the interests of future generations, who are neither voters nor consumers, have not been adequately taken into account by our political and economic systems.

Interdependence also exists between genetic resources and biotechnology. In general, genetic resources provide the raw material for biotechnologies. New and more powerful biotechnologies drastically increase the potential of using genetic resources, but, in some instances, they can also raise new risks for the environment, and socioeconomic concerns, if the only purpose of the users is immediate profit without ethical considerations. Regulatory mechanisms need to be developed to maximize the potentials and minimize the risks. It is important to ensure that both the new and traditional biotechnologies contribute together to the sustainable and efficient utilization of biological diversity, and to the fair and equitable distribution of the benefits derived from their use.

The industrialized world has developed legal-economic mechanisms, such as intellectual property rights (e.g., patents and plant breeders' rights) to provide incentives for the development of new biotechnologies and to compensate their inventors. However, there are no economic or legal mechanisms to compensate or provide incentives for the developers of the raw material, the genetic resources themselves. An important step in this direction has been the unanimous recognition by Food and Agriculture Organization of the United Nations (FAO) Member Countries of Farmers' Rights, whereby farmers are recognized as donors of genetic resources, as a counterweight to plant breeders' rights (rights of the donors of technology). Unlike plant breeders' rights, however, Farmers' Rights are not yet operative.

It is the inescapable responsibility of our generation to develop ethical solutions to the problems and issues raised above, within a political framework that allows an equitable sharing of benefits for all countries, and ensures food and agriculture for future generations. The United Nations, as a universal intergovernmental forum, has a fundamental role to play in the facilitation of the necessary inter-governmental negotiations to accomplish this task.

In the 1970s, action to address the global management of plant genetic resources began within the FAO resulting in the establishment, in 1983, of the first permanent intergovernmental forum on this subject: the Commission on Genetic Resources for Food and Agriculture (CGRFA), which is currently composed of 164 Member Countries and the European Community. In the 1980s this forum made possible the negotiation and development of an International Undertaking on Plant Genetic Resources, which recognized Farmers' Rights as being complementary to plant breeders' rights. Farmers' Rights were adopted by the FAO Commission in 1989 as "rights arising from the past, present and future contributions of farmers in conserving, improving, and making available plant genetic resources, particularly those in the centers of origin/diversity", with the aim of allowing "farmers, their communities, and countries in all regions, to participate fully in the benefits derived, at present and in the future, from the improved use of plant genetic resources." The members of the Commission then negotiated a revision of the International Undertaking in harmony with the Convention on Biological Diversity (CBD), which allows the regulation of access to genetic resources for food and agriculture, the fair and equitable sharing of the benefits derived from their use and the realization of Farmers' Rights. This international agreement, which is binding, was adopted by consensus by the FAO Member Countries on November 3, 2001, with the name of International Treaty on Plant Genetic Resources for Food and Agriculture (ITPGRFA). It entered into force on June 29, 2004.

The FAO/CGRFA also negotiates other international agreements related to the conservation and sustainable utilization of plant genetic resources for food and agriculture (PGRFA). In 1993, it adopted the International Code of Conduct for Plant Germplasm Collecting and Transfer; and it is currently discussing a Code of Conduct on Biotechnology as it relates to genetic resources for food and agriculture.

This chapter provides a description of some of the key economic, technical, and legal issues that arose in the negotiations and ultimately the design of the ITPGRFA, together with a description of the main provisions of the ITPGRFA. This is followed by a brief description of the motivation and key provisions of the two other international agreements negotiated in the FAO/CGRFA.

2. THE INTERNATIONAL TREATY ON PLANT GENETIC RESOURCES FOR FOOD AND AGRICULTURE

2.1. Background: The uniqueness of PGRFA

In the long negotiating process to develop and agree on the ITPGRFA, a key consideration was the fact that PGRFA differs substantially from other plant genetic resources and, therefore, specific solutions were needed for their conservation and development and the fair and equitable sharing of the benefits derived from their use, which would not necessarily be similar to those required for other kinds of biodiversity.

Unique features of PGRFA include:

(i) They are essentially *man-made*, that is, biological diversity devel-oped and consciously selected by farmers since the origins of agriculture, who have guided the evolution and development of these plants for over 10,000 years. In recent times, scientific plant breeders have built upon this rich inheritance. Much of the genetic diversity of cultivated plants can only survive through continued human conservation and maintenance.

(ii) They are not randomly distributed over the world, but rather concentrated in the so-called "centers of origin and diversity" of cultivated plants and their wild relatives, which are largely located in the tropical and sub-tropical areas (see *Appendix 1*).

(iii) Because of the diffusion of agriculture all over the world, over the last 10,000 years, and because of the association of major crops with the spread of civilizations, many crop genes, genotypes, and populations have spread, and continue to develop, all over the planet. Moreover, PGRFA have been systematically and freely collected and exchanged for over two hundred years, and a large proportion have been incorporated in *ex situ* collections.[2]

(iv) There is much greater inter-dependence among countries for PGRFA than for any other kind of biodiversity (see *Appendixes 1 and 2*).[3] Continued agricultural progress implies the need for continued access to the global stock of PGRFA. No region can afford to be isolated, or isolate itself, from the germplasm of other parts of the world.

For such reasons, the second session of the Conference of the Parties to the CBD, in 1995, adopted decision II/15, "recognizing the special nature of agricultural biodiversity, its distinctive features and problems needing distinctive solutions." The Conference of the Parties also supported the negotiations for the ITPGRFA, in order to provide such solutions.

2.2. Economic, technical, and legal issues involved in a multilateral system to regulate the conservation and sustainable use of PGRFA

During the negotiations for the ITPGRFA, a number of complex economic, technical, and legal issues needed to be examined and understood in order to develop, negotiate, and reach consensus on innovative concepts and provisions, based on an interdisciplinary approach. To facilitate this negotiation, the secretariat of the negotiating body, the Commission on Genetic Resources for Food and Agriculture commissioned a number of technical papers, as part of its series "Background Study Papers" (available on the internet at http://). Many of the concepts discussed in this section were presented and developed in such papers.

2.2.1. Economic issues

Wild and weedy crop-relatives and landraces provide the foundation-breeding materials for crop improvement and sustainable agriculture. They allow value to be added or provide a "value of use" in breeding and farming activities. This value is realized through the use of germplasm from *in situ* conditions, as well as material in *ex situ* collections.

Besides the current use-value of plant genetic resources, there are a variety of other values that can be derived from plant genetic resources. The

---

[2] These collections were made before the entry into force of, and hence outside the CBD, as Resolution 3 of the Nairobi Conference for the Adoption of the Agreed Text of the CBD recognized.

[3] More information on countries' dependence for its major crops on genetic resources that originated abroad is given on a country-by-country basis in Flores (1997).

portfolio value is the value of retaining a relatively wide range of assets within biological production systems, to smooth yield fluctuations. The *option value* is the value of retaining a wide range of known agro-biodiversity across time, as a source of currently unknown potential usefulness. The *exploration value* is the value retaining unexplored biodiversity, for the same reasons. Another way of grouping those values is to see them as *insurance values* (diversity acts as an insurance against crop yield fluctuations) and *information values*,[4] (specific information coded in the germplasm may later prove to be of concrete value). (See chapters 4, 5, 6 and 19 for alternative discussions of the value of conserving plant genetic resources.)

It is clear that the conservation of PGRFA generates a use-value. A further question is how an exchange-value may arise, that is, how is it possible to set a price, or determine an appropriate level of economic remuneration, for the exchange of these resources? An understanding of this matter is necessary in attempting to identify effective incentives for the conservation and sustainable use of PGRFA.

Traditional farmers, their communities, and countries maintain agro-diversity *in situ*, and thereby conserve and further develop the diversity contained in their landraces and related materials. A problem arises, however, in that they often have an economic incentive to replace their heterogeneous landraces by homogeneous modern varieties, as these frequently offer higher yields and productivity, and thus, higher incomes. While this process of conversion (the replacement of landraces by modern varieties) may be a rational decision on the part of an individual farmer, increasing conversion means a continuous and irreversible loss of diversity, which is not in the global interest.[5] (See chapters 5, 6, and 19 for other discussion on public and private values of *in situ* conservation.)

---

[4] Swanson *et al.*, 1994 (supplemented by personal communication with Swanson), consider that there are two parts to the information value of biodiversity: one part is unappropriable under all known mechanisms, while one part (*exploration value*) is appropriable under current conditions. They believe that the returns earned by plant breeders and seed companies, when they market a new variety over which they have any form of exclusive marketing right, include this value.

[5] An example may be given of how fast the conversion process is. Tarwi (*Lupinus mutabilis*) is one of the Andean crops that have formed the staple diet of the area for thousands of years, as a protein source. These landraces were selected by farmers over many generations for the quantity (as much as 40%) and quality of their proteins. Although of lesser interest to the farmer, tarwi also has a high fat content (as much as 26%). There is, however, a negative correlation between productivity and the oil content of tarwi seeds. In 1977, in a foreign assistance project to industrialize the crop, an experimental factory for the extraction of tarwi oil, was established south of Lima. The commercial production of new varieties of this crop, which had been selected to offer better characteristics for oil extraction, was encouraged, and farmers replaced their very heterogeneous and protein-rich landraces with the new, uniform, oil-rich but protein-poor varieties. The experiment failed, and the factory was closed in 1979. Farmers found

The question of the realization of an exchange-value for PGFRA is complex, because the farmers and communities developing and cultivating landraces, and other related genetic resources in their farming systems are, in fact, creating a global economic value, much of which they are unable to appropriate. In other words, they have no mechanism for obtaining a price, or other form of compensation, for the valuable germplasm they generate and conserve. It is the germplasm, which they have developed within their farming systems, that is the world's main source of PGRFA (whether it is still maintained in the fields, or in *ex situ* collections). This germplasm is, however, mostly available at no cost.

Traditional farmers thereby generate externalities, as providers of a "public good" (that is, a good that cannot be appropriated by its producers, and which may be used by many without exhausting it, and without adding cost). To the extent that traditional farmers, and their communities and countries, are not able to appropriate the values that they generate, they lack economic incentives to continue developing and conserving the diverse PGRFA, on which agricultural development will continue to depend. That is, they lack economic incentives to maintain this biodiversity, rather than converting to improved varieties.

In more general terms, where public goods are created, the investments for producing or preserving them necessarily tend to be suboptimal, because their producers are unable to fully benefit from the rents such goods may generate. This is a typical market failure, and is also often found in areas such as the funding of basic science.

The public nature of the goods generated by traditional farming does not mean, however, that other agents do not appropriate and benefit from these values, at a later point of the development and production process. Plant breeders and seed companies do, for example, capture at least part of the rents generated by the farmers' germplasm which they have incorporated in their varieties, especially when these are protected by plant breeders' rights, or other forms of intellectual property right. But this value is not appropriated at the correct point in the production cycle to provide the necessary incentives to promote *ex* and *in situ* conservation.

If it is in the global interest to maintain landraces and other diverse PGRFA, it is necessary that farmers and communities, who develop and conserve diversity, and their countries, either appropriate the value of maintaining diversity directly, or are compensated for the costs of conserving diversity, including the potential benefits that they forego by not

---

themselves without seeds of their old, more nutritious, landraces, the useful genes of which would have been lost forever, had not some samples previously been collected and kept viable through storage. In situations like this, a few years of the substitution of landraces by modern varieties are often enough to cause the permanent loss of germplasm that has been selected in traditional farming systems over thousands of years (Esquinas-Alcázar, 1983).

converting to modern varieties. A major difficulty arises with agro-biodiversity in the design of incentive mechanisms, as values are both difficult to estimate and to appropriate. In fact, an essential part of these values, specifically those of global nature, cannot be appropriated.

Economic analysis suggests that, for an agreement to be economically effective, it should be forward-looking and include structural incentives to favor and reward conservation in a clear, transparent manner. These incentives must be greater than the benefits foregone by renouncing conversion to specialized agriculture. If necessary, they could be linked to conservation for precise periods of time. The implementation of such incentives would require international arrangements, within the framework of an overarching multilateral agreement. Such a system might, in principle, be based on market mechanisms (for example using intellectual property rights or contracts), on nonmarket mechanisms (such as an international fund), or on a mixture or combination of mechanisms (such as a system of payments from countries on the basis of the commercial benefits derived from the use of foreign PGRFA to an international fund, and utilized to pay countries and farming communities maintaining diverse PGRFA, for making specific commitments).[6] These three possible mechanisms— especially the latter two—provide ways in which Farmers' Rights and benefit-sharing could be implemented.

The design of such mechanisms, in turn, raises a number of technical and legal questions, which condition their feasibility and enforceability; and these are discussed in the following sections.

A further approach, which avoids the burden of many separate bilateral agreements and the need to trace material in use with all the technical difficulties involved, is the development of a multilateral system shaped to the needs of agriculture, and which would not compensate individual farmers but facilitate access generally and, as benefits, mobilize finance for projects, programs, and activities for internationally agreed priorities, which promote conservation and sustainable use of PGRFA and, in particular, support small farmers holding biological diversity in their farming systems.

2.2.2. Technical issues

For the design and implementation of mechanisms for the appropriation of, or compensation for, values generated by PGRFA, the identity and origin of material must be identifiable, at least when bilateral instruments are concerned. A major question is how far this is possible, that is, whether it is possible to identify and track the geographic origin and distribution of plant genetic resources in use, over time. A document commissioned by FAO

---

[6] Editors note: Chapters 9 and 10 describe some of the issues in designing such a fund and suggests some approaches to its implementation.

reviews the capabilities and limitations of genetic fingerprinting, and related modern techniques, in identifying PGRFA, and establishing their geographical origin.[7]

In this analysis, a distinction is made between an original accession, the population from which that accession was sampled, a single genotype from that accession, and a particular gene from an accession. While any individual organism appears as a phenotype,[8] genetic fingerprinting and related techniques help to analyze the genotype and the particular combination of genes and gene variants (that is, alleles) it contains, independently of the environment in which it may be expressed. Diverse populations can be described in terms of genotype and allele frequencies.

It must also be noted that there are important differences in the genetic structure, as well as the genetic variability contained in landraces, when compared with the modern varieties that are the subject of plant breeders' rights. Current plant breeders' rights legislation applies only to propagating materials that are distinct, uniform, and stable, and can thus easily be identified, that is, to modern varieties. These contain much less variation than is usually present in a landrace. A landrace is the product, at a particular moment of time, of continuous, changing evolutionary processes that result in great variability in the gene pool, but which also provide the capacity to adapt to changing human needs (expressed through selection by farmers) and environmental conditions (expressed through evolutionary pressure). It is these characteristics that give landraces their high value as sources of plant germplasm. However, these same dynamics mean that the identification of a landrace is much more difficult than the identification of a modern variety.

Genetically inherited traits, such as flower color, growth habits, and disease resistance, can be used to identify PGRFA. More precise identification can also be obtained at the level of biochemical and molecular composition, especially through proteins and DNA-sequences.

The examples given in the Hardon, Vosman, and Van Hintum (1994) review for FAO show that, in specific instances, a number of techniques have been used to distinguish varieties and accessions. However, it is unlikely that such techniques can be routinely used to prove the identity of specific genotypes, or gene sequences, and even less the origin of unknown genetic material. There are several reasons for this:

- (i) The high costs of some of the techniques, particularly sequencing and restriction fragment length polymorphisms (RFLPs).
- (ii) The same, or similar, genetic material may exist, and be detected, in more than one place, especially in neighboring countries.
- (iii) Different methods of analysis may give different genetic estimates for the same accessions, which may lead to disputes.

---

7 Hardon et al. (1994).

8 The expression of a particular genotype in a particular environment.

(iv) The complex pedigrees of most improved varieties resulting from a plant-breeding program complicate attempts to trace specific genes, and to infer their possible relative values.

In addition, it must be borne in mind that, on the rare occasions when the ultimate geographical origin can be identified, it may not necessarily benefit the country or region of origin, since this might not be the provider of the accession, which, in line with the CBD, will usually be the subject of any rights.[9]

All these were strong arguments in favor of multilateralism as the preferred option to deal with PGRFA.

2.2.3. Legal issues

There is a need to establish a clear distinction between sovereign rights and property rights, as well as between physical and intangible property. The recognition of sovereign rights over PGRFA is not equivalent to the attribution, or existence, of property rights over such resources: sovereignty only means that the State may, within the limits imposed by the nature of such resources, determine what type and modalities of property rights, if any, are recognized.

The values of PGRFA are derived from the genetic information contained in their germplasm. It is from this point of view that intellectual property rights become relevant. Intellectual property rights cover the intangible content of processes or goods: In the case of living forms, for instance, they may govern knowledge of the information contained in genes, or other subcellular components, in cells propagating materials or plants. However, the existence of intellectual property rights over such information is not equivalent to property rights over the individual organism that carries such information, but is the right to exclude third parties from producing or selling such organisms without prior agreement.

Intellectual property rights (in particular, patents and breeders' rights) cannot currently apply to crop landraces and farmers' varieties. An important question is whether it is technically sound, and legally feasible, to extend such rights, possibly in a modified, *sui generis*, form to cover such heterogeneous populations, and whether this would create adequate incentives for the conservation of landraces. The issue raises a number of complex legal problems. These include the definition of the subject of such rights, requirements for protection, who may become titleholders, the territorial validity and administration of the system, and the actual enforceability of rights. A proposal to extend intellectual property rights to landraces, if feasible, would also have to consider the transaction costs involved in the establishment and operation of the system.

---

[9] Article 2 of the CBD.

In certain cases, the value of plant genetic resources may also be appropriated by contractual arrangements, whereby the suppliers of germplasm are remunerated, or otherwise ensured an equitable sharing in the benefits of their exploitation. Most contracts concluded until now relate to genetic resources of specific pharmaceutical or industrial value under a bilateral arrangement, rather than PGRFA, where multilateral approaches are likely to be more efficient.

Under either a multilateral or a bilateral approach, "material transfer agreements" (a form of contract) may be useful in regulating the transfer of material. Material transfer agreements typically regulate the use of the materials by the receiver, issues relating to intellectual property rights, and economic compensation to the supplying source.

Figure 1 below shows the relationship between access to genetic resources and related biotechnologies for food and agriculture, and between plant breeder's rights and Farmers' Rights, as well as the possible role of *sui generis* systems in ensuring harmony, coordination, and synergy among the various related international agreements in all relevant sectors, especially agriculture, environment, and trade. Intellectual property rights allow individuals and plant breeders to appropriate the benefit of applying biotechnologies to make commercial products. These rights are individual and have already been established for many types of biotechnologies. In contrast, farmers' rights and other forms of communal rights are the means by which benefits from the genetic resource inputs to the development of commercialized products can be obtained. These rights are collective and yet to be operationalized. The institutions governing intellectual property rights include the World Intellectual Property Organization (WIPO), the World Trade Organization/Trade Related Intellectual Property Agreement: (WTO/TRIPS) and the Union for the Protection of New Varieties of Plants (UPOV). Institutions governing farmers' rights and other types of communal rights are the FAO ITPGRFA and the CBD Article 8j.

National legislation for countries that are members of these international binding agreements needs to comply with international legal provisions contained in the agreements noted on both sides of Figure 1. This can be done through the establishment of *sui generis* systems of rights as referred to in Article 27.3.b of the TRIPs agreement and the recognition of rights through national legislation as ways in which to balance the appropriation of values from genetic resources and the biotechnologies that use them. One example is recently enacted Indian legislation, "The Protection of Plant Varieties and Farmers' Rights Act".

2.2.4 Implications for the design of the ITPGR

Many of these economic, technical, and legal issues have been overcome during the negotiation process of the ITPGRFA by innovative provisions. A key outcome was the development of a multilateral system shaped to the needs of agriculture, and which would not compensate individual farmers but facilitate access generally and share benefits, by mobilizing finance for projects, programs, and activities for internationally agreed priorities, which promote conservation and sustainable use of PGRFA, and, in particular, support small farmers holding biological diversity in their farming systems. This approach was developed to overcome the difficulties that arise from the frequent lack of knowledge of the origin of specific germplasm contributions; the difficulty of attributing value; the fact that the same diversity may be found in *in situ* conditions in a number of countries; and the onerous transaction costs of bilateralism.

Under the ITPGRFA, Farmers' Rights are the responsibility of national governments, and separate from the concept of the sharing of the monetary and other benefits of commercialization. The latter are expected to contribute to the implementation of a rolling Global Plan of Action periodically agreed by the governing body of the ITPGRFA.

In order to identify needs and priorities, national and regional plans of actions were developed in a country-driven process that culminated with the negotiation and adoption by 150 countries of the Global Plan of Action for the Conservation and Sustainable Utilization of Plant Genetic Resources for Food and Agriculture, at the Fourth FAO International Technical Conference on Plant Genetic Resources for Food and Agriculture in Leipzig (Germany) in 1996.[10] This first Global Plan of Action identified 20 priority activities on PGRFA and estimated the funds necessary for their implementation.

2.3. Relevant provisions of the International Treaty on Plant Genetic Resources for Food and Agriculture

On November 3, 2001, the 31st Session of the Conference of the FAO adopted, by consensus and as a binding international agreement, the ITPGRFA, which entered into force on June 29, 2004.

The text of the Treaty and related information are available on the Internet at http://www.fao.org/ag/cgrfa.

The Treaty's *objectives as stated in Article 1,* are "the conservation and sustainable use of PGRFA and the fair and equitable sharing of the benefits arising out of their use, in harmony with the CBD, for sustainable agriculture and food security." Article 3 establishes that its *scope* "relates to plant genetic resources for food and agriculture." For the first time in a binding international agreement, the Treaty makes provision for *Farmers' Rights*, in recognition of the collective innovation on which agriculture is based.

[10] It is available on the Internet at ftp://ext-ftp.fao.org/waicent/pub/cgrfa8/GS/gpaE.pdf.

Under Article 9, contracting parties to the Treaty are called upon to take measures to protect and promote Farmers' Rights. Specific measures called for include the protection of traditional knowledge, the right to equitably participate in sharing benefits arising from the utilization of PGRFA; and the right to participate in making decisions, at the national level, on matters related to the conservation and sustainable use of PGRFA.

A further important and innovative part of the Treaty is contained in Articles 10 to 13 (Part IV), which establish a *Multilateral System of Access and Benefit-Sharing*, which applies to a list of 64 crops, selected according to criteria of food security and interdependence. The crops in question cover about 80% of the world's food calorie intake from plants. Under this part of the Treaty, contracting parties agreed to include the plant genetic resources of these crops that are under their management and control and in the public domain into the multilateral system. In addition, they will encourage natural and legal persons within their jurisdiction to include the plant genetic resources they hold into the multilateral system. The *ex situ* collections of the International Agricultural Research Centres of the Consultative Group on International Agricultural Research (CGIAR) will also be brought into the Multilateral System, through agreements with the Governing Body.

The multilateral system is designed to provide access to PGRFA solely for the purpose of utilization and conservation for research, breeding and training for food and agriculture, provided that "such purpose does not include chemical, pharmaceutical and/or other nonfood/feed industrial uses" (Article 12.3a). Recipients of materials from the multilateral system cannot claim any intellectual property or other rights that would limit access to the resource or their genetic parts and components, in the form that it is received from the system. Access to the materials is to be provided under a standard material transfer agreement, which has yet to be developed by the Governing Body of the Treaty.

Benefit-sharing takes several forms under the multilateral system. One benefit which is shared among all contracting parties is facilitated access to the resources. Other benefits provided are information sharing, as well as access and transfer of technology. The latter are to be shared through measures such as:

> "the establishment and maintenance of, and participation in, crop-based thematic groups on utilization of plant genetic resources for food and agriculture, all types of partnership in research and development and in commercial joint ventures relating to the material received, human resource development, and effective access to research facilities" (Article 13.2.b).

Monetary benefits from the commercialization of resources accessed under the multilateral system are also to be shared through payments of an equitable share of the overall benefit to an international funding mechanism which will support conservation and sustainable utilization of PGRFA. Agreements on how an equitable share of the benefits is to be calculated and how the funding mechanism is to be managed are yet to be developed by the Governing Body of the Treaty.

Following its adoption by the FAO Conference, other universal fora have expressed unanimous support for the ITPGRFA:

The Sixth Meeting of the Conference of the Parties to the CBD (April 7-19, 2002, The Hague), in its Ministerial Declaration, which was agreed by the delegations of 176 countries, including some 130 Ministers, "urged all States to ratify and fully implement [...] the International Treaty on Plant Genetic Resources for Food and Agriculture." The Declaration adopted by the World Food Summit five years later (10-13 June 2002, Rome) recognizes "the importance of the International Treaty on Plant Genetic Resources for Food and Agriculture, in support of food security objective," and calls "on all countries that have not yet done so to consider signing and ratifying the International Treaty on Plant Genetic Resources for Food and Agriculture, in order that it enter into force as soon as possible." In the Johannesburg Declaration on Sustainable Development, countries which were represented at the highest level in the World Summit on Sustainable Development (August 26–September 4, 2002) stated that they committed themselves to the Johannesburg Plan of Implementation, in which they "invite countries that have not done so to ratify the international Treaty on Plant Genetic Resources for Food and Agriculture."

The adoption of the Treaty marks a milestone in international cooperation. It entered into force after 40 governments ratified it. Governments that have ratified it make up its Governing Body. At its first meeting, likely to be held in 2005, this Governing Body will address important questions such as the level, form, and manner of monetary payments on commercialization, a standard Material Transfer Agreement for plant genetic resources, mechanisms to promote compliance with the Treaty, and the funding strategy.

An up-to-date listing of governments that have signed and ratified the Treaty is available on the Internet at http://www.fao.org/Legal/treaties/treaty-e.htm.

3. OTHER INTERNATIONAL INSTRUMENTS DEVELOPED OR BEING DEVELOPED BY THE FAO COMMISSION ON GENETIC RESOURCES FOR FOOD AND AGRICULTURE

The International Code of Conduct for Plant Germplasm Collecting and Transfer was negotiated by the Commission on Genetic Resources for Food and Agriculture and adopted by the FAO Conference at its 27th session, in November, 1993. A Code of Conduct on Biotechnology, as it relates to Genetic Resources for Food and Agriculture, is currently under development.

3.1. The International Code of Conduct for Plant Germplasm Collecting and Transfer[11]

This Code of Conduct aims to promote the rational collection and sustainable use of genetic resources, to prevent genetic erosion, and to protect the interests of both donors and collectors of germplasm. The Code, a voluntary one, has been developed by FAO and negotiated by its Member Nations through the FAO Commission on Genetic Resources for Food and Agriculture.[12]

The Code proposes procedures to request and to issue licenses for collecting missions, provides guidelines for collectors themselves, and extends responsibilities and obligations to the sponsors of missions, the curators of gene banks, and the users of genetic material. It calls for the participation of farmers and local institutions in collecting missions and proposes that users of germplasm share the benefits derived from the use of plant genetic resources with the host country and its farmers.

The primary function of the Code is to serve as a point of reference until such time as individual countries establish their own codes of regulations for germplasm exploration and collection, conservation, exchange, and utilization.

3.2. Towards a Code of Conduct on Biotechnology as It Relates to Genetic Resources for Food and Agriculture[13]

Following the proposal of the Commission on Genetic Resources for Food and Agriculture in the 1990s, a survey of more than 400 international experts in plant genetics identified four major areas that should be covered in the Code of Conduct: biosafety, intellectual property rights, the substitution of traditional agricultural products and the development of biotechnologies appropriate for developing countries.

Biosafety: Public interest groups are concerned over the possible environmental and health risks resulting from biotechnology, especially from the field testing and release of genetically engineered organisms and plants in the area of food and agriculture. They argue that there is a lack of

---

[11] FAO (1994).

[12] At that stage, still the Commission on Plant Genetic Resources.

[13] FAO (1993).

scientific data to evaluate such risks and only the rudiments of a valid risk assessment procedure are currently possible. In the absence of an international agreement, countries that neglect to adopt adequate regulatory policies may become attractive as test sites for genetically modified organisms and plants in ways forbidden in other countries. Once released, however, organisms modified by biotechnology will not be limited by political boundaries. It is critical that means of regulation be developed at the international level. It was therefore suggested that the Code could set international standards for testing and release of such organisms.

Intellectual property rights: Some developed countries have extended legislation on intellectual property rights to cover biotechnology processes and products, in order to stimulate and protect research. But often such measures tend to restrict the exchange of germplasm, scientific information, and technologies. It was suggested that the Code lay the foundations of an international system of cooperation, through a set of agreed principles which, while being in harmony with existing international agreements, would promote research and transfer of technologies, and prevent the appropriation of existing genetic resources, traditional knowledge, and local technologies.

Substitution of traditional agricultural products: Biotechnology offers future possibilities for developing substitutes for existing crops, such as laboratory-produced vanilline flavor substituting for vanilla, which provides the livelihood of 70,000 farmers in Madagascar alone. Cocoa and sugar are two other crops threatened by substitutes. Current international economic equilibrium could dramatically shift if biotechnologies displace workers and markets in developing countries. It was proposed that the Code of Conduct offer options to minimize such effects, resulting in less drastic economic change.

Development of appropriate biotechnologies to fit the needs of developing countries: Biotechnology research is expensive, and thus tends to concentrate on cash crops and commodities of major economic interest. Unless suitable provisions are taken, crops of local and social importance in developing countries could be neglected. It was proposed that the Code promote economic incentives and suitable institutional arrangements needed to stimulate research for developing biotechnologies more appropriate for the needs of developing countries.

A first draft Code of Conduct, with four modules corresponding to the above-mentioned issues, was discussed by the FAO intergovernmental Commission on Genetic Resources for Food and Agriculture in 1993. The draft Code is aimed at biotechnologies insofar as they "affect the conservation and utilization of plant genetic resources." It recognizes that the new biotechnologies have tremendous possibilities both for improving the conservation of plant genetic resources and for stimulating throughout the world the creation of improvement programs. It further recognizes the risks inherent in these technologies, as well as how their application could have a negative effect, in particular in developing countries. The purpose of the Code is to enhance the positive effects of these new biotechnologies, and mitigate the negative effects foreseen. More information on this issue and the draft Code itself can be found on the Internet at http://www.fao.org/ag/cgrfa/biocode.htm.

Subsequently, and taking into account that the CBD was developing a Biosafety protocol, the Commission recommended that the module on biosafety in relation to PGRFA be sent to the Executive Secretary of the Convention as FAO's contribution to the development of the protocol. In 1995, the Commission examined a report on recent international developments of interest to the draft Code of Conduct, and recommended that its further development and the negotiation be postponed until the negotiations for the ITPGRFA had been completed.

Following the adoption of the ITPGRFA by the FAO Conference in 2001, the Secretariat of the Commission carried out a new survey among FAO Member Nations and a large number of stakeholders to revise the possible components for the Code on the line of recent biotechnology developments. Subjects suggested by countries and stakeholders, and currently being considered as possible components of this Code, include access to and transfer of biotechnology, capacity-building, biosafety and environmental concerns, public awareness, development of appropriate biotechnologies for poor farmers and developing countries, ethical questions regarding new biotechnologies, genetic use restriction technologies ("terminator" technology), GMOs, gene flow and the question of liability, voluntary certification schemes, and possible FAO universal declarations on plant and animal genomes. The Ninth Regular Session of the Commission (October, 2002) discussed a working paper on the subject, and requested the Secretariat to prepare a study covering all the issues raised in the survey which would identify what is being done in other forums and what remains to be done on this issue so it would help the Commission to identify the issues on which it should concentrate in the future, with respect to a Code, guidelines, or other courses of action.

4. FINAL REMARKS

The Treaty is the outcome of many years of intense negotiations in FAO's intergovernmental Commission on Genetic Resources for Food and Agriculture, to revise the voluntary International Undertaking on Plant Genetic Resources for Food and Agriculture. As the 30th Session of the Conference of the FAO noted, these negotiations were at the meeting point between agriculture, the environment and commerce. The Conference agreed that there should be consistency and synergy in the agreements being developed in these different The innovative provisions of the Treaty provide for facilitated access to PGFRA and an agreed way of benefit-sharing, without deriving these benefits from individual negotiations, on a case-by-case basis, between the provider and the user of these resources. They provide for both access and benefit-sharing to be through multilateral arrangements. This avoids the high transaction costs that such bilateral contracts involve, which are hard to justify in the context of plant breeding, which has for thousands of years been characterized by repetitive exchange, crossing, selection, and local adaptation of the intraspecific genetic resources of crops, within and between countries and regions. This ensures that plant breeders, in both the public and private sectors, can have access to the widest possible range of the resources crucial for world food security. This will benefit consumers, by providing a stream of improved and varied agricultural products, and it will benefit the seed and biotechnology industries, by providing an agreed international framework, within which to plan their investments. It also provides a firm international framework for IARCs of the CGIAR and other international organizations, whereby they hold PGRFA in trust, under the IT.

The IT and other relevant international agreements need to be fully enforced at the national level. The development of national legislation for implementation of their provisions will be essential in deterring genetic erosion, protecting indigenous germplasm and Farmers' Rights, facilitating access to genetic resources for food and agriculture, and ensuring benefit sharing.

Political and economic support for implementation of these agreements can be stimulated, if the public is informed about the importance of genetic diversity and the dangers of its depletion, and encouraged to act to stop genetic erosion. It should not be forgotten, however, that genetic erosion is but one consequence of man's abusive exploitation of the planet's natural resources, which has broken the balance of many ecosystems and brought about an increasing degradation of the biosphere. Safeguarding genetic resources by protecting them *ex situ* or *in situ* is crucial, if the process that has been unleashed is to be reversed, or controlled, at all. The fundamental problem remains man's lack of respect for the rest of nature, and any lasting solution will have to involve establishing a new relationship with our small planet, in full understanding and recognition of its limitations and fragility. If humanity is to have a future, it is imperative that children learn this in primary schools, and that adults make it part of their life.

REFERENCES

Brush, S. B., 1994, Providing Farmers' Rights through *in situ* conservation of crop genetic resources, Background Study Paper No. 3, CGRFA-FAO.

Chevassis, S., Weisell, R., 2001, Nutritional value of some of the crops under discussion in the development of a Multilateral System, Background Study Paper No. 11, CGRFA-FAO.

Correa, C.M., 1999, Access to plant genetic resources and intellectual property rights, Background Study Paper No. 8, CGRFA-FAO.

Correa, C.M., 1994, Sovereign and property rights over plant genetic resources, Background Study Paper No. 2, CGRFA-FAO.

Cunningham, E.P., 1999, Recent developments in biotechnology as they relate to animal genetic resources for food and agriculture, Background Study Paper No. 10, CGRFA-FAO.

Esquinas-Alcázar, J. T., 1993, Plant genetic resources, in *Plant Breeding: Principles and Prospects*, M. D. Hayward, N. O. Bosemark, and I. Romagosa, eds., Chapman and Hall, London.

Esquinas-Alcázar, J. T., 1983, Los Recursos Fitogenéticos: Una Inversión Segura Para el Futur, INIA, Madrid.

FAO, 1991, Biotechnology and plant genetic resources and elements of a code of conduct for biotechnology, CPGR 91/12, FAO.

FAO, 1993, Towards an international code of conduct for plant biotechnology as is affects the conservation and utilization of plant genetic resources, CPGR 93/9, FAO.

FAO, 1994, International code of conduct for plant germplasm collecting and transfer.

FAO, 1995, Recent international developments of relevance to the draft Code of Conduct for Plant Biotechnology, CPGR 6/95/15, FAO.

FAO, 1996, The Global Plan of Action for the Conservation and Sustainable Utilization of Plant Genetic Resources for Food and Agriculture.

FAO, 2001, International Treaty on Plant Genetic Resources for Food and Agriculture.

Flores, X., 1997, Contribution to the estimation of countries' interdependence in the area of plant genetic resources, Background Study Paper No. 7 (Rev.1), CGRFA-FAO.

Hardon, J. J., Vosman B., Van Hintum, T. J. L., 1994, Identifying genetic resources and their origin: The capabilities and limitations of modern biochemical and legal systems, Background Study Paper No. 4, CGRFA-FAO.

IPGRI, 2001, Crops proposed for the Multilateral System: centres of diversity, locations of *ex situ* collections, and major producing countries, Background Study Paper No. 12, CGRFA-FAO.

Kloppenburg J. R., and Kleinman, D. L., 1988, Seeds of controversy: National property versus common heritage, in: *Seeds and Sovereignty*, J. R. Kloppenburg, Jr., ed., Duke University Press, Durham and London.

Spillane, C., 1999, Recent developments in biotechnology as they relate to plant genetic resources for food and agriculture, Background Study Paper No. 9, CGRFA-FAO.

Swanson, T., Pearce, D. W., and Cervigni, R., 1994, The appropriation of the benefits of plant genetic resources for agriculture: An economic analysis of the alternative mechanisms for biodiversity conservation, Background Study Paper No. 1, CGRFA-FAO.

Visser, B., Eaton, D., Louwaars, N., and Engels, J. (GFAR), 2001, Transaction costs of germplasm exchange under bilateral agreements, Background Study Paper No. 14, CGRFA-FAO.

***Appendix 1*: Cultivated plants and their regions of diversity**[14]

1. Chinese-Japanese Region:
   - Proso millet, foxtail millet, naked oat
   - Soybean, adzuki bean
   - Leafy mustard
   - Orange/*citrus*, peach, apricot, litchi
   - Bamboo, ramie, tung oil tree, tea

2. Indochinese-Indonesian Region:
   - Rice
   - Rice bean, winged bean
   - Cucurbits/ash gourd
   - Mango, banana, rambutan, durian, bread fruit, *citrus*/lime, grapefruit
   - Bamboos, nutmeg, clove, sago palm, ginger, taros and yams, betel nut, coconut

3. Australian Region:
   - *Eucalyptus, acacia, macadamia* nut

[14] Esquinas-Alcázar (1993; based on Zeven and Zhukovsk, 1975, and Zeven and de Wet, 1982).

4. Hindustani Region:
   - Rice, little millet
   - Black gram, green gram, moth bean, rice bean, *Dolichos* bean, pigeon pea, cowpea, chickpea, horse gram, jute
   - Eggplant, okra, cucumber, leafy mustard, rat's tail radish, taros and yams
   - *Citrus*, banana, mango, sunn hemp, tree cotton
   - Sesame, ginger, turmeric, cardamom, Arecanut, sugarcane, black pepper, indigo

5. Central Asian Region:
   - Wheat (bread/club/shot), rye
   - *Allium*/onion, garlic, spinach, peas, beetroot, faba bean
   - Lentil, chickpea
   - Apricot, plum, pear, apple, walnut, almond, pistachio, melon, grape, carrot, radish
   - Hemp/cannabis, sesame, flax, safflower

6. Near Eastern Region:
   - Wheat (Einkorn, durum, poulard, bread), barley, rye/*Secale*
   - Faba bean, chickpea, French bean, lentil, pea
   - *Brassica oleracea, allium*, melon, grape, plum, pear, apple, apricot, pistachio, fig, pomegranate, almond
   - Safflower, sesame, flax
   - Lupin, medics

7. Mediterranean Region:
   - Wheat (durum, *turgidum*), oats
   - *Brassica oleracea*, lettuce, beetroot, colza
   - Faba bean, radish
   - Olive, *Trifolium*/berseem, lupin, *Crocus*, grape, fennel, cumin, celery, linseed

8. African Region:
   - Wheat, (durum, emmer, poulard, bread)
   - African rice, sorghum, pearl millet, finger millet, teff
   - Cowpea, bottle gourd, okra, yams, cucumber
   - Castor bean, sesame, niger, oil palm, safflower, flax
   - Cotton, kenaf, coffee
   - Kola, Bambara groundnut, date palm, Ensete, melons

9. European-Siberian Region:
   - Peach, pear, plum, apricot, apple, almond, walnut, pistachio, cherry
   - Cannabis, mustard (black), chicory, hops, lettuce

10. South American Region:
    - Potato, sweet potato, *xanthosoma*
    - Lima bean, amaranth, *chenopodium, cucurbita*, tomato, tobacco, lupin
    - Papaya, pineapple
    - Groundnut, sea island cotton
    - Cassava, cacao, rubber tree, passion fruit

11. Central American and Mexican Region:
    - Maize, French bean, potato, *cucurbita*, pepper/chili, amaranth, *chenopodium,* tobacco, sisal hemp, upland cotton

12. North American Region:
    - Jerusalem artichoke, sunflower, plum, raspberry, strawberry

Appendix 2: Percentages of regional food crop production[15] accounted for by crops associated with different regions of diversity

Regions of diversity

Regions of Production	**Chino-Japanese**	**Indo-Chinese**	**Australian**	**Hindus-tanean**	**West Central**	**Mediter-ranean**	**African**	**Euro-Siberian**	**Latin American**	**North American**	**Total Dependence**
Chino-Japanese	37.2	0.0	0.0	0.0	16.4	2.3	3.1	0.3	40.7	0.0	62.8
Indochinese	0.9	66.8	0.0	0.0	0.0	0.0	0.2	0.0	31.9	0.0	62.8
Australian	1.7	0.9	0.0	0.5	82.1	0.3	2.9	7.0	4.6	0.0	100.0
Hindustanean	0.8	4.5	0.0	51.4	18.6	0.2	12.8	0.0	11.5	0.0	48.6
West Central Asiatic	4.9	3.2	0.0	3.0	69.2	0.7	1.2	0.8	17.0	0.0	30.8
Mediterranean	8.5	1.4	0.0	0.9	46.4	1.8	0.7	1.2	39.0	0.0	98.2
African	2.4	22.3	0.0	1.5	4.9	0.3	12.3	0.1	56.3	0.0	87.7
Euro-Siberian	0.4	0.1	0.0	0.1	51.7	2.6	0.4	9.2	35.5	0.0	90.8
Latin American	18.7	12.5	0.0	2.3	13.3	0.4	7.8	0.5	44.4	0.0	55.6
North American	15.8	0.4	0.0	0.4	36.1	0.5	3.6	2.8	40.3	0.0	100.0
World	**12.9**	**7.5**	**0.0**	**5.7**	**30.0**	**1.4**	**4.0**	**2.9**	**35.6**	**0.0**	

Reading horizontally, the figures can be interpreted as measures of the extent (in percentages) to which a given region of production depends upon each of the regions of diversity. The column labeled "total dependence" shows the percentage of a given region's production that is accounted for by crops associated with nonindigenous regions of diversity. (This is the total of the figures in the row, except for the percentage of auto-dependence.) Due to rounding, the figures in each row do not always sum exactly to 100.

[15] Extracted from Kloppenburg and Kleinman (1988). The figures are based on the 20 food crops of current economic importance that lead global production in tonnage. These are: wheat, maize, rice, potato, barley, cassava, sweet potato, soybean, grape, sorghum, tomato, oats, banana, orange, apple, cabbage, coconut, rye, millet, and yam.

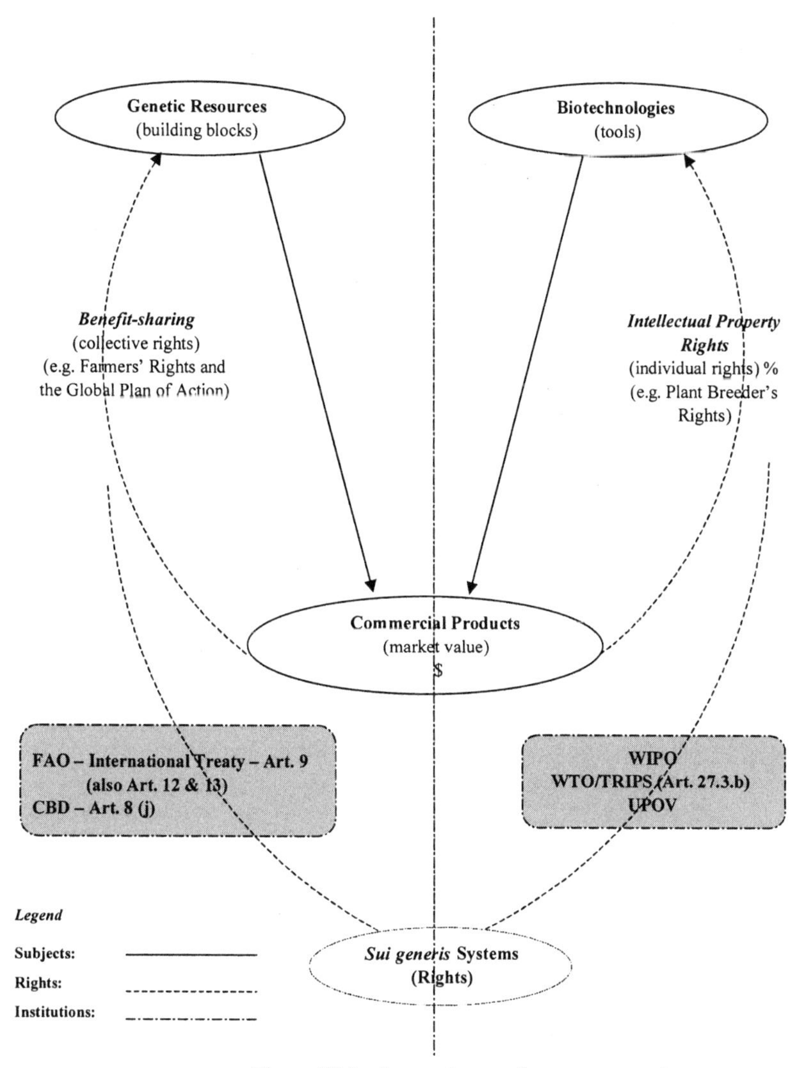

Figure 20.1. Access to genetic resources and

Chapter 21

SYNTHESIS CHAPTER: MANAGING PLANT GENETIC DIVERSITY AND AGRICULTURAL BIOTECHNOLOGY FOR DEVELOPMENT

Leslie Lipper,[1] Joseph C. Cooper,[2] and David Zilberman[3]

[1]Economist, Agricultural and Development Economic Analysis Division, Food and Agriculture Organization of the U.N., Viale delle Terme di Caracalla 00100, Rome, Italy; [2]Deputy Director, Resource Economics Division, Economic Research Service (United States Department of Agriculture), 1800 M Street, NW, Washington, DC 20036-5831;[3]Professor, Department of Agricultural and Resource Economics, University of California, 207 Giannini Hall, University of California, Berkeley, CA 94720

Abstract: This chapter synthesizes the arguments presented in 20 different chapters on various aspects of agricultural biodiversity conservation, managing biotechnology for development, equity issues in the management of plant genetic resources, and the policy implications associated with the respective analyses. Overall, the analyses in this book indicate that agricultural biodiversity and biotechnology are co-evolving, with a number of different points of intersection. Recognition of the inter-dependency between biotechnology and biodiversity is critical to the achievement of sound policy design for the management of agricultural biotechnology and biodiversity in the context of economic development. The analyses suggest that on efficiency as well as equity grounds, direct beneficiaries from agricultural biodiversity conservation should be made to reward the providers of the benefits, based both on actual and expected gains. However, benefit-sharing mechanisms must be designed to recognize the significant benefits associated with maintaining a free flow of genetic resources. Several directions for future research were identified including: the economic assessment of the impacts of adoption of various types of agricultural biotechnologies as they evolve, identification and assessment of the risks associated with biotechnology adoption relative to potential benefits, designing institutions for monitoring the environmental impacts of agricultural biotechnology, assessment of the contribution and value of various forms of genetic resources and the costs associated with their loss, including the local and global public good values of diversity in terms of

---

* The views contained herein are those of the authors and do not necessarily represent policies or views of the Economic Research Service or United States Department of Agriculture.

reducing vulnerabilities to pests and diseases, the value of maintaining diversity as an input to agricultural breeding programs. Combining research on valuation and costs could be a highly useful guide to developing countries on targeting strategies for conservation. Markets, due to their increasing importance as a mechanism for the allocation of resources, need to be analyzed in terms of their role in providing incentives and disincentives for conservation. Finally, an important area for further research is the equity implications of alternative management schemes for plant genetic diversity conservation and agricultural biotechnology. Designing incentives for *in situ* conservation, which address not only current but also future opportunity costs associated with conservation in the presence of economic development, is another important equity issue where the analysis in the book indicates the need for more economic research.

Key words: agricultural biodiversity; agricultural biotechnology; benefit-sharing; developing countries *ex situ* conservation; *in situ* conservation; intellectual property; market institutions; plant genetic resources; technology adoption; technology diffusion; technology transfer.

1. KEY POLICY CHALLENGES

In this volume we have brought together a unique set of analyses on managing agricultural biodiversity and biotechnology in the context of development. One of the key features of the book is using the common thread of plant genetic resource management as a point of departure in analyzing the potential for, and barriers to, jointly managing agro-biodiversity and biotechnology to achieve a range of development-related goals. These include increasing agricultural productivity and sustainability, reducing poverty, and improving the conservation of genetic diversity. Policies governing the management of both genetic diversity and biotechnologies have the potential to affect the ability of countries to achieve any of these objectives, depending on prevailing socioeconomic and environmental circumstances. One of the strengths of the approach taken in this volume is that it allows for an assessment of where there are overlaps, synergies, and contradictions in policy approaches to managing biodiversity and bio-technology. Just as importantly, the approach helps to identify where there is not likely to be any interaction between biodiversity and biotechnology management, and if not, thc types of policy intervention needed in each separate arena to achieve desired outcomes.

Increasing the productivity of agricultural production systems through the development and dissemination of improved genetic resources is a primary means of accelerating economic growth and addressing the problems of food insecurity and poverty in developing countries. Bio-technologies provide one important vehicle to achieve this improvement, albeit with several caveats. Both institutional and technological obstacles need to be overcome to realize its potential. Since molecular biotechnology is a major innovation, the full ramifications of its impacts are still unknown, and safeguards for preventing undesirable consequences are necessary, but difficult to formulate in the presence of uncertainty. Advances in biotechnology have also generated radical institutional changes in plant breeding with a major shift of funds and control from the public to private sector. Harnessing the benefits of biotechnology for developing countries and poor farmers thus requires new institutions and new ways of managing existing institutions.

There are other means of improving genetic resource productivity in agriculture besides biotechnology, which may be more effective, particularly in marginal production areas. These include approaches such as production ecology (Chapter 17), participatory plant breeding (Chapter 5), or reducing the costs of accessing a diverse set of genetic resources by increasing the supply of diversity (Chapter 19). However, these strategies face their own set of constraints, including cost effectiveness. Biotechnology-based approaches to improving genetic resource productivity are not mutually exclusive with alternative approaches: In fact, enhancing the productivity of conventional approaches is likely to be one of the most valuable contributions of the technology to development. Regardless of the approach taken to improve the access to and performance of genetic resources, greater focus on improvements in the development and delivery of genetic resources that meet the specific production and consumption constraints of the poor is necessary in order to achieve poverty-reduction goals.

The conservation and sustainable utilization of agricultural biodiversity is another important policy objective in developing countries, which is separate but linked to that of increasing the productivity of genetic resources for food and agriculture. Agricultural biodiversity is a broader concept than genetic diversity, encompassing human knowledge and ecosystem functions as well as the genetic variability of plant, animal and microorganisms. Its conservation generates benefits that are realized globally as well as nationally and locally. As signatories to the International Treaty on Plant Genetic Resources and the Convention on Biological Diversity, many developing countries have assumed obligations to promote the conservation and sustainable utilization of agricultural biodiversity. The relevant decisions that face developing country policymakers are how to optimize the benefits from agricultural biodiversity management, including the potential for increasing productivity and sustainability of agricultural production, as well as receiving compensation for providing public environmental goods and services.

Much of the world's valuable plant genetic diversity is located in developing countries and the conservation of this resource generates both private and public goods (Chapters 5, 9, 19). To the extent that maintaining genetic diversity results in private benefits to the farmers who provide it through their planting decisions, incentives to conserve exist, although oftentimes rapidly eroding under processes of social and economic change. Maintenance of the public good aspects of diversity conservation (e.g., reduced vulnerability to pest and diseases, and options for future genetic inputs to plant-breeding efforts) requires some type of policy intervention. One way to generate a socially desirable level of conservation is by setting up mechanisms to allow for flows of payments from the beneficiaries to the providers (e.g., Chapters 9, 10, 11, 19, and 20). However, two major problems arise with such mechanisms. First, there are concerns that establishing property rights and rights to compensation for diversity conservation will lead to a reduction in the free exchange in genetic materials among crop breeders and farmers that has prevailed thus far, thus reducing their capacity to generate new varieties, and ultimately reducing farmer access to genetic resources in developing countries (Chapter 9). Second, even if payments are desirable, payment mechanisms are difficult to design due to difficulties in valuing the benefits associated with conservation (Chapters 4, 5, 9, and 20). This debate over compensation and benefit sharing is part of a bigger discussion about the ownership of genetic materials and the benefits that farmers, breeders, and other groups obtain from conserving agrobiodiversity, which have been the focus of the International Treaty on Plant Genetic Resources and for Food and Agriculture whose implementation mechanisms have yet to be designed (Chapter 20).

The challenges outlined in the paragraphs above are being faced by developing country policymakers in a rapidly changing and high-stakes environment, where current policy choices may have significant consequences for current and future generations. This book has been designed to provide insight into the key problems of managing agro-biodiversity and biotechnology efficiently and equitably in the context of economic development. It is structured in an incremental fashion, first looking at the key forces and factors which are shaping the overall "rules of the game" under which biotechnology and biodiversity can be managed, and the impact of these on the ability to achieve efficient and equitable management regimes. Next, specific considerations of biodiversity conservation, biotechnology development and dissemination, and sharing benefits from genetic resource management are addressed. The last part includes a series of policy-oriented chapters drawing upon the analyses in earlier sections.

In the first part of the book, the chapters describe a series of major changes in global economic and environmental settings. On the environmental side, there is a decline in the global natural capital asset base of plant genetic diversity, together with a rising appreciation of their value and institutions designed to promote them. On the economic side, increased integration of global agricultural markets gives rise to changes in the structure of production and marketing, resulting in the expansion of markets and the economic opportunities associated with them. However these same changes may also result in increased barriers to market participation, particularly among small-scale and low-income producers (Chapter 2).

Three key lessons can be summarized from the chapters in the first part. The first is that markets are increasingly important as mechanisms for transmitting incentives for production and consumption decisions, as they are expanding in terms of participants on both the supply and demand side for a wide range of agricultural input and output products as well as for environmental goods and services (Chapters 2 and 3). Secondly, demand and supply are increasingly determined at supra-national levels—e.g., consumers and suppliers beyond national borders have increasing impact on market signals at a national and subnational level. One example is the rise of environmental concerns in developed countries leading to increased willingness to pay for agricultural products grown under specific environmental conditions, e.g., organic, no genetically modified organisms (GMOs), etc., affecting the production decisions of farmers in developing countries who supply international markets (Chapter 2). Another is the potential impact of the privatization and commercialization of genetic resources in developed countries on the cost of accessing these resources in developing countries (Chapters 3 and 17).

The third, and perhaps most key point, is that we cannot rely on market forces alone to generate socially desirable levels of poverty alleviation, agricultural biodiversity conservation and biotechnology development (Chapters 3, 5, 15, 16, 17, and 19). In some cases this is because markets are non-existent for the socially desirable goods and services, as is the case with agricultural biodiversity conservation. The value and sources of diversity are difficult to identify and quantify and, thus, so is the establishment of market-based payment mechanisms. In other cases the problem arises from distortions and poorly functioning markets. Examples here include the inability of poor farmers to express their demand for improved genetic resources in commercial seed markets due to their limited purchasing power and poorly functioning seed, credit, and other markets. Increased reliance on a market-based research and development system in this era of increased privatization would bypass the needs of the poor farmers in technology development. Another example is concentration and vertical integration in agricultural input and output markets that lead to noncompetitive and inefficient markets. Finally, markets are a means of achieving efficient, but not necessarily equitable, allocations of resources and thus interventions may be required to achieve socially desirable levels of equity.

The tension between the increasing importance of markets as a means of improving genetic resource management for both conservation and development on the one hand, contrasted with the increasing recognition of a need for policy interventions to either improve, supplement, or substitute for markets on the other hand, is a theme that recurs throughout the analyses presented in this volume.

2. TOWARDS EFFICIENT AND EQUITABLE STRATEGIES FOR CONSERVING AGRI-CULTURAL BIODIVERSITY

Agricultural biodiversity is a major and valuable form of natural and human capital, comprised of several components, including plant genetic diversity, which has been the focus in this book. The gains from the conservation and enhancement of agricultural biodiversity spread far beyond the location such activities take place. Agricultural biodiversity has strong public good properties, but its benefits are uncertain and vary across locations and over time. Thus, market forces by themselves will lead to a socially undesirable rate of loss, and global collective action and cooperation are required to efficiently manage agricultural biodiversity. Several means of attaining conservation have been identified in this volume, associated with varying costs and benefits. The socially desirable levels of activities in conservation and enhancement of crop biodiversities may be most efficiently achieved through compensation in exchange for these activities. Ideally, conservation funds should be allocated across the portfolio of potential activities, in order to maximize the expected benefits of agricultural biodiversity conservation, and coordination between the various forms of conservation promoted in order to enhance cost effectiveness. In reality this is difficult to achieve, due to several issues raised in the chapters of this volume. These include the following: the valuation of conservation benefits, the identification of criteria for establishing efficient conservation programs, the design of mechanisms to provide incentives to developing countries for conservation and developing means of incorporating diversity conservation into overall agricultural and economic development concerns and strategies.

A key question which arises in this design of effective policies for conservation, is just how exactly should diversity be defined—what is it that we are trying to conserve? The answer is complex, depending on the type of value focused upon, as well as assumptions about how best to generate or maintain it. Chapter 3 discusses the controversies over defining genetic narrowing in crop genetic diversity, noting several relevant dimensions, including spatial vs. temporal diversity, variation within vs. among varieties, and variation within landrace vs. modern varieties. Chapter 5 states that the biological diversity of crops encompasses phenotypic as well as genotypic variations, resulting in differences in the perception of crop genetic diversity between farmers and plant breeders. The author also notes the importance of conserving rare alleles in centers of crop origin, which requires a different type of conservation strategy than one targeted at maintaining high levels of varietal heterogeneity. Chapter 19 describes several forms of agricultural diversity, including species and varieties, ecosystems and human knowledge, all of which, the authors argue, are important to consider in conservation programs.

Clearly agricultural biodiversity conservation generates several types of goods and services and conservation programs will vary depending on which are of key concern. However, there is considerable uncertainty about the most effective means of generating goods and services from conservation, as well as uncertainty about the relative values of these services. Thus one of the biggest problems in designing effective conservation programs is defining what should be conserved and where.

Several chapters in the book provide insight into where and how the conservation of agricultural biodiversity in general, and plant genetic diversity specifically, can be most effective (Chapters 5, 6, 7, and 19). A variety of conservation methods exist, ranging from *ex situ* gene collections to *in situ* farm-based diversity management. However the high degree of uncertainty associated with both the private and public values of diversity, as well as a lack of information about the actual and opportunity costs involved, means that conservation efforts are often not efficient.

The private benefits associated with plant genetic diversity conservation are realized by farmers whose maintenance of diverse cropping systems can be thought of as the outcome of a constrained utility maximization problem. These values are described and analyzed in some detail in Chapters 5, 6 and 7. Chapter 5 summarizes the results of several studies where risk management, responsiveness to highly heterogeneous production conditions, labor management, and preferred consumption characteristics have all been found to be important determinants of on-farm diversity. These private values of diversity are determined by agroecology, population density, and the level of commercial market development. Chapter 7 adds another important determinant of the private values of crop genetic diversity: the seed system, which affects the availability and accessibility of genetic resources and information at the farm level. The incidence of natural disaster and political strife that can disrupt supply systems can also be important determinants the private value of maintaining crop genetic diversity (Chapters 6 and 7).

The market failure in diversity conservation arises from the fact that conservation generates several types of public goods. One is in the form of reduced vulnerability to pests and disease incidence, which occurs mostly as a local public good, but also with potentially wider benefits (Chapters 5, 6 and 19). Much of the use benefits associated with diversity conservation have not yet been realized and, as such, remain as potential. In such cases, the benefits of biodiversity conservation are primarily in the form of an option value as is discussed in Chapters 4 and 10. Chapter 4 presents one approach to the measurement of this value, using an empirical example from teak breeding. This chapter concludes that the value of increasing the number of potential parents for breeding is actually quite low at the margin, measured in terms of changes in the consumer and producer surplus. This chapter raises the important question of how much effort and cost should be made in maintaining crop genetic diversity as a source of input to future breeding efforts. The question is still open to considerable debate, and is likely to vary considerably among crops. Rausser and Small (2000) argue that the option values of agricultural genetic resources are likely to be sufficiently high to support market-based bioprospecting activities, since researchers have prior knowledge about where the most promising leads are likely to be found. Chapter 19 discusses various criteria for assessing option values, and discuss the disincentives to plant breeders in broadening the genetic base of their breeding lines. Public sector interventions to promote genetically diverse "pre-breeding" activities could lead to higher option values for crop genetic resources.

Moving towards consideration of criteria for designing conservation programs, one approach identified is minimizing associated costs. Chapter 6 examines this issue in detail in the context of *ex situ* conservation, which is the term applied to all conservation methods in which the species or varieties are taken out of their traditional ecosystems and are kept in an environment managed by humans. An estimated 6.2 million accessions of 80 different crops are stored in 1,320 gene banks and related facilities in 131 countries at local, national, and international levels (FAO, 1998). Chapter 6 also discusses the inefficient management of these facilities, finding significant differences in the degree of national commitment and expenditures on PGR conservation, which are not necessarily tied to the level and value of domestic genetic diversity. The chapter concludes that better collaborative relationships are the primary vehicle for reducing costs and improving the management of *ex situ* sites at a regional level, between public and private entities, and within the multilateral system.

The primary costs associated with *in situ* conservation are opportunity costs, which are addressed in Chapters 5, 6, and 19. Chapter 5 provides a conceptual framework for assessing public/private tradeoffs in maintaining *in situ* conservation, differentiating between situations where the private and public values of diversity maintenance coincide, versus come into conflict. Ostensibly, situations where they coincide require no intervention to maintain desired levels of conservation. This implies that the least-cost means of *in situ* conservation is to focus on areas where private values of diversity are high and, thus, opportunity costs of conservation are low. However, in a dynamic setting, problems arise as high private values of diversity conservation are often negatively associated with processes of economic development, particularly increasing integration of farmers into markets. Assessing the future opportunity costs farmers may face in maintaining diversity given efforts to promote economic development thus becomes a critical issue. Reducing the future opportunity costs farmers may face in maintaining on-farm diversity, and thus providing incentives for its maintenance, can be achieved by either addressing the change in conditioning factors that reduce the value of diversity or through compensation programs.

A key strategy for reducing future opportunity costs of conservation is to increase the supply of diversity and reduce access costs. This strategy includes increasing the supply of a diverse range of improved crop varieties (that is, more diversity in modern, i.e., genetically uniform, varieties) as well as enhancements to existing varieties and populations that encompass a high range of diversity (e.g., enhance the performance from varieties that encompass genetically diverse populations, landraces, and seed lots). Increasing diversity supply is an issue which is addressed throughout the book, with various pathways identified. The chapters in Part II analyze the potential for changes to traditional and conventional breeding systems that may lead to higher levels of diversity supply. Participatory plant breeding, broadening the genetic base of conventional breeding programs, and the establishment of community seed banks and registers are all examples of programs that fit here.

A major problem identified with these programs is their cost effectiveness. The inability of such programs to cover costs does not mean they are undesirable. However, some level of public support will be required to achieve the desired objective of increasing the supply of genetic diversity and thus increase the provision of both private and public goods associated with conservation.

Chapter 19 argues that plant genetic diversity conservation requires consideration of the costs and benefits of all available options. The chapter analyzes a variety of mechanisms which may be appropriate for promoting efficient conservation in both *in situ* and *ex situ* situations, ranging from direct approaches such as payments to farmers for growing diverse crop varieties and royalty payments on genetic resource inputs to commercialized products, to more indirect methods such as provision of access to biotechnologies and other forms of technology and institutional support. Agricultural research and development and plant-breeding management, seed regulation, input and output market development, information transmission, and seed provision under disaster conditions all have implications for the costs and values of *in situ* conservation, and these have not been well researched to date.

3. LINKAGES BETWEEN BIOTECHNOLOGY AND PLANT GENETIC DIVERSITY

Advances in biotechnology will have a significant impact on both the demand for, and supply of, plant genetic diversity conservation, and several chapters in the book address this issue. In this discussion clear definitions are critical: Within both biodiversity and biotechnology, there is a range of meanings, and the relationship between the two depends on which specific aspect is being considered.

Improving information about the nature, source, and value of biodiversity is one critical function biotechnologies offer to the improvement of genetic diversity conservation. As raised in several points throughout the book, lack of information is a serious problem hampering effective agricultural biodiversity conservation efforts. In Chapter 8, Virchow suggests that the value of genetic collections is reduced by the uncertainty regarding the properties and impacts of genetic materials stored in specific seed varieties. The existing and emerging tools of biotechnology will expand the capacity to utilize the information stored within *in situ* and *ex situ* collections, allowing analyses of the genetic content and potential of stored seeds. Emerging techniques of molecular and cell biology, and in particular the tools of computational genomics, allow for rigorous classification and documentation of genetic materials and, thus, a reduction in the cost of accessing and utilizing genetic materials stored in various collections (Chapters 8 and 12). This improves the ability of researchers to identify promising genetic materials for incorporation into breeding products, which is likely to increase their marginal value and demand for conservation.

The production and dissemination of GMOs are another aspect of biotechnology development likely to have significant impacts on agricultural biodiversity in general, and plant genetic diversity specifically. The introduction of GMOs may affect the number as well as genetic content of new varieties available for adoption in developing countries. Adoption patterns will affect both spatial and temporal patterns of diversity through two processes: The replacement of one type of germplasm for another, and the integration of new genetic materials into existing gene pools through gene flows. The first is a human-driven process, dependent on the supply of and demand for GMOs. The second is governed by the natural process of gene flow and integration. Ultimately, the impact on genetic diversity depends on (1) a series of forces which drive supply and demand patterns, (2) the baseline situation with regard to crop genetic diversity, (3) vulnerability of the crop to geneflow (reproductive characteristics, presence of weedy relatives), and (4) the way in which diversity is defined.

Chapter 16 looks at factors that determine the supply of GMOs in developing countries. They argue that the capacity to adapt biotechnologies to local materials is a critical determinant of the potential benefits of GMOs in agricultural development as well as impacts on crop genetic diversity. The strength of intellectual property rights (IPRs) over plant genetic resources and their enforcement within a country, together with the level of competence in the plant-breeding sector and the level of transactions costs associated with accessing biotechnologies, are identified as the key determinants of the numbers and genetic content of GMO varieties likely to be supplied in developing countries. Countries with strong IPRs, advanced breeding capacity, and relatively low transactions costs are most likely to develop a wider range of GMOs for any one crop, as the marginal costs of adding a transgenic trait to an increasing number of varieties of a sexually propagated species is smaller than the marginal benefits. In addition, the degree of local materials incorporated and thus conserved into GMO varieties is likely to be higher under these conditions. The authors argue that GMO development under these conditions can lead to an increase in crop genetic diversity, as incentives exist to modify local materials with improved traits and generate several varieties, resulting in a higher number of improved varieties, with a higher content of local materials preserved. The impacts on diversity also depend on what the GMO varieties are replacing; the implications are quite different if they are replacing a few conventionally bred modern varieties versus landrace populations.

The introduction of GMO varieties may also affect diversity through gene flows from transgenics to other planted varieties (Chapter 12). Managing undesired gene flows is an important aspect of biosafety regulation, but the degree to which gene flows pose a risk to biodiversity conservation and the degree to which regulations will be effective in managing such risks are still unknown.

Apart from the technology and products of biotechnology per se, several authors raised concerns about the impacts of the institutional changes accompanying biotechnology on diversity conservation. Chapter 3 argues that biotechnology-induced changes in IPR regimes increase the privatization of knowledge and could increase the costs of accessing breeding materials. Therefore, stringent IPR regimes may well reduce the capacity of breeders in developing countries and the CGIAR centers to access new materials and technologies. They also note that the absence of transparent and well-functioning biosafety regulations are likely to restrict access, as suppliers of the technology may be unwilling to enter such markets. Public sector access to genetic materials is a critical concern since it is this sector that will be focused on crops of most importance to the poor, which in many cases are not commercially attractive. IPRs have also been associated with increases in the number of new varieties developed. Chapter 15 argues that IPRs were a crucial stimulant to the development of private sector research and development in canola, leading to an explosion in the number of new varieties developed. However, Graff and Zilberman describe the current situation with IPRs in agricultural biotechnology as an *anti-commons climate*, restricting both public and private sector access to technologies and thus development of new varieties. Chapter 2 discusses the implications of changing IPRs under impetus from the TRIPS agreement of the World Trade Organization on agricultural biodiversity, finding the potential for both positive and negative impacts. These chapters indicate that the numbers, genetic content, and accessibility of improved varieties are changing in response to institutional changes associated with biotechnology; however, assessing the impacts on plant genetic diversity conservation is again a function of how diversity is defined.

Overall, the analyses in this book indicate that agricultural biodiversity and biotechnology are co-evolving, with a number of different points of intersection. The adoption of transgenic products may harm or enhance crop biodiversity. The new tools of biotechnology improve our capacity to interpret and utilize agricultural biodiversity. Improvements in the conservation of plant genetic diversity are likely to increase the productivity and value of agricultural biotechnology. The analyses in this book suggest that recognition of the interdependency between biotechnology and biodiversity is critical to the achievement of sound policy design for the management of agricultural biotechnology and biodiversity in the context of economic development.

4. EQUITY ISSUES IN THE MANAGEMENT OF PLANT GENETIC RESOURCES

Sharing the benefits (and costs) of plant genetic diversity conservation and maintaining access to genetic resources and biotechnologies for low-income groups are critical concerns addressed throughout this volume, but particularly in Part III and Chapter 20. Equity (and efficiency) criteria would suggest that since much of the natural capital embodied in agricultural biodiversity is in developing countries, companies and nations in the North, which are potential beneficiaries of this conservation, should contribute to crop biodiversity conservation funds.

Developed countries tend to be in regions whose original genetic endowment in the major agricultural crops was lower than in biodiversity hotspot areas. As Tables 10-1 and 10-2 demonstrate, primary centers of agricultural genetic diversity are mostly in developing countries. Thus, private breeders largely from developed countries develop and market varieties that rely on genetic materials that originated at some point (perhaps many generations ago) from the developing world. Many less-developed countries (LDCs) or associated interest groups claim that these breeders are benefiting from utilization of their native landraces without compensating the farmers responsible for their maintenance. Furthermore, they assert that developed countries are benefiting more from the utilization of plant genetic resources for food and agriculture (PGRFA) from developing countries than do the LDCs themselves and that these LDCs are not being compensated in return for using these resources. The issue has become particularly acute with the development of biotechnology and privatization of agricultural research and development. This perspective leads to active demand for compensation of farmers and others in LDCs for past conservation efforts.

However, at least from the economics standpoint, there is some difference between biodiversity funds that aim to compensate for past conservation and funds that aim to encourage future conservation. Paying LDC farmers for past conservation efforts is largely an equity issue given that insufficient data are available to establish compensation payments based on the economic value of conservation efforts in the past and, as such, one must appeal to equity, even though it is a weak mechanism for allocating funds (Chapters 9 and 11). Paying for current and future conservation activities can have more potential to be made using notions of economic efficiency (i.e., making conservation payments such that the marginal benefit of conservation effort equals its marginal cost). Chapter 11, for example, demonstrates a proxy measure for economic value that can at least be used as a rough mechanism for distributing conservation funds to world regions with an eye on increasing the economic benefits to society of conservation efforts.

The analyses suggest that on efficiency as well as equity grounds, direct beneficiaries from agricultural biodiversity conservation would be made to reward the providers of the benefits, based both on actual and expected gains. However, there are also significant benefits to maintaining a free flow of genetic resources among breeders and other researchers, and this is a difficult issue to address in the design of compensation and incentive mechanisms. On the one hand, improved property rights over genetic resources and their embodied values would facilitate the establishment of exchange and compensation mechanisms. However, at the same time, economic efficiency and equity criteria suggest that the continued sharing of the benefits associated with these goods be promoted. While this book suggests some possibilities for cost-sharing mechanisms, their exact design still needs further research.

Chapter 21 describes in detail how issues of equity and benefit sharing have been incorporated into the design of the International Treaty on Plant Genetic Resources for Food and Agriculture. The chapter discusses the economic, technical, and legal reasons for the establishment of a multilateral system to facilitate access to and sharing of benefits from the utilization of plant genetic resources. The chapter also discusses the role of various forms of property rights, including intellectual property rights and farmers' rights and how the two systems can complement each other to ensure that incentives to innovate are maintained, while at the same time ensuring the capacity of rural communities to benefit from their conservation of plant genetic resources.

5. BIOTECHNOLOGY: MAXIMIZING THE BENEFITS AND MINIMIZING THE COSTS

One of the major points made about biotechnology in this volume is that it is much more than just a tool for genetically modifying crop varieties. Aside from the crop sector, livestock, fisheries, and forestry biotechnology products of relevance to the poor are currently under development (Chapters 12 and 13). As noted above, one of the key benefits of biotechnology is through increasing information on genomics and, thus, values of biodiversity, which are necessary for developing effective conservation and compensation strategies and programs. The main focus of the potential benefits of biotechnology in economic development has been on the increased potential to generate breeding materials that are specifically relevant to the production and consumption conditions in developing countries, and in a much more targeted fashion and shorter time frame than is possible with conventional breeding methods.

Several chapters in the book describe the experience that has already been seen with biotechnology adoption in developing country agriculture. Chapters 12, 13, 14, and 16) note the dominance of tissue culture technologies in developing countries, and their importance in generating disease-free plants. Other chapters focus on the experience with GMOs in both developed and developing countries. Transgenics are in the early stages of their development, yet GMOs that control pests have high adoption rates for major crops in Latin America and China. Nevertheless, the adoption of transgenics in the majority of developing countries has been minimal, and no GMOs have been introduced for several major staples consumed by the poor (rice, wheat, cassava) in developing countries. At this point it is not possible to draw firm conclusions about the potential impacts of the adoption of transgenics in developing countries. However, the current evidence provides valuable insights, including:

- Adoption patterns and impacts of GMOs vary over different economic and agronomic circumstances. Chapter 14 cites evidence on how differences in pest incidence, land quality, and credit availability affect adoption rates. The availability, effectiveness, and prior use of pesticides determine the extent to which GMOs reduce chemical use and affect output levels, and GMOs may increase agricultural production where other approaches have not been effective in controlling pest damage at lower environmental costs.
- By reducing the variability of crop yields, GMOs can serve as an insurance strategy allowing the farmer to cope with the randomness

of pest infestation within and between seasons. The benefits of GMOs consist both of their average yield effect and yield risk-reducing effects.

- The yield gains from the adoption of GMOs are likely to be smaller if the modified varieties are generic, as opposed to those based on local materials adapted to the local conditions. Chapter 14 suggests that using GMOs not adapted to local conditions is likely to introduce new sources of yield risks.
- Whether transgenics increase yields or reduce pest-control costs, they tend to increase the overall supply of the crops. Chapter 17 suggests that this may lead to reduction in the prices of the modified commodity, which will benefit consumers including urban population, the rural landless poor, and net-consuming farm households. However, lower prices may harm the nonadopting farmers.
- Transgenics are a highly divisible technology with low fixed costs and low management requirements—e.g., they have limited requirements for human capital inputs (Chapters 14 and 17). These characteristics make GMOs accessible and attractive to small- and low-income producers. Nonetheless, the traditional constraints to technology adoption among the poor—such as lack of credit, poorly developed input and output markets, and the presence of risk—are likely to impede adoption among smallholders.
- The adoption of GMOs may generate environmental and human health benefits through the reduction of pesticide use, and yield effects may lead to reduction of land conversion to agricultural use and thus reduce deforestation and land degradation (Chapter 13). These benefits have to be weighed against the risks that may be introduced with GMOs, such as irreversible changes in genetic populations through geneflow.

The substantial rates of adoption of agricultural biotechnologies in some developed and developing countries and their realized net benefits suggest that these technologies are likely to play a significant role in global agriculture as they evolve (Chapters 13 and 16). Chapters 12 and 13 highlight the applications of agricultural biotechnology currently available and in the development pipeline, which could be highly beneficial to low-income farmers in particular and to developing countries in general. However, the degree to which these potential benefits of biotechnology are realized by poor farmers and developing countries is likely to be determined more at a macro than at a microlevel (Chapters 14, 15, and 16). The benefits of biotechnology to farmers and the poor will depend on the degree to which biotechnology innovations address production and consumption constraints, and are affordable and accessible to farmers (Chapters 15, 16, and 17).

Farmer access to biotechnology is determined by the type, amount, and cost of technologies produced by plant breeders—either nationally or internationally. As argued in Chapter 14, these factors are, in turn, driven by the combination of intellectual property (IP) regimes, local breeding capacity, commercial seed industry development, and biosafety regulation regimes. Transactions costs (affected by IPRs and biosafety regulations) associated with obtaining breeding materials will determine the degree to which private sector materials would be available to local breeders, while local breeding capacity will determine the costs of adapting them to local conditions, and the development of the commercial seed sector drives the degree to which such innovations could be disseminated to farmers (Chapter 16). High transactions costs in obtaining breeding materials and local breeding capacity are the two most critical determinants of potential beneficial effects of biotechnology in developing countries (Chapters 13-18), which can be addressed by institutional reforms at both the national and international levels.

Chapter 18 gives one example of such an institutional reform, arguing that the transaction costs can be significantly reduced by establishing clearinghouses for IP, which will provide crop breeders with information on the status of IP over various crops and technologies and assist them to obtain access to it. The recently established Public Intellectual Property Resources for Agriculture (PIPRA) is one example of a clearinghouse that aims to reduce the transaction cost constraints of agricultural biotechnology.[4] Chapters 16 and 18 also argue that reducing registration requirements for agricultural biotechnology will reduce transaction costs and lead to a more diversified portfolio of modified varieties. A clear example of a policy that reduces transactions costs is the requirement of registration and safety testing only for new biotechnology events (such as development of a parent GMV, which through back crossing can lead to insertion of the modification from the parent into all the varieties of the crop) rather than for every modified variety. Chapters 3 and 15 argue that the CGIAR centers have an important role to play in filling the gap created by a lack of local breeding capacity in many developing countries, as well as greater integration of NARS research work over agroecological regions

In many developing countries, agricultural biotechnology may not be the least cost or most efficient means of improving agricultural productivity (Chapter 17). The national breeding capacity, type of farming systems present, and constraints to increases in agricultural productivity are key determinants of the degree to which developing countries will benefit from agricultural biotechnology (Chapters 3, 16, and 17).

---

[4] For an example of a knowledge clearinghouse for agricultural biotechnology, see www.ers.usda.gov/data/AgBiotechIP/.

The future of agricultural biotechnology and its impacts on economic development will be affected by the management of the human health and environmental risks associated with it (Chapter 17). Continuous research and monitoring of the potential risks involved are clearly critically important. The efficiency of the regulation of biotechnology applications would increase significantly with greater quantification and definition of the risks associated with these applications. However, the high degree of uncertainty and the lack of information on the risks prevent precise estimation. Building this uncertainty into biosafety regulatory structures would provide more meaningful information than that associated with simply providing mean measures of risks. One way to overcome the lack of information at the initial stage may be to quantify the potential risks under plausible pessimistic scenarios, and assess their costs relative to the expected economic and environmental benefits of the technology. It is important to recognize that, beyond a certain stage, estimates of outcomes and the technology itself will not improve significantly without field experience, which implies that the efficiency of assessment and regulation of technologies will be increased if they can incorporate adaptive learning and through taking advantage of findings in the laboratory as well as outcomes in the field. Poorly designed biosafety regulations that lead to excessive delay in the introduction of biotechnologies may generate significant economic costs in terms of foregone opportunities for technological development, including learning by doing, and improvements in agricultural productivity.

An important reference point for the development of biosafety regulations in the context of agricultural biotechnology is the Draft Code of Conduct on Biotechnology as it relates to Genetic Resources for Food and Agriculture. (Chapter 20) The objective of the code is to maximize the positive effects and minimize the possible negative effects, of biotechnology (http://www.fao.org/ag/cgrfa/biocode.htm). The draft Code is based on the results of two major surveys of stakeholders in 1993 and 2001, which identified the key issues of concern. Issues currently being considered as possible components of the Code include access to and transfer of biotechnology, capacity-building, biosafety and environmental concerns, public awareness, development of appropriate biotechnologies for poor farmers and developing countries, ethical questions regarding new biotechnologies, genetic use restriction technologies ("terminator" technology), GMOs, gene flow and the question of liability, voluntary certification schemes, and possible FAO universal declarations on plant and animal genomes.

A clear message that emerges from the analyses in this volume is that appropriately designed policies and institutions are essential for enabling agricultural biotechnology to fulfill its promise for developing countries. One policy implication arising from the analyses presented is the potential benefits to be reaped from strengthening of the capacity of developing country agricultural research and development and seed sectors to introduce desired traits into local varieties, rather than relying upon imports of generic transgenic varieties. A second policy implication that emerges is the need for regulations to manage the risks associated with the new technology as well as the importance of including cost considerations—particularly the costs of foregoing opportunities to improve productivity—when designing such regulations. Thirdly, barriers to access the intellectual property needed for the development of transgenic crops for developing countries should be reduced through institutional arrangements for technology transfer and sharing of knowledge about IPR and technology management.

A clear message that emerges from the analyses in this volume is that designing appropriate policies and institutions is essential for enabling agricultural biotechnology to fulfill its promise for developing countries. First, benefits to developing countries will be greater if the capacity of the seed sector in these countries is enhanced to allow introduction of desired traits into local varieties than with simply importing generic transgenic varieties. Second, economic efficiency suggests that the level of regulation of new varieties to allow control against risks has to be balanced against cost considerations—particularly the costs of foregoing opportunities to improve productivity—when designing such regulations. Third, the most efficient way to reduce barriers to access the IP needed for the development of transgenic crops for developing countries would be likely be through institutional arrangements for technology transfer and sharing of knowledge about IPR and technology management.

6. DIRECTIONS FOR FUTURE RESEARCH

Several areas where new research is needed on issues related to managing plant genetic diversity and agricultural biotechnology for economic development have been identified throughout this volume.

With regards to biotechnology, it is important to continue to assess the economic impacts of adoption of various types of agricultural biotechnologies as they evolve. To best assess these impacts, we need quantitative understanding on how the features of various technologies, the economic and environmental conditions in various locations, the institutional setup in general, and the policies associated with the new technologies affect their impacts in terms of pricing and welfare of various groups. This research will allow identification of the countries and situations where investments in agricultural biotechnology are likely to generate significant returns in terms of agricultural productivity increases and poverty alleviation, relative to other potential strategies.

The potential environmental side effects of agricultural biotechnology are a continuous source of controversy that will affect the future of this technology. Identifying and assessing the risks associated with biotechnology adoption relative to potential benefits is a major priority and, more importantly, designing institutions for monitoring the environmental impacts of agricultural biotechnology and effectively regulating to control potential risks is a major policy challenge. It has also emerged as an area where more research is urgently needed.

We also need to identify features of biotechnology products that are especially desirable from the perspective of the developing world and identify mechanisms that will help developing countries gain access to them, especially if they will not be pursued as part of the agenda of the private sector. For example, it is important to understand to what extent can biotechnology enhance the micronutrient content of food consumed in developing countries and to what extent the biotechnology innovations that serve this purpose will be pursued by the private sector and, if they are not pursued privately, whether and how to provide the incentives for their introduction.

We need a better understanding of the role and effects of regulatory regimes, including environmental, IPR, and market structure regulations on the evolution and adoption of new biotechnology products and their impact on the environment. As new institutions for the management and regulation of biotechnology are introduced, we need research that assesses their performance and suggests design modification and reform. Specifically, more work on policy and institutional reforms necessary to facilitate the potential benefits of biotechnology to the poor is necessary—particularly in reducing the transactions costs associated with access under increasingly restrictive property rights for genetic materials and associated technologies.

On the topic of genetic diversity conservation, first we need to have a better handle on the contribution and value of various forms of genetic resources and the costs associated with their loss. One could use emerging information technologies to collect data on use of various genetic collections and analyze it statistically. It is crucial to understand how improved capabilities affect the usage and productivity of biodiversity in order to better their storage and distribution, an understanding which requires interdisciplinary research cooperation. Determining how to optimize the value of both *in situ* and *ex situ* conservation to developing countries requires better information on what these values are, as well as the costs associated with obtaining them, considered in the dynamic context of economic development. Some of this valuation work must be inferred indirectly from greater understanding of the improved economic value derived by bioresources. Valuation work on plant genetic diversity has focused at the farm level in looking at household decision-making over a portfolio of crops and varieties. More work is needed on the local and global public good values of diversity in terms of reducing vulnerabilities to pests and diseases. In addition, further work on the value of maintaining diversity as an input to agricultural breeding programs is needed, following up and expanding on the work of Simpson (Chapter 5) and others (Rausser and Small 2000). Combining research on valuation and costs could be a highly useful guide to developing countries on targeting strategies for conservation.

Together with an assessment of the most efficient conservation opportunities, there is a need for analysis of the most effective and equitable mechanisms for providing incentives for conservation. Markets, due to their increasing importance as a mechanism for the allocation of resources, need to be analyzed in terms of their role in providing incentives and disincentives for conservation. Here markets are taken in the widest sense, ranging from local commodity exchanges up to global markets for environmental goods. The efficiency of markets in allocating plant genetic resources and the implications this has for diversity at the farm and local level are areas where more research is needed. The efficiency and optimal design of market-based mechanisms for maximizing global public good values associated with diversity conservation are other areas where gaps in the economic literature exist. However, market forces are not the only drivers of interest in assessing conservation incentives: The impact of nonmarket forces, particularly government regulations in the agricultural and seed sectors, is also a critical area for further research. Regulations of interest range from biosafety, to seed certification and release procedures, to agricultural pricing interventions.

Finally, an important area for further research is the equity implications of alternative management schemes for plant genetic diversity conservation and agricultural biotechnology. Designing mechanisms to compensate farm communities for their past services in conserving and providing genetic resources to the formal breeding sector, which do not create new barriers to exchanges and thus reduces access, is a challenging area where more work is needed. Designing incentives for *in situ* conservation, which address not only current but also future opportunity costs associated with conservation in the presence of economic development, is another important equity issue where the analysis in the book indicates the need for more economic research. Finally, further analyses of the distribution of benefits and costs to agricultural biotechnology investment and adoption and the impact, particularly on low productivity agricultural populations relative to other means of productivity increases, is a highly important area of research both from an equity and efficiency standpoint.

REFERENCES

FAO, 1998, *The State of the World's Plant Genetic Resources for Food and Agriculture*, FAO, Rome.

Rausser, G. C., Small, A. A., 2000, Valuing research leads: bioprospecting and the conservation of genetic resources, *J. of Polit. Econ.* **108**(1):173-206.

AUTHOR INDEX

SUBJECT INDEX